BIOLOGY OF THE
MAMMARY GLAND

# ADVANCES IN EXPERIMENTAL MEDICINE AND BIOLOGY



---

Recent Volumes in this Series

Volume 474
HYPOXIA: Into the Next Millennium
Edited by Robert C. Roach, Peter D. Wagner, and Peter H. Hackett

Volume 475
OXYGEN SENSING: Molecule to Man
Edited by Sukhamay Lahiri, Nanduri R. Prabhakar, and Robert E. Forster, II

Volume 476
ANGIOGENESIS: From the Molecular to Integrative Pharmacology
Edited by Michael E. Maragoudakis

Volume 477
CELLULAR PEPTIDASES IN IMMUNE FUNCTIONS AND DISEASES 2
Edited by Jürgen Langner and Siegfried Ansorge

Volume 478
SHORT AND LONG TERM EFFECTS OF BREAST FEEDING ON CHILD HEALTH
Edited by Berthold Koletzko, Olle Hernell, and Kim Fleischer Michaelsen

Volume 479
THE BIOLOGY AND PATHOLOGY OF INNATE IMMUNITY MECHANISMS
Edited by Yona Keisari and Itzhak Ofek

Volume 480
BIOLOGY OF THE MAMMARY GLAND
Edited by Jan A. Mol and Roger A. Clegg

Volume 481
ELASTIC FILAMENTS OF THE CELL
Edited by Henk L. Granzier and Gerald H. Pollack

Volume 482
CHROMOGRANINS: Functional and Clinical Aspects
Edited by Karen B. Helle and Dominique Aunis

Volume 483
TAURINE 4: Taurine and Excitable Tissues
Edited by Laura Della Corte, Ryan H. Huxtable, Giampetro Sgaragli, and Keith F. Tipton

Volume 484
PHYLOGENETIC PERSPECTIVES ON THE VERTEBRATE IMMUNE SYSTEM
Edited by Gregory Beck, Manickam Sugumaran, and Edwin L. Cooper

---

A Continuation Order Plan is available for this series. A continuation order will bring delivery of each new volume immediately upon publication. Volumes are billed only upon actual shipment. For further information please contact the publisher.

# BIOLOGY OF THE MAMMARY GLAND

Edited by

Jan A. Mol
Utrecht University
Utrecht, The Netherlands

and

Roger A. Clegg
Hannah Research Institute
Ayr, Scotland, United Kingdom

KLUWER ACADEMIC / PLENUM PUBLISHERS
New York, Boston, Dordrecht, London, Moscow

Library of Congress Cataloging-in-Publication Data

Biology of the mammary gland / edited by Jan A. Mol and Roger A. Clegg.
p. cm. -- (Advances in experimental medicine and biology ; v. 480)
Papers from the First International Conference on the Biology of the Mammary Gland, held in Tours, France, Sept. 16-18, 1999.
Includes bibliographical references.
ISBN 0-306-46414-4
1. Mammary glands--Physiology--Congresses. 2. Mammary glands--Molecular aspects--Congresses. I. Mol, Jan A. II. Clegg, Roger A. III. International Conference on the Biology of the Mammary Gland (1st : 1999 : Tours, France) IV. Series.

QP188.M3 B56 2000
573.6'79--dc21

00-033102

Proceedings of the European Cooperation in the Field of Scientific and Technical Research (COST 825) Symposium on Mammary Gland Biology, held September 16–18, 1999, in Tours, France

ISBN 0-306-46414-4


233 Spring Street, New York, New York 10013

http://www.wkap.nl/

10 9 8 7 6 5 4 3 2 1

A C.I.P. record for this book is available from the Library of Congress



Printed in the United States of America

# Preface

It is difficult to overstate the evolutionary and functional significance of mammary tissue in biology. Substantial progress has been made by researchers in various disciplines, particularly over the last fifteen years, towards realizing the potential of this tissue to yield powerful experimental models for morphogenesis and tissue development; for cellular differentiation; for the biosynthesis and secretion of proteins, lipids, small molecules and inorganic salts; and for the coordination and regulation of these processes. More recently, the possibility of exploiting the secretory epithelial cells of mammary tissue as "cell factories" has become a reality and the recombinant production by lactating animals of an increasing number of proteins, valuable both in the pharmaceutical and "nutraceutical" fields, is in progress or under development. The fable of the goose that laid the golden egg has given way to the reality of Dolly the sheep and latter-day Dollys yielding milks of staggeringly high value. Alongside this high-profile biotechnology, the traditional role for mankind of lactating ruminant animals remains, as a "renewable" source of foodstuffs for human consumption. Also in this sphere of agricultural production, genetic as well as nutritional technologies are under investigation and exploitation to optimize milk composition for various end-uses - for instance in food process and manufacture. The possibilities of deriving health benefit from the bioactive properties of some of the minor constituents of milk are emerging to counter the highly-publicized negative health impact of excessive consumption of saturated animal fats. In human nutrition and medicine, the mammary gland is both a source of nutrition to the neonate and of potential health threat to the adult female - breast cancer remains the major single cause of female mortality in most developed countries.

For 3 days in the autumn of 1999 (September 16th-18th) a group of around 175 biologists, all sharing a common interest in the mammary gland, gathered together in Tours, France. Individual researchers within the group owed allegiance to many scientific traditions including those of cellular and developmental biology, biochemistry and molecular biology, nutrition, genetics, biotechnology and medicine. In all probability, this was the greatest number of mammary gland biologists ever to have assembled in a single location within Europe - perhaps, the world. The occasion for such a concentration of mammary tissue specialism was the first International Conference on the Biology of the Mammary Gland. The undoubted success of this conference, against a background of such disciplinary diversity, stands as a tribute to the success of the European networking project that gave rise to it.

To trace back to the origin of the idea for the Tours conference it is necessary to explore the genesis of the COST Action (#825) on Mammary Gland Biology, since the two were conceptually interconnected from the beginning, although the location in Tours only emerged much later. For the benefit of readers not familiar with the European Union systems, structures and programmes for the Europe-wide support of science, COST is one of those acronymic programmes (for cooperation in the field of scientific and technical research). COST exists to nurture and support networks of scientists with the objective of stimulating scientific collaboration, cooperation and communication across, and somewhat beyond, the EU. The COST 825 seed was sown paradoxically far from Europe, in New Hampshire, in 1991. In the spartan New England conditions of a Gordon Conference on Mammary Gland Biology, conversation among a group of scientists turned (certainly not for the first time in the distinguished history of the Gordon Conferences) to a consideration of an enigma - why do European scientists in this field apparently have to cross the Atlantic ocean to find the opportunity to meet together and to discuss shared interests? Michèle Ollivier-Bousquet was in the midst of this debate, and during the remaining days of the Gordon Conference, began discretely to speak with colleagues about her vision of a European Mammary Gland Biology group. The seed grew, and in 1994 a proposal from a consortium of European mammary gland biologists, led by Ollivier-Bousquet, was submitted to the EU COST New Actions Committee. In due course, Action #825 - Mammary Gland Biology - was officially inaugurated in June 1995, with initial support for 5 years.

A core aim of the COST 825 network has been to bring together the different scientific traditions which, while sharing an interest in mammary gland biology, have historically remained separate and even remote from one another. The medical (breast cancer) and the agricultural (lactation and animal production) traditions represent the extremes of this separation. As the end of the century approached, all the infrastructure of a mature network was well-established, under the umbrella of COST, through meetings, workshops, a newsletter, a web-site, mailing-lists, and a searchable database of workers in the field. It seemed a good moment to draw together the varied scientific threads of the network and to join with colleagues worldwide in attempting to define just where we had reached in the field of mammary gland biology and the directions in which we were heading. Thus, although the Tours Conference owed its existence to the COST European network, the contributions of scientists from many countries both inside and outside the EU/COST orbit, ensured that the dimensions of this defining moment were truly international.

Collected together in this volume, the invited communications presented at the Tours International Conference on the Biology of the Mammary Gland give a unique glimpse of our understanding, at the cutting edge of a variety of disciplines, of this versatile and extraordinary tissue, at the birth of the twenty-first century.

Jan Mol
Roger Clegg
*April 2000*

# Acknowledgements

The chapters collected together here give an account of the proceedings of the first International Conference on the Biology of the Mammary Gland. The generous financial support for this Conference by the following not-for-profit bodies is acknowledged: Conseil Régional de la Région Centre, France; European Union COST programme; International Union of Biochemistry and Molecular Biology; National Institute of Agricultural Research (INRA), France. In addition, the Conference enjoyed commercial sponsorship from Atelier Coperta, Gent, Belgium; Intervet International B.V., The Netherlands; Lipha Santé, France; Merck Sharp & Dohme-Chibret, France; Monsanto Services Int SA, Belgium; Rhône Poulenc Animal Nutrition, France; Roche Pharma, France.

Numerous individual members of the COST Action for Mammary Gland Biology contributed to the success of the Conference; their inputs were ably co-ordinated by an Organizing Committee consisting of Antonella Baldi (Italy), Paul Edwards (UK), Anne-Marie Massart-Leen (Belgium), Michèle Ollivier-Bousquet (France), Armand Sanchez (Spain) and Dieter Schams (Germany). Finally, it was the job of the local organizer, Charles Couet, to translate the will of the Organizing Committee into the reality of a Conference - a task that he undertook with great zeal, efficiency and good humour, and with valuable support from the Université François Rabelais, Tours. The staff at the Centre Vinci, Tours, not only ensured that the day-to-day activities and facilities of the Conference ran without any hitches but also facilitated the collection of many of the manuscripts leading to this publication.

The editors express their thanks to the contributors of individual chapters for their cooperation and to Joanna Lawrence of the London Office staff at Kluwer Academic/Plenum Publishers for her help and support in bringing this volume to publication.

# Contents

1.

# Fibroblast Growth Factor Signalling and Cyclin D1 Function are Necessary for Normal Mammary Gland Development during Pregnancy

*A transgenic mouse approach*

Vera Fantl, Anna Creer, Christian Dillon, Janine Bresnick, David Jackson, [2]Paul Edwards, Ian Rosewell and Clive Dickson

*Imperial Cancer Research Fund, Lincoln's Inn Fields, London , UK, [2]Dept. of Pathology, University of Cambridge, UK.*

Key words: Fibroblast growth factor, fibroblast growth factor receptor, cyclin D1, mammary, dominant negative receptor

Abstract: A number of growth factors, growth factor receptors and cell cycle regulatory proteins have been implicated in the genesis of mammary carcinomas both in animal models as well as in human breast tumour samples. Studies on the development of the mammary gland has revealed that several of the proto-oncogenes, or their closely related gene-family members, have a function in the normal growth and differentiation of the gland. In this review the role of fibroblast growth factor signalling and the critical requirement for the cell cycle regulator, cyclin D1 is discussed with respect to their normal function in mammary gland development and abnormal role in mammary carcinogenesis.

## 1. INTRODUCTION

The search for genetic alterations that give rise to mammary cancer, in animal models or humans, has revealed several growth factors, growth factor receptors, as well as cell cycle regulators, that can act as proto-oncogenes when they are over-expressed or constitutively activated through somatic mutation. In the mouse, where mouse mammary tumour virus has been shown to act as a powerful insertional mutagen, there is a preferential activation of dominant acting growth and differentiation factors, such as fibroblast growth factors (FGFs) or members of the *wnt* gene family (reviewed[1,2]). In human breast cancer, the inappropriate activation of cell cycle genes such as *c-myc* and *cyclin D1* often in conjunction with loss of tumour suppressor genes appears to be more common, although mutation or over-expression of the epidermal growth factor-tyrosine receptor gene family also plays a significant role (reviewed[3]).

During the last decade, there has been a massive expansion in understanding of the genes involved in normal mammalian development which has implicated many proto-oncogenes, as important signalling molecules in pattern formation during organogenesis[4]. The mammary gland is unusual in that a large proportion of its development occurs after birth, through puberty and during pregnancy and lactation. A key question was therefore whether, when correctly regulated, the oncogenes implicated in mammary cancer are involved in normal mammary gland development. We have examined this question with regard to *FGF-3* (originally called *int-2*) which is inappropriately expressed in many MMTV induced mouse mammary tumours, and cyclin D1 which is abnormally elevated in 40% of human breast cancers.

## 2. A ROLE FOR FIBROBLAST GROWTH FACTORS IN LOBULOALVEOLAR DEVELOPMENT

Fibroblast growth factors are pleiotropic cell to cell signalling molecules that can act as broad spectrum mitogens, promote cell migration or modulate cellular differentiation depending on their context (reviewed[5,6]). These secreted ligands signal by binding to high affinity cell surface receptors, inducing their dimerization and subsequent activation of their cytoplasmic tyrosine kinase (reviewed[6,7]). *FGF*-receptors contain two or three immunoglobin-like loops in the extracellular domain, a transmembrane element as well as the cytoplasmic tyrosine kinase. There are four FGF receptor genes (*FGFR-1* to *FGFR-4*), in mammals, although alternative splicing of *FGFR-1*, *FGFR-2* and*FGFR-3*, yields seven proto-type receptors with different FGF binding specificities and tissue distributions[8,9,10]. The two membrane proximal Ig-like domains constitute the ligand binding site, and it is part of the third Ig-loop which is encoded by an alternative exon (designated IIIb or IIIc) that defines FGF binding specificity.

As FGF-3 is not found in the normal mammary gland, it seemed likely that an FGF with a receptor binding specificity overlapping that of FGF-3 was important for gland development. To determine the role such an FGF might have, we sought to compromise FGF signalling in the mammary epithelium and assess the effect this had during puberty and pregnancy.

FGF-3 binds to the IIIb isoform of both *FGFR*-1 and FGFR-2, and the latter receptor was selected for expression as a dominant negative receptor in the mammary epithelium of transgenic mice[11]. The dominant negative receptor gene was constructed by removing the cytoplasmic tyrosine kinase domain from an appropriate mouse cDNA, leaving the extracellular and transmembrane coding domains intact. Hence, ligand mediated dimerization of the truncated receptor protein potentiates the sequestration of wild-type endogenous receptors as inactive heterodimers. Expression in the mammary epithelium was achieved using the MMTV promoter, which has a particularly high expression in this cell lineage, especially during pregnancy.

Whole mount preparations and histological examination of the mammary glands from adult virgin females showed no discernible abnormalities associated with transgene expression[12]. However, by mid-pregnancy, the mammary glands of transgenic mice showed a distinct reduction in lobuloalveolar development that was maintained into lactation, although those alveolar lobules that developed showed a normal histology, and were filled with milk (Fig. 1). These findings indicate that FGF signalling is an important facet of lobuloalveolar development during pregnancy. Moreover, it explains why inappropriate and deregulated FGF-3 expression in this tissue leads to mammary hyperplasia since this would institute autocrine stimulation of epithelial cell growth.

## 3. CYCLIN D1 IS REQUIRED FOR MAMMARY GLAND GROWTH AND DIFFERENTIATION

The *FGF-3* locus maps to human chromosome 11 band q13 and is amplified in approximately 15% of breast tumour DNA samples[12], although the gene is very rarely expressed in these tumours. This suggested that amplification of *FGF3* was fortuitous, and that another gene on the amplicon was driving tumour development. The cyclin D1 gene (*CCND1*) was identified as a likely candidate oncogene since it was closely linked to and always co-amplified with *FGF-3*[14]. Furthermore, *CCND1* was involved in chromosomal translocations for a subset of lymphomas and parathyroid adenomas further supporting its potential as an oncogene[15]. Moreover, immunohistochemical examination of breast cancer samples revealed that cyclin D1 is expressed at elevated levels in 40% of tumour samples, considerably more than the subset accounted for by gene amplification[16,17].

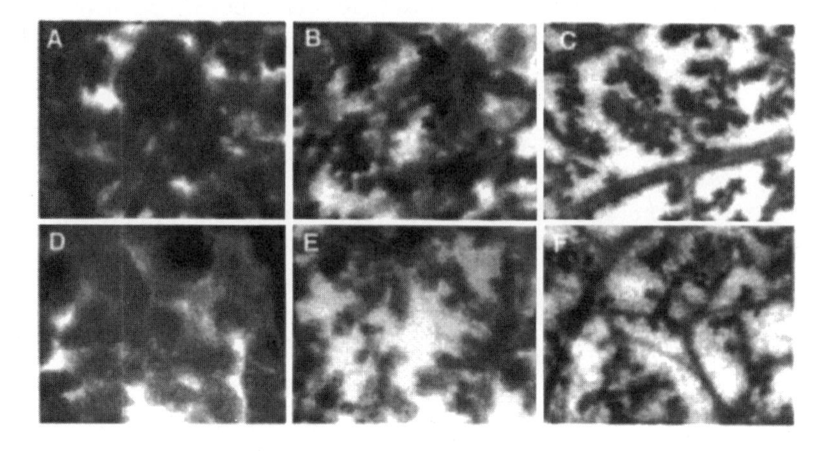


Figure 1. Whole mount preparations of mammary glands from control and transgenic mice. The glands were taken at 17.5 days of pregnancy (A, B and C) and 1 day post partum (D, E and F) from control (A and D), DN-FGFR-2 (IIIb) (B and E), and *Cyl-1*$^{-/-}$ mice (C and F). A reduced lobuloalveolar density is apparent in the mammary glands from pregnant DN-FGFR-2(IIIb) and *Cyl-1*$^{-/-}$ mice. However, in the 1 day post partum samples the alveoli in the mammary gland of the DN-FGFR-2(IIIb) mice are distended with milk as are the controls, but those of the *Cyl-1*$^{-/-}$ gland remain contracted.

Other studies have shown that inappropriate expression of cyclin D1 as a transgene in the mouse mammary epithelium caused a late but significant incidence of mammary hyperplasia, strengthening its potential involvement in mammary cancer[18].

Cyclin D1 is one of three closely related proteins that act as regulatory subunits for the cyclin dependent kinases (CDK) CDK4 and CDK6 (reviewed[19]). The CDK/cyclinD complexes are activated during the G1 phase of the cell division cycle and the primary targets of the kinase complexes are the retinoblastoma (Rb) family of proteins (reviewed[20,21]). Phosphorylation of Rb related proteins appear to relieve their repressive effect on a number of transcription factors that regulate genes necessary for DNA synthesis and subsequent entry into cell division. Thus an increase in cyclin D1 expression could perturb the negative regulation on cell division imposed by Rb, which in turn could result in abnormal cell proliferation, one of the early changes on the path to tumour formation.

To investigate the role of cyclin D1 in vivo, and assess the level of redundancy among the three D-type cyclin family members, mutant mice lacking the protein have been generated[22,23]. These *Cyl-1*$^{-/-}$ mice are viable and fertile but have a retinopathy and, demonstrate defective mammary gland development during pregnancy. While the mammary glands of virgin cyclin D1 deficient mice are comparable to normals, by 7.5 days of pregnancy there is a distinct retardation of lobuloalveolar development, which continues to term (Fig. 1). Moreover, there is also a very poor lactogenic response, with a delayed appearance of milk proteins that are present in reduced amounts compared with control glands. Transplantation of mammary epithelium from mutant mice into the empty fat pad of histocompatible wild-type mice showed the same mammary gland phenotype as the *Cyl-1*$^{-/-}$ mice demonstrating that the defect in mammary gland development is a property of its epithelial component[24].

# 4. CONCLUSIONS

Two oncogenes identified in mammary cancer studies have now been shown to have crucial functions in the growth and differentiation of the normal mammary gland. The results clearly demonstrate a need for FGF signalling to produce the normal density of lobuloalveolar growth during pregnancy. However, the alveoli that do form when FGF signalling is deficient appear to function normally, and mothers are able to suckle their young. However, the litters produced are runted in appearance due to an insufficient supply of milk. The mammary glands of *Cyl-1*$^{-/-}$ mice also showed a reduction of lobuloalveolar development, but the alveoli also gave a poor lactogenic response, and although milk proteins can be detected biochemically, histologically the alveoli appear to be contracted with little luminal milk apparent (Fig. 1). Moreover, *Cyl-1*$^{-/-}$ mothers fail to nurture their young.

These studies demonstrate that signalling and cell cycle regulatory molecules involved in normal mammary gland growth and differentiation during pregnancy can also act as oncogenes when inappropriately expressed, reinforcing the strong links between tumour and developmental biology.

# 5. REFERENCES

1. Nusse, R. Insertional mutagenesis in mouse mammary tumorigenesis. Curr. Topics Microbiol. Immunol. 171, 44-65. 1991.

2. Peters, G. Inappropriate expression of growth factor genes in tumors induced by mouse mammary tumor virus. Sem. Virol. 2, 319-328. 1991.
3. Callahan, R. and Campbell, G. Mutations in human breast cancer: an overview. J. Natl. Cancer Inst. 81, 1780-1786. 1989.
4. Yamaguchi, T. P. and Rossant, J. Fibroblast growth-factors in mammalian development. Curr Opin Genet Dev. 5, 485-491. 1995.
5. Basilico, C. and Moscatelli, D. The FGF family of growth-factors and oncogenes. Adv Cancer Res. 59, 115-165. 1992.
6. McKeehan, W. L., Wang, F. and Kan, M. The heparan-sulfate fibroblast growth-factor family - diversity of structure and function. Prog Nucleic Acid Res Mol Biol. 59, 1998.
7. Johnson, D. and Williams, L. Structural and functional diversity in the FGF receptor multigene family. Adv Cancer Res. 60, 1-41. 1993.
8. Ornitz, D., Xu, J., Colvin, J., McEwen, D., MacArthur, C., Coulier, F., Gao, G. and Goldfarb, M. Receptor specificity of the fibroblast growth-factor family. J Biol Chem. 271, 15292-15297. 1996.
9. Orr-Urtreger, A., Bedford, M., Burakova, T., Arman, E., Zimmer, Y., Yayon, A., Givol, D. and Lonai, P. Developmental localization of the splicing alternatives of fibroblast growth-factor receptor-2 (FGFR2). Dev Biol. 158, 475-486. 1993.
10. Peters, K., Werner, S., Chen, G. and Williams, L. Two FGF receptor genes are differentially expressed in epithelial and mesenchymal tissues during limb formation and organogenesis in the mouse. Development. 114, 233-243. 1992.
11. Mathieu, M., Chatelain, E., Ornitz, D., Bresnick, J., Mason, I., Kiefer, P. and Dickson, C. Receptor-binding and mitogenic properties of mouse fibroblast-growth-factor-3 - modulation of response by heparin. J Biol Chem. 270, 24197-24203. 1995.
12. Jackson, D., Bresnick, J., Rosewell, I., Crafton, T., Poulson, R., Stamp, G. and Dickson, C. Fibroblast growth factor signalling has a role in lobuloalveolar development of the mammary gland. J. Cell Science. 110, 1261-1268. 1997.
13. Fantl, V., Richards, M., Smith, R., Lammie, G., Johnstone, G., Allen, D., Gregory, W., Peters, G., Dickson, C. and Barnes, D. Gene amplification on chromosome band 11q13 and oestrogen receptor status in breast cancer. Eur. J. Cancer. 26, 423-429. 1990.
14. Lammie, G. A., Fantl, V., Smith, R., Schuuring, E., Brookes, S., Michalides, R., Dickson, C., Arnold, A. and Peters, G. D11S287, a putative oncogene on chromosome 11q13, is amplified and expressed in squamous cell and mammary carcinomas and linked to BCL-1. Oncogene. 6, 439-444. 1991.
15. Lammie, G. A. and Peters, G. Chromosome 11q13 abnormalities in human cancer. Cancer Cells. 3, 413-420. 1991.
16. Bartkova, J., Lukas, J., Muller, H., Lutzhoft, D., Strauss, M. and Bartek, J. Cyclin D1 protein expression and function in human breast-cancer. Int J Cancer. 57, 353-361. 1994.
17. Gillett, C., Fantl, V., Smith, R., Fisher, C., Bartek, J., Dickson, C., Barnes, D. and Peters, G. Amplification and overexpression of cyclin-D1 in breast-cancer detected by immunohistochemical staining. Cancer Res. 54, 1812-1817. 1994.
18. Wang, T. C., Cardiff, R. D., Zukerberg, L., Lees, E., Arnold, A. and Schmidt, E. V. Mammary hyperplasia and carcinoma in MMTV-cyclin D1 transgenic mice. Nature. 369, 669-671. 1994.
19. Sherr, C. D-type cyclins. TIBS. 20, 187-190. 1995.
20. Sherr, C. Cancer cell cycles. Science. 274, 1672-1677. 1996.
21. Weinberg, R. The retinoblastoma protein and cell cycle control. Cell. 81, 323-330. 1995.

22. Fantl, V., Stamp, G., Andrews, A., Rosewell, I. and Dickson, C. Mice lacking cyclin D1 are small and show defects in eye and mammary-gland development. Genes Dev. 9, 2364-2372. 1995.
23. Sicinski, P., Donaher, J. L., Parker, S. B., Li, T. S., Gardner, H., Haslam, S. Z., Bronson, R. T., Elledge, S. J. and Weinberg, R. A. Cyclin D1 provides a link between development and oncogenesis in the retina and breast. Cell. 82, 621-630. 1995.
24. Fantl, V., Edwards, P., Steel, J., Vonderhaar, B. and Dickson, C. Impaired Mammary Gland Development in Cyl-1-/- Mice During Pregnancy and Lactation is Epithelial Cell Autonomous. Devel. Biol. 212, 1-11. 1999.

2.

# Hepatocyte Growth Factor and Neuregulin in Mammary Gland Cell Morphogenesis

Catherin Niemann, Volker Brinkmann, Walter Birchmeier
*Imperial Cancer Research Fund, Lincolns Inn Fields, London, UK and Max Delbrueck Center for Molecular Medicine, Berlin, Germany*

Key words: Hepatocyte Growth Factor, neuregulin, mammary epithelial cells, receptor tyrosine kinase signalling

Abstract: Organ culture and transplantation experiments in the early 1960s and 1970s have demonstrated that growth and morphogenesis of the epithelium of the mammary gland are controlled by mesenchymal-epithelial interactions. The identification of molecules that provide the essential signals exchanged in mesenchymal-epithelial interactions is an area of active research. Recent evidence suggests that morphogenic programs of epithelia can be triggered by mesenchymal factors that signal via tyrosine kinase receptors. This review concentrates on the effects of two mesenchymal factors, Hepatocyte Growth Factor/Scatter Factor and neuregulin, on morphogenesis and differentiation of mammary epithelial cells in vitro and signalling pathways involved during morphogenesis of mammary epithelial cells.

## 1. INTRODUCTION

During development, differentiation and morphogenesis of epithelial organs are essential processes and lead to an astounding array of different structures. Signals for growth and differentiation of epithelia are frequently provided by neighbouring mesenchymal cells. The molecular basis for signals exchanged during mesenchymal-epithelial interaction is provided by paracrine signalling systems consisting of epithelial tyrosine kinase receptors and their mesenchymal ligands. Hepatocyte Growth Factor/ Scatter Factor (HGF/SF) and the c-met receptor, neuregulin and the members of the c-erbB receptor family, Glial cell line Derived Neurotrophic factor (GDNF) and the c-ret receptor as well as the Fibroblast Growth Factors (FGFs) and their receptors constitute such paracrine signalling systems. During recent years, the analysis of mice that carry targeted mutations or express transdominant receptors have demonstrated an important role for epithelial tyrosine kinase receptors and their mesenchymal ligands for epithelial development[1].

We have previously identified an important function for HGF/SF and neuregulin in the development of the mammary gland by using the whole organ culture system.

In this review we will concentrate on the effects HGF/SF and neuregulin exert on mammary epithelial cells cultured on matrigel. We observe that these two growth factors elicit fundamentally distinct morphogenic responses. HGF/SF induces the formation of tubular structures, whereas neuregulin evokes the formation of alveolar structures. The structures generated in cell culture by these two growth factors resemble the ones which are formed during mammary gland development in vivo. We used this cell culture system to further analyse the signalling pathways that become activated by HGF/SF or neuregulin and lead to the different morphogenic responses of mammary epithelial cells.

## 2. THE EFFECT OF HGF/SF AND NEUREGULIN ON MORPHOGENESIS AND DIFFERENTIATION OF MAMMARY EPITHELIAL CELLS

During mouse mammary gland development HGF/SF and neuregulin are expressed by the mesenchym that surrounds the epithelial cells which produce receptors for both ligands. Interestingly, HGF/SF is highly expressed in the mammary gland during puberty whereas neuregulin is specifically highly expressed during pregnancy of the mouse[2]. In experiments using whole organ cultures of mouse mammary glands HGF/SF promotes branching of ductal trees and inhibits terminal differentiation, as assessed by the expression of milk proteins. Neuregulin stimulates lobulo-alveolar differentiation in this culture system and promotes the expression of milk components such as β-casein[2]. These results indicate that different growth factors elicit distinct responses in organ culture of the mammary gland and raise the questions of whether different cellular populations of the mammary epithelium are affected by these factors and how these different responses are evoked on a molecular level. On one hand, the organ culture system provides a good model for the analysis of morphogenic events but on the other hand, its complex cellular composition precludes a biochemical analysis of signalling cascades activated by morphogenic factors.

In the search for an appropriate way to examine signalling pathways responsible for morphogenic changes, we found that mammary epithelial cells (EpH4) grown on matrigel are affected by two growth factors, HGF/SF and neuregulin.

EpH4/K6 mouse mammary epithelial cells[3] cultured on matrigel in the presence of hormones (insulin, prolactin and hydrocortisone) form small spheroids and secrete milk proteins[4-6]. These aggregates consist mainly of 3 to 7 single-layered epithelial cells surrounding a small lumen. When HGF/SF was added, EpH4/K6 cells started to grow faster and built long tubular structures that could reach a length of up to several millimeters. Histological analysis revealed that the tubular structures induced by HGF/SF consist of several layers of cells lining the elongated lumina. In contrast, neuregulin induced an entirely different response in EpH4/K6 cells grown on matrigel: large alveolar-like structures or lobulo-alveolar-like aggregates are formed. Histological analysis demonstrated that neuregulin-induced alveoli and lobulo-alveoli consist of single-layered cells surrounding large lumina. We also tested other growth factors for their morphogenic effect on this mammary epithelial cells grown on matrigel: epidermal growth factor (EGF) stimulated growth, whereas keratinocyte growth factor (KGF) had only a moderate effect on proliferation. However, neither growth factor, EGF or KGF, elicited any particular morphogenic response: tubular structures or large alveoli were not induced[7].

We went on to analyse the distribution of the cell adhesion molecule E-cadherin and the production of the differentiation marker β-casein by immunofluorescence-staining (Fig. 1). In control aggregates, β-casein production was high but E-cadherin was evenly distributed along all cell surfaces, suggesting only moderate polarization of the cells (Fig 1A). HGF/SF reduced β-casein expression in the tubular structures as assessed by the reduction of immunofluorescence-staining intensity (corroborated by a decrease of mRNA expression for casein, observed by northern blotting). E-cadherin was largely distributed along the whole cell surface (Fig 1B). However, in accordance with the more pronounced polarization of cells in neuregulin-induced alveoli, E-cadherin was predominantly located at lateral membranes. β-Casein was strongly expressed in these cells as assessed by the staining intensity observed by immunofluorescence analysis (Fig 1C)[7].

Thus, EpH4 mammary gland epithelial cells respond in a distinct manner when stimulated with HGF/SF or neuregulin. While HGF/SF induces branched multilayered tubes and inhibits the production of milk proteins, neuregulin causes the formation of alveoli and lobulo-alveolar-like aggregates which often consist of a monolayered epithelium and possess a higher degree of polarization and differentiation. Both growth factors can induce complex morphogenic programs in mammary gland epithelial cells, which resemble those observed in the mammary gland during postnatal development.

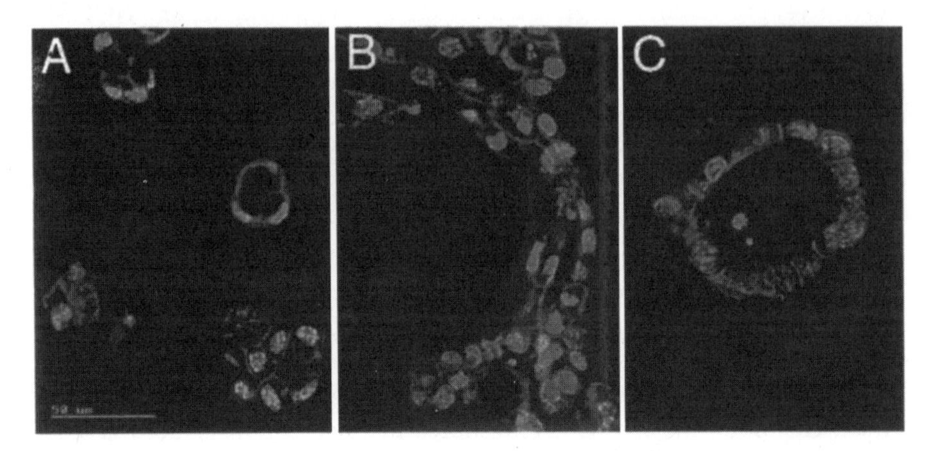


Figure 1. Scanning confocal micrograph of EpH4/K6 aggregates treated with HGF/SF (B) or neuregulin (C); controls are shown in A.

## 3. SIGNALLING PATHWAYS ACTIVATED BY HGF/SF AND NEUREGULIN IN MAMMARY EPITHELIAL CELLS

We used this cell culture system to further analyse components of the signalling cascades activated by HGF/SF and its receptor c-met as well as cascades regulated by neuregulin and the members of the c-erbB family of receptor tyrosine kinases. EpH4/K6 cells were transfected with the cDNA of the recently identified substrate of the c-met receptor, Gab1 [8]. When grown on matrigel, the transfected cells produced tubular structures, even in the absence of HGF/SF. These structures were not observed with cells transfected with a control plasmid only. The histology of the Gab1-induced structures was identical to those induced by HGF/SF.

In order to investigate which member of the c-erbB receptor family or what receptor combination is responsible for the neuregulin-induced phenotype of mammary epithelial cells, we performed transfection experiments with hybrid receptor molecules. Neuregulin signals are mediated by direct interaction with the high affinity receptors c-erbB3 or c-erbB4. In addition, c-erbB2 acts as an important coreceptor for the transmission of neuregulin[9-11]. We transfected EpH4/K6 cells with cDNAs for hybrid receptors consisting of extracellular portion of trk (the receptor for nerve growth factor, NGF) and intracellular c-erbBs. Remarkably, clones of EpH4 cells which stably expressed the hybrid trk/c-erbB2 receptor produced alveolar-like structures in the presence of NGF. These structures were found to be histologically identical to the neuregulin-induced alveoli. Control transfectants expressing pSVneo, trk or an trk/KGF hybrid receptor did not show this morphogenic response in matrigel. Similarly, stable cell clones expressing a trk/c-erbB4 hybrid receptor did not form alveolar structures in the presence of NGF.

We then examined EpH4/K6 cells expressing mutants of trk/c-erbB2 to identify essential docking sites in the c-erbB2 receptor that elicit this morphogenic response. Recently, five phosphorylated tyrosine residues in the COOH-terminal substrate-binding region have been identified as important docking sites for signalling substrates acting downstream of the c-erbB2 [12,13]. We found that a hybrid receptor trk/c-erbB2 containing only the first four tyrosines was sufficient to induce alveolar structures when overexpressed in EpH4 cells and activated by NGF. In contrast, a hybrid receptor of trk/c-erbB2 containing only the last one of these five tyrosines was not capable of inducing alveolar morphogenesis. These findings demonstrate that an activated trk/c-erbB2 is sufficient to elicit a specific morphogenic response in mammary epithelial cells, the formation of alveoli and lobulo-alveolar structures, even in the absence of active c-erbB3 and c-erbB4 coreceptors. Furthermore, our results indicate that only the first four (Y1028, Y1144, Y1201 and Y1226/27) but not the last tyrosine (Y1253) are important for mediating the neuregulin-induced morphogenic response.

In order to examine which of the known signalling cascades mediates the morphogenic responses evoked by HGF/SF and neuregulin we used inhibitors that interfere with specific signalling pathways. Addition of the PI3 kinase inhibitors, wortmannin and LY294002, to EpH4/K6 cells reduced the formation of elongated structures, whereas the MAPK kinase inhibitor PD98059 had no effect on the stimulation of elongated aggregates. In contrast, PD98059 quenched the induction of larger alveoli, whereas wortmannin was without effect on the formation of larger alveolar structures[7].

These results suggest that the formation of tubular structures induced by HGF/SF requires pathways involving Gab1 and PI3 kinase, whereas for the induction of alveoli and lobulo-alveolar structures by neuregulin, an activated c-erbB2 receptor is sufficient and most likely the activity of MAPK kinase is required.

# 4. DISCUSSION

Morphogenic activities of various growth factors on mammary gland development have been examined previously; for instance it has been shown that EGF (or TGF-α) and TGF-β can influence ductal as well as alveolar development[14-21]. These factors are predominantly expressed in an autocrine fashion by epithelial cells and are produced throughout mammary gland postnatal development. Mesenchymal-epithelial interactions are essential for pre- and postnatal development of the mammary gland[22-26], so we have focused on the functional analysis of mesenchymal ligands, HGF/SF and neuregulin, and their epithelial tyrosine kinase, c-met and the c-erbBs. Interestingly, both growth factors are highly expressed within the mammary gland during specific developmental phases: HGF/SF is high expressed during puberty, whereas neuregulin is strongly expressed only during pregnancy[2].

It has already been shown that epithelial cells isolated from kidney, breast and other organs respond to HGF/SF by morphological differentiation, the formation of branched tubules when grown in a collagen matrix[27-32]. Here we used matrigel as a substrate to investigate effects of HGF/SF and neuregulin on functional differentiation, that is, expression of milk proteins. HGF/SF inhibits the expression of β-casein, indicating that a morphogenic program without concomitant functional differentiation is activated. This is consistent with developmental processes in vivo: milk production is blocked during branching of the ductal tree during puberty. In the whole organ culture system of the mouse mammary gland, HGF/SF inhibits the expression of milk components whereas neuregulin stimulates the production of milk components[2]. We did not observe an increase of β-casein expression by neuregulin in our cell culture system. It seems possible that neuregulin promotes an already programmed differentiation process leading to alveolar morphogenesis and concomitant production of milk components in the whole organ cultures of the mammary gland whereas such a complex differentiation program does not exist in our EpH4 cell system.

Interestingly, neuregulin induced a redistribution of the cell adhesion molecule E-cadherin to the lateral membranes of EpH4/K6 cells. Recently, this redistribution of cadherins has also been reported for other epithelial cell lines[33]. In addition, it has been demonstrated that neuregulin can promote the formation of ring-shaped multicellular structures in various cancer cell lines cultured on plastic[33].

Several tyrosine kinase receptors have been reported to affect epithelial cells (e.g. trk, c-ros, KGFR) but only c-met was found to induce tubulogenesis[34]. Although a variety of substrates were found to bind to tyrosine phosphorylation sites in the COOH-terminus of c-met[8,35,36], a substrate which can mediate the signal responsible for branching morphogenesis, Gab1, has only recently been identified[37,38]. It has previously demonstrated that Gab1 is the predominant protein phosphorylated following activation of the c-met receptor in epithelial cells[39]. The c-met receptor can signal through ras and PI3 kinase[35,40,41]. Signalling through ras, resulting in cellular growth, requires the substrate Grb2, which binds to tyrosine residue Y1254 of c-met (i.e. the second tyrosine residue of the bidentate docking site[35,36,39]. Motility and morphogenic responses evoked by an active c-met receptor require Gab1, which binds strongly to Y1347 (the first residue of the bidentate docking site[37,39]. Gab1 can also bind to Y1354, but in the presence of Grb2 binding to Y1347 is preferred and stabilised[39]. Furthermore, it has recently been shown that the pleckstrin homology domain of Gab1 is crucial for its subcellular localization to cell-cell contacts as well as for the morphological response in MDCK cells[38]. Gab1 harbours three potential PI3 kinase-binding sites. Indeed, we found that inhibitors of PI3 kinase, wortmannin and LY294002, specifically reduce HGF/SF-induced formation of elongated structures[7]. Recently it has been shown that LY294002 could inhibit the localization of Gab1 to areas of cell-cell contacts[38]. These data indicate that tubular and branching morphogenesis depends on pathways involving PI3 kinase. In contrast, the neuregulin-induced formation of larger alveoli could be inhibited by a MAPK kinase inhibitor, PD98059. Thus, components of the ras/MAPK kinase pathway are important for alveolar development of mammary epithelial cells. It will now be important to identify which substrates or substrate combinations elicit the alveolar morphogenesis through the activated c-erbB2 receptor.

# 5. REFERENCES

1. Birchmeier C, Birchmeier W. Molecular aspects of mesenchymal-epithelial interactions. Annu Rev Cell Biol 9, 511-540, 1993.
2. Yang Y, Spitzer E, Meyer D, Sachs M, Niemann C, Hartmann G, Weidner KM, Birchmeier C, Birchmeier W. Sequential requirement of hepatocyte growth factor and neuregulin in the morphogenesis and differentiation of the mammary gland. J Cell Biol 131, 215-226, 1995.
3. Reichmann E, Ball R, Groner B, Friis RR. New mammary epithelial and fibroblastic cell clones in coculture form structures competent to differentiate functionally. J Cell Biol 108, 1127-1138, 1989.
4. Wicha MS, Lowrie G, Kohn E, Bagavandoss P, Mahn T. Extracellular matrix promotes mammary epithelial growth and differentiation in vitro. Proc Natl Acad Sci USA 79, 3213-3217, 1982.
5. Lin CQ, Bissell MJ. Multi-faceted regulation of cell differentiation by extracellular matrix. FASEB J 7, 737-743, 1993.

6. Streuli CH, Schmidhauser C, Bailey N, Yurchenco P, Skubitz AP, Roskelley C, Bissell MJ. Laminin mediates tissue-specific gene expression in mammary epithelia. J Cell Biol 129, 591-603, 1995.
7. Niemann C, Brinkmann V, Spitzer E, Hartmann G, Sachs M, Naundorf H, Birchmeier W. Reconstitution of mammary gland development in vitro: Requirement of c-met and c-erbB2 signalling for branching and alveolar morphogenesis. J Cell Biol 143, 533-545, 1998.
8. Weidner KM, Sachs M, Riethmacher D, Birchmeier W. Mutation of juxtamembrane tyrosine residue 1001 suppresses loss-of-function mutations of the met receptor in epithelial cells. Proc Natl Acad Sci. USA 92, 2597-2601, 1995.
9. Plowman GD, Green JM, Culouscou JM, Carlton GW, Rothwell VM, Buckley S. Heregulin induces tyrosine phosphorylation of HER4/p180erbB4. Nature. 366, 473-475, 1993.
10. Carraway K3, Cantley LC. A neu acquaintance for erbB3 and erbB4: a role for receptor heterodimerization in growth signalling. Cell 78, 5-8, 1994.
11. Carraway K3, Burden SJ. Neuregulins and their receptors. Curr Opin Neurobiol 5, 606-612, 1995.
12. Ben-Levy R, Paterson HF, Marshall CJ, and Yarden Y. A single autophosphorylation site confers oncogenicity to the Neu/ErbB-2 receptor and enables coupling to the MAP kinase pathway. EMBO J 13:3302-3311, 1994.
13. Dankort DL, Wang Z, Blackmore V, Moran MF, Muller WJ. Distinct tyrosine autophosphorylation sites negatively and positively modulate neu-mediated transformation. Mol Cell Biol 17, 5410-5425, 1997.
14. Silberstein GB, Daniel CW. Reversible inhibition of mammary gland growth by transforming growth factor-beta. Science 237, 291-293, 1987.
15. Vonderhaar BK. Local effects of EGF, alpha-TGF, and EGF-like growth factors on lobuloalveolar development of the mouse mammary gland in vivo. J Cell Physiol 132, 581-584, 1987.
16. Coleman S, Silberstein GB, Daniel CW. Ductal morphogenesis in the mouse mammary gland: evidence supporting a role for epidermal growth factor. Dev Biol 127, 304-315, 1988.
17. Jhappan C, Stahle C, Harkins RN, Fausto N, Smith GH, Merlino GT. TGF alpha overexpression in transgenic mice induces liver neoplasia and abnormal development of the mammary gland and pancreas. Cell 61, 1137-1146, 1990.
18. Jhappan C, Geiser AG, Kordon EC, Bagheri D, Hennighausen L, Roberts AB, Smith GH, Merlino G. Targeting expression of a transforming growth factor beta 1 transgene to the pregnant mammary gland inhibits alveolar development and lactation. EMBO J 12, 1835-1845, 1993.
19. Robinson SD, Silberstein GB, Roberts AB, Flanders KC, Daniel CW. Regulated expression and growth inhibitory effects of transforming growth factor-beta isoforms in mouse mammary gland development. Development 113, 867-878, 1991.
20. Snedeker SM, Brown CF, DiAugustine RP. Expression and functional properties of transforming growth factor alpha and epidermal growth factor during mouse mammary gland ductal morphogenesis. Proc Natl Acad Sci USA 88, 276-280, 1991.
21. Pierce D, Johnson MD, Matsui Y, Robinson SD, Gold LI, Purchio AF, Daniel CW, Hogan BL, Moses HL. Inhibition of mammary duct development but not alveolar outgrowth during pregnancy in transgenic mice expressing active TGF-beta 1. Genes Dev 7, 2308-2317, 1993.

22. Kratochwil K. Tissue combination and organ culture studies in the development of the embryonic mammary gland. In Developmental Biology: A Comprehensive Synthesis, Gwatkin RBL, editor. Plenum Press, NY, 315-334, 1987.
23. Sakakura T, Nishizuka Y, Dawe CJ. Mesenchyme-dependent morphogenesis and epithelium-specific cytodifferentiation in mouse mammary gland. Science 194, 1439-1441, 1976.
24. Durnberger H, Kratochwil K. Specificity of tissue interaction and origin of mesenchymal cells in the androgen response of the embryonic mammary gland. Cell 19, 465-471, 1980.
25. Cunha GR, Young P, Hamamoto S, Guzman R, Nandi S. Developmental response of adult mammary epithelial cells to various fetal and neonatal mesenchymes. Epithelial Cell Biol 1, 105-118, 1992.
26. Cunha GR, Hom YK. Role of mesenchymal-epithelial interactions in mammary gland development. J Mamm Gl Biol Neopl 1, 21-35, 1996.
27. Montesano R, Matsumoto K, Nakamura T, Orci L. Identification of a fibroblast-derived epithelial morphogen as hepatocyte growth factor. Cell 67, 901-908, 1991.
28. Montesano R, Schaller G, Orci L. Induction of epithelial tubular morphogenesis in vitro by fibroblast-derived soluble factors. Cell 66, 697-711, 1991.
29. Weidner KM, Sachs M, Birchmeier W. The Met receptor tyrosine kinase transduces motility, proliferation, and morphogenic signals of scatter factor/hepatocyte growth factor in epithelial cells. J Cell Biol 121, 145-154, 1993.
30. Berdichevsky F, Alford D, D'Souza B, Taylor-Papadimitriou J. Branching morphogenesis of human mammary epithelial cells in collagen gels. J Cell Sci 107, 3557-3568, 1994.
31. Brinkmann V, Foroutan H, Sachs M, Weidner KM, Birchmeier W. Hepatocyte growth factor/scatter factor induces a variety of tissue-specific morphogenic programs in epithelial cells. J Cell Biol 131, 1573-1586, 1995.
32. Soriano JV, Pepper MS, Nakamura T, Orci L, Montesano R. Hepatocyte growth factor stimulates extensive development of branching duct-like structures by cloned mammary gland epithelial cells. J Cell Sci 108, 413-430, 1995.
33. Chausovsky A, Tsarfaty I, Kam Z, Yarden Y, Geiger B, Bershadsky AD. Morphogenic effects of neuregulin (Neu Differentiation Factor) in cultured epithelial cells. Mol Biol Cell 9, 3195-3209, 1998.
34. Sachs M, Weidner KM, Brinkmann V, Walther I, Obermeier A, Ullrich A, Birchmeier W. Motogenic and morphogenic activity of epithelial receptor tyrosine kinases. J Cell Biol 133, 1095-1107, 1996.
35. Ponzetto C, Bardelli A, Zhen Z, Maina F, Dalla-Zonca P, Giordano S, Graziani A, Panayotou G, Comoglio PM. A multifunctional docking site mediates signalling and transformation by the hepatocyte growth factor/scatter factor receptor family. Cell 77, 261-27, 1994.
36. Fixman ED, Holgado-Madruga M, Nguyen L, Kamikura DM, Fournier TM, Wong AJ, Park M. Efficient cellular transformation by the Met oncoprotein requires a functional Grb2 binding site and correlates with phosphorylation of the Grb2-associated proteins, Cbl and Gab1. J Biol Chem 272, 20167-20172, 1997.
37. Weidner KM, Di Cesare S, Sachs M, Brinkmann V, Behrens J, Birchmeier W. Interaction between Gab1 and the c-Met receptor tyrosine kinase is responsible for epithelial morphogenesis. Nature 384, 173-176, 1996.
38. Maroun CR, Holgado-Madruga M, Royal I, Naujokas MN, Fournier TM, Wong AJ, Park M. The Gab1 PH domain is required for localization of Gab1 at sites of cell-cell contact

and epithelial morphogenesis downstream from the met receptor tyrosine kinase. Mol Cell Biol 19, 1784-1799, 1999.

39. Nguyen L, Holgado-Madruga M, Maroun C, Fixman ED, Kamikura D, Fournier T, Charest A, Tremblay ML, Wong AJ, Park M. Association of the multisubstrate docking protein Gab1 with the hepatocyte growth factor receptor requires a functional Grb2 binding site involving tyrosine 1356. J Biol Chem 272, 20811-20819, 1997.
40. Hartmann G, Weidner KM, Schwarz H, Birchmeier W. The motility signal of scatter factor/hepatocyte growth factor mediated through the receptor tyrosine kinase met requires intracellular action of Ras. J Biol Chem 269, 21936-21939, 1994.
41. Royal I, Fournier TM, Park M. Differential requirement of Grb2 and PI3-kinase in HGF/SF-induced cell motility and tubulogenesis. J Cell Physiol 173, 196-201, 1997.

3.

# Expression and Localization of Growth Factors during Mammary Gland Development

[1]Fred Sinowatz, [2]Dieter Schams, [2]Annette Plath, [1]Sabine Kölle
*[1]Institute of Veterinary Anatomy, University of Munich, Germany, [2]Institute of Physiology, Technical University of Munich, Weihenstephan, Germany*

Key words: Bovine, mammary gland, growth factors

Abstract: Growth and differentiation of the mammary gland during development and lactation are controlled by complex hormonal mechanisms. Additionally growth factors are supposed to act as local mediators of the hormonally controlled developmental processes. Mammary tissue for this study was obtained from non pregnant control heifers, primigravid heifers (second part of pregnancy), around parturition, during lactation (early and late) and from dry cows. Using RT-PCR and ribonuclease protections assay (RPA) the expression of the following growth factors was studied in the different phases bovine mammary gland development : Insulin-like growth factor I (IGF-I), insulin-like growth factor II (IGF-II), fibroblast growth factor 1 (FGF-1), fibroblast growth factor 2 (FGF-2), transforming growth factor α (TGF-α). Additionally the expression of fibroblast growth factor receptor (FGFR) and growth hormone receptor (GHR) was investigated. The cellular distribution pattern of several of these growth factors and GHR was obtained using Immunocytochemical techniques. The detailed expression and localization pattern of these growth factors are presented and their role in the local regulation of the bovine mammary gland is briefly discussed

## 1. INTRODUCTION

The profound effects of systemic hormones on all aspects of pre- and postnatal development, differentiation and secretory activity of mammary gland are well documented[1,2]. The cyclic events of mammary growth and differentiation that occur during pregnancy and lactation are controlled by various steroid and peptide hormones[3]. At the cellular and tissue level the action of many of these hormones is mediated by growth factors in an intracrine, autocrine, juxtacrine or paracrine manner[1,4]. In the present communication we report on the expression of various growth factors (IGF-I, IGF-II, FGF-1, FGF-2, TGF-α) in the bovine mammary gland during well defined stages of mammogenesis, lactation and involution. Additionally the expression pattern of GHR during these stages will be discussed.

## 2. MATERIAL AND METHODS

### 2.1 Animals, tissue sampling and preparation.

Mammary tissue was obtained from 27 German Brown Swiss heifers and cows at distinct periods of mammary gland development immediately after slaughter. The material studied comprised mammary tissue from non-pregnant and primigravid heifers, cows during lactogenesis, early and late lactation and dry, non pregnant cows. The following classification of animals was established:

I Mammogenesis
- I.1 Mammogenesis-ductal growth (non pregnant heifers)
- I.2 Mammogenesis-lobuloalveolar development during first pregnancy
  - I.2.1: days 194-213 of pregnancy
  - I.2.2: days 255-272 of pregnancy

II Lactogenesis: days 5-11 post partum

III Galactopoiesis
- III.1 peak lactation
- III.2 late lactation

IV Involution (3-4 weeks dried off, non pregnant)

For histology and immunohistochemistry, tissue samples (approximately 15 mm long and 5 mm thick) were fixed in Bouin's solution or in methanol/glacial acid (ratio 1:1, w/v) for 24 h, dehydrated in a graded series of ethanol, cleared in xylene and embedded in paraffin. Serial sections (5 μm) were cut on a Leitz microtome and mounted on glass slides. Following deparaffinization, the presence of growth factors (IGF-I, IGF-II, TGF-α, FGF-1, FGF-2, and receptors (FGFR and GHR) was demonstrated immunohistochemically by the streptavidin-biotin horseradish peroxidase complex (ABC) technique[5]. Controls were performed by a) omission of the primary antibody; b) replacing the antiserum against the growth factors or FGFR and GHR by normal mouse serum (Sigma, Munich) in different concentrations (dilution 1:5; 1:10; 1:100). c) Preabsorption of the primary antibody. It was performed by preincubation of the primary antibody for 24 h at 40°C with the respective recombinant growth factors in a siliconized polypyrene tube before incubation on the slides.

## 2.2 RNA extraction

Total RNA was isolated from bovine tissues using a specifically adapted guanidium thiocyanate/phenol procedure[6]. The RNA pellet was washed in 75% ethanol and diluted in distilled, DEPC treated water. Concentration of RNA was measured by photometry. For RT-PCR of GHR, total RNA was prepared from snap frozen bovine mammary gland tissue using Tripure Reagent (Sigma, Deisenhofen, Germany) according to the manufacturer's instructions.

## 2.3 Reverse transcription polymerase chain reaction (RT-PCR)

For RT-PCR, total cellular RNA was denatured at 65°C for 10 min, quick-cooled on ice and reverse transcribed in a final volume of 20 ml. When RNA had been prepared using snap-frozen tissue and Tripure Reagent, 5 mg RNA were used for the RT reaction. The RT reaction mix included 1x Expand™ Reverse Transcriptase buffer (50 mM Tris-HCl, 40 mM $MgCl_2$, 0.5% Tween 20 (v/v), pH 8.3), 10 mM DTT, 1.6 mg oligo dT, 0.75 mM dNTP, 3 mM $MgCl_2$, 30 units RNase inhibitor and 50 units Expand™ reverse transcriptase. All reagents were purchased from Boehringer Mannheim, Germany. The reaction mix contained 1x reaction buffer (10 mM Tris, 50 mM KCI, 1.5 mM $MgCl_2$, pH 8.3), 0.4 mM dNTP, 4 units Taq polymerase (Boehringer, Mannheim, Germany) and 100 pmol of each primer. Controls were performed by using RNA instead of cDNA in the PCR reaction. The primers used for RT-PCR for the different growth factors and for GHR as well as the specific conditions of amplification are described previously[4,7,8].

## 2.4 RNase protection assay

The total RNA (30 mg) was introduced into a commercial RNase protection assay (RPA) (Ambion, Texas, USA) and performed as previously described[4]. Precast 10% polyacrylamide gels supplied with 7 M urea (Cleangels, Pharmacia, Freiburg, Germany) were used. Sense and $^{32}P$-antisense riboprobes were generated by subcloning of the desired homologues gene products according to the manufacturer's instructions (PCR-Cloning and RNA-Transcription Kit; Stratagene, La Jolla, CA, USA). To quantify the mRNA of the growth factors by RPA, increasing amounts of in vivo synthesized sense RNA were hybridised with the respective $^{32}$P-antisense riboprobes and compared with the mammary RNA sample using densitometry. There was a linear increase in the abundance of RNase-protected sense/antisense fragments between 0.25 and 5 pg. Individual samples were analysed to verify RT-PCR results and afterwards pooled RNA was examined to obtain the expression pattern throughout the different stages of mammary gland development.

## 3. RESULTS AND DISCUSSION

Semiquantitative densitometric data of the RPA for different growth factors are given in table 1

Table I: Ribonuclease protection assay of growth factors

| | I.1<br>Mam. | I.2.1<br>Mam. | I.2.2<br>Mam. | II<br>Lac. | III.1<br>Galac. | III.2<br>Galac. | IV<br>Invol. |
|---|---|---|---|---|---|---|---|
| IGF-1 | +++ | + | ++ | - | - | - | + - ++ |
| TGFa | ++ | + | + | - | - | - | ++ |
| FGF-1 | +++ | ++ | ++ | - | - | - | + |
| FGF-2 | ++ | + | +++ | + | + | - | + |
| FGFR | ++ | + | ++ | + | + | ± | + |

A general tendency of the expression pattern during the different stages of mammary gland development can be seen. All growth factors studied displayed a relatively high expression in virgin heifers (I.1), reduced (IGF-I and TGF-α) or equal levels (FGF-1 and FGF-2) during pregnancy (I.2.1 and I.2.2), negative or weak expression during lactogenesis (II) and galacto-poiesis (III.1 and III.2) and again increased levels during involution (IV). A similar distribution pattern was found with RT-PCR. Additionally, using RT-PCR, the growth factor IGF-II was demonstrated during all stages of mammary gland development but the tendency for a stronger expression during mammogenesis, as seen for the other growth factors, could not be proven for IGF-II.

The immunohistochemical distribution pattern was studied for IGF-I, IGF-II, FGF-1 and FGF-2. The results for immunostaining of IGF-I are summarized in table 2. Only a weak staining for IGF-I occurred in the epithelium of the ducts during mammogenesis. The epithelium of the alveoli were negative during mammogenesis, lactogenesis and galactopoiesis but displayed distinct IGF-I activity during involution. In the stroma a distinct staining of the cytoplasm of adipocytes and of vascular smooth muscle cells could be observed. A certain percentage of fibroblasts (usually 20 to 30 %) were also immunopositive. Macrophages which occurred in various numbers during the different stages of mammary gland development (comparatively many during mammogenesis and involution, few during lactopoiesis and galactopoiesis) always showed a strong reaction for immunoreactive IGF-I. A similar pattern of immunoreactivity was found for FGF-1. Whereas mRNA for IGF-II could be demonstrated during all stages of mammary gland development using RT-PCR, no immunostaining was seen for this growth factor.

Table 2: IGF-I Immunohistochemistry in bovine mammary gland

| | I.1 | I.2.1 | I.2.2 | II | III.1 | III.2 | IV |
|---|---|---|---|---|---|---|---|
| Epithelium | | | | | | | |
| Ducts | + | + | ± | ± | ± | + | ++ |
| Alveoli | No | - | - | ± | - | + | ++ |
| Myoepithelium | - | - | - | - | - | - | - |
| Stroma | | | | | | | |
| Fibroblasts | + | + | ± | + few | + few | + few | + |
| Adipocytes | ++ | ++ | ++ | ++ | no | no | ++ |
| Endothelium | - | - | - | - | - | - | - |
| Smooth muscle | ++ | ++ | + - ++ | + | + | + - ++ | |
| Macrophages | +++ | +++ | +++ | +++ (few) | +++ (few) | +++ | +++ (many) |

Immunoreactive FGF-2 was observed in virgin heifers in endothelial cells, ductal epithelial cells and myoepithelial cells. During mammogenesis positive immunostaining occurred additionally in the epithelium of some alveoli. During lactogenesis only endothelial and myoepithelial cells were immunopositive. During involution positive staining was again observed in vascular cells, fibroblasts, smooth muscle cells and myoepithelial cells. The pronounced localization of FGF-2 in the endothelial cells of the vascular system of the mammary gland suggests an important role of FGF-2 in the changes of vascularization during the mammogenesis, lactation and involution.

The physiological role of GH for mammogenesis and lactation is not fully known and especially the role of GHR is controversial discussed. Using RT-PCR, non radioactive in situ hybridisation and immunohistochemistry we could demonstrate a characteristic pattern of GHR expression in the epithelial and stromal compartments of the bovine mammary gland (Table 3)

The ductular epithelium showed distinct staining during most stages of development. The secretory epithelium of the alveoli contained a moderate amount of GHR during pregnancy which significantly increased during lactation and galactopoiesis. In dry cows the immunostaining for GHR in the alveoli was only weak or negative. Contrary to the GHR protein, the amount of mRNA encoding GHR appeared to be relatively constant during mammogenesis and lactation and with in situ hybridization a distinct signal for GHR was found in the epithelial cells of ducts and alveoli during this stages.

Table 3: GHR-immunohistochemistry in bovine mammary gland

| | I.1 | I.2.1 | I.2.2 | II | III.1 | III.2 | IV |
|---|---|---|---|---|---|---|---|
| Epithelium | | | | | | | |
| Ducts | ++ | + - ++ | + - ++ | + - ++ | ++ | ++ | +/- |
| Alveoli | No | + | + | + - ++ | ++ | ++ | +/- |
| Myoepithelium | +/- | - | - | - | - | - | - |
| Stroma | | | | | | | |
| Fibroblasts | 10-20% + | 10-20% + | + | + | + | + | + (few) |
| Adipocytes | ++ | + - ++ | + - ++ | + - ++ | no | no | + |
| Endothelium | + | + - ++ | + - ++ | + | + | + | + |
| Smooth muscle | ++ | ++ | ++ | ++ | ++ | ++ | ++ |
| Macrophages | +++ | +++ | +++ | +++ (few) | +++ (few) | +++ | +++ (many) |

Even though an important role for GH in maintaining milk yield in ruminants is well established the precise mechanisms of its actions in the mammary gland is unclear as GHR appears to lack in this organ[9]. Proposals for GH action includes homeostatic mechanisms such as changes in energy partitioning and suggestions that GH increases milk production indirectly via stimulation of IGF-I production, either in the liver or locally in the mammary gland, which in turn increases milk production. Our results clearly point to a direct role of GH via its receptor. Using different techniques (RT-PCR, in situ hybridisation, immunocytochemistry) we could demonstrate a distinct amount of mRNA encoding GHR and its receptor protein during lactation and galactopoiesis whereas at the same time. the amount of IGF-I appeared to be low or negative in the glandular epithelium. and only weakly expressed in the stroma.

## 4. REFERENCES

1. Woodward TL, Dumont N, OConnor-McCourt M. Turner JD, Philip A. Characterisation of transforming growth factor, growth regulatory effects and receptors on bovine mammary cells. J Cell Physiol 165: 339-348. 1995.

2. Knabel M, Kölle S, Sinowatz F. Expression of growth hormone receptor in the bovine mammary gland during prenatal development. Anat Embryol 198: 163-169, 1998.
3. Schams D, Amselgruber W, Sinowatz F. Immunolocalisation of insulin-like growth factors and basic fibroblast growth factor in the bovine mammary gland. In: Intercellular signalling in the mammary gland, New York and London, Plenum Press. 1995, pp. 97-98.
4. Plath A. Zur Expression von Wachstumsfaktoren in der Rindermilchdruse. Thesis. University of Munich. 1996.
5. Hsu SH, Raine L, Fanger H. Use of avidin-biotin peroxidase complex (ABC) in immunoperoxidase techniques. A comparison between ABC and unlabelled antibody (PAP) procedures. J Histochem Cytochem 29: 577580, 1981.
6. Plath A. Einspanier R. Peters F, Sinowatz F. Schams D. Expression of transforming growth factors alpha and beta-l messenger RNNA in the bovine mammary gland during different stages of development and lactation. J Endocrinol 155: 501-511, 1997.
7. Plath A, Einspanier R, Gabler C, Peters F, Sinowatz F. Gospodarowicz D. Schams D. Expression and localization of members of the fibroblast growth factor family in the bovine mammary gland. J Dairy Sci 81: 2604-2613, 1998.
8. Kölle S. Sinowatz F, Boie G, Lincoln D. Developmental changes in the expression of the growth hormone receptor messenger ribonucleic acid and protein in the bovine ovary. Biol Reprod 59: 836-842, 1998.
9. Flint DJ. Regulation of milk secretion and composition by growth hormone and prolactin. In: Intercellular signalling in the mammary gland, New York and London, Plenum Press, 1995, pp. 131.151.

4.

# Involvement of Growth Factors in the Regulation of Pubertal Mammary Growth in Cattle

Purup, S., Vestergaard, M. and Sejrsen, K.
*Dept. of Animal Nutrition and Physiology, Danish Institute of Agricultural Sciences, Research Centre Foulum, DK-8830 Tjele, Denmark*

Key words: Bovine, mammary gland, growth factors

Abstract: Pubertal mammary growth in heifers is dependent on interactions of many hormones and growth factors of which some are stimulatory while others are inhibitory. Although estrogen and growth hormone (GH) are of primary importance, more recent studies have suggested a role for both systemic and mammary tissue-specific growth factors. Growth factors may act as mediators of estrogen and GH or through specific effects of their own. These growth factors include insulin (INS), IGFs (IGF-I and IGF-II), epidermal growth factor (EGF), FGFs (FGF-1 and FGF-2), TGFs (TGF-α and TGF-ß's, amphiregulin (AR), platelets derived growth factor (PDGF), and mammary derived growth factor-1 (MDGF-1). Using mammary epithelial cells derived from prepubertal heifers and cultured in three-dimensional collagen gels as an in vitro model, we have investigated the mitogenic effects of a number of different growth factors (IGF-I, des(1-3) IGF-I, IGF-II, INS, EGF, TGF-α, AR, FGF-1, FGF-2, and TGF-ß1). As expected, IGF-I, des(1-3)IGF-I, IGF-II and INS all stimulated proliferation of mammary cells with des(1-3)IGF-I being the most potent and INS the least potent. The mitogenic effect of IGF-I could be inhibited by both IGFBP-2 and IGFBP-3 showing that these binding proteins modulate the bioactivity of IGF-I in the mammary gland at the cellular level. Regulation of IGF availability by IGFBPs in the extracellular environment therefore is critical for IGF action in the mammary gland. Proliferation of mammary epithelial cells was also stimulated by growth factors of the EGF family, i.e. EGF, TGF-α and AR, however, not as much as growth factors from the IGF family. Members of the fibroblast growth factor family showed various mitogenic activities. FGF-1 stimulated DNA synthesis while FGF-2 in concentrations above 10 ng/ml inhibited DNA synthesis. TGF-ß1 at very low concentrations stimulated proliferation slightly whereas higher concentrations strongly inhibited proliferation of mammary epithelial cells and inhibited mitogenesis induced by growth factors of both the

EGF- and IGF family. This shows that TGF-ß1 is a very potent regulator of pubertal mammary growth.

# 1. INTRODUCTION

Growth and development of the mammary glands in pubertal heifers is important for milk yield potential at maturity[1]. The regulation of mammary growth involves complex interactions of many hormones and growth factors, of which some are growth-stimulating and others growth-inhibitory. Although the systemic importance of the hormones estrogen and growth hormone (GH), for regulation of growth of the mammary gland has been well documented, these hormones have no or very little stimulating effect on mammary epithelial cell growth in vitro (for review[2,3]). Thus other factors of either systemic or local origin are of importance for growth and development of the mammary gland. In recent years, a number of growth factors and binding proteins have been discovered that may be involved in mediating the effects of estrogen and GH or have specific effects of their own. In the present paper, we present data from studies in which we have investigated the mitogenic effect of specific growth factors and binding proteins using an in vitro model of epithelial cells from 8-9 mo heifers. The factors studied include growth factors of the insulin-like growth factor (IGF) family, the epidermal growth factor (EGF) family, the fibroblast growth factor (FGF) family, the transforming growth factor-ß (TGF-ß) family, and two IGF binding proteins. Since factors are studied in the same cell culture system, it is possible to evaluate their relative importance in regulation of growth and development of the pubertal mammary gland.

# 2. PRIMARY CELL CULTURES OF MAMMARY EPITHELIUM

Mitogenic effects of growth factors have been investigated in different cell culture systems with mammary cells from different species at different developmental stages. Unfortunately, species specificity exists such that findings from laboratory animals cannot automatically be applied to ruminants. Therefore, comparisons of the observed growth-stimulatory or growth-inhibitory effects are difficult. To study the importance of growth factors for growth and development of the pubertal heifer mammary glands, mammary epithelial cells were isolated from heifers with body weights between 200 and 250 kg and 8-9 mo of age. At this body weight first estrus has normally not yet occurred and the heifers have significant amounts of mammary tissue showing allometric growth (it is growing at a faster rate than the body). Therefore heifers at this development stage are ideal for studying regulation of mammary growth.

Mammary epithelial cells were cultured in three-dimensional collagen gels as described previously[4]. Briefly, tissue pieces excised aseptically form mammary parenchyma were digested in basal medium supplemented with collagenase, hyaluronidase, DNase and insulin. Organoids were isolated by filtration, centrifugation and precipitation, and subsequently stored in liquid nitrogen until use in cell culture experiments. Experimental designs always included 3 replicates per treatment. Cells were cultured for 24 h in basal serum-free medium 199 containing insulin (10 ng/ml), followed by an additional 4 d in treatment media containing growth factors and/or binding proteins. Culture media were changed every 2 d and 1 μCi [methyl-$^{3}$H]thymidine was added for the last 24 h of the culture period. DNA synthesis was measured as thymidine incorporation as described previously[4].

## 3. INSULIN AND INSULIN-LIKE GROWTH FACTORS (IGFS)

Insulin and insulin-like growth factors are structurally homologues peptides that are involved in many biological processes[5]. The biological effect of IGFs is influenced by association with IGF binding proteins (IGFBPs)[5]. IGFs and IGFBPs are detected in many biological fluids including serum, colostrum, milk, lymph and follicular fluid. IGFs and IGFBPs are mainly synthesized in the liver but also locally in many tissues including mammary tissue.

In ruminants, the stimulatory effects of insulin, IGF-I, IGF-II and the natural variant of IGF-I, des-(1-3) IGF-I, on DNA synthesis in ruminant mammary tissue in cultures has been documented in a number of reports. These studies include undifferentiated mammary epithelial cells from 4-6 mo old heifers[6,7], mammary epithelial cells prepared from pregnant sheep[8] and pregnant, nonlactating hcifcrs[9,10], and prepartum and lactating cows[11]. In undifferentiated mammary epithelial cells, des-(1-3) IGF-I was found to be as potent as IGF-I, while IGF-II was significantly less active (Peri et al., 1992). In 8-9 mo old heifers, we found des-(1-3) IGF-I to be more potent than IGF-I, IGF-II and insulin (Fig. 1). Des-(1-3) attained half-maximal activity at a concentration of less than 1.5 ng/ml, native IGF-I at 1.5-3 ng/ml, whereas IGF-II was required at 3-6 ng/ml before reaching half-maximal activity. However, the maximal proliferative activity of des-(1-3) IGF-I, native IGF-I and IGF-II were not different, while insulin exhibited lower total activity and did not attain the maximal activity of the IGFs even in a concentration of 1000 ng/ml (this concentration not shown).

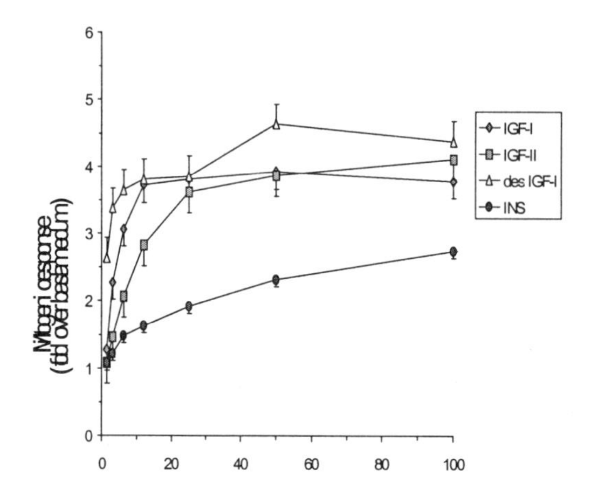


Fig. 1 Effect of IGF-I, IGF-II, des-(1-3) IGF-I and insulin on DNA synthesis, measured by incorporation of [methyl-$^{3}$H]thymidine, into mammary epithelial cells prepared from 8-9 mo old prepubertal heifers and grown in 3-dimensional collagen gels. Values are LSmeans ± SEM of values obtained from cultures with triplicate samples.

These results show that IGFs are very potent factors for mammary epithelial cell growth. Differences in potency of IGF-I, des-(1-3) IGF-I and IGF-II in different studies may depend on variation in affinity and number of IGF membrane receptors. Two types of IGF receptors (type 1 and 2) are known. We have confirmed the presence of type 1 IGF receptors in heifers at this developmental stage in ligand-binding assays with crude mammary membranes[12]. The potency of the IGFs in different culture systems may also depend on the extent of IGF-binding protein (IGFBP) accumulation in the particular culture system since IGF bioactivity is modulated by these specific, high-affinity binding proteins. Mammary cells derived from pregnant, nonlactating heifers and grown in collagen gels show IGF-I inducible secretion of IGFBP-2 and IGFBP-3[10]. We have recently shown the abundance of mRNA and protein for IGFBP-1, -2, -3 and -4 in mammary tissue from heifers[13] suggesting the involvement of the binding proteins in regulating growth of mammary epithelial cells. We therefore studied the effect of recombinant IGFBP-2 at the cellular level by adding increasing concentrations to cell culture medium with (Fig. 2b) or without (Fig. 2a) IGF-I. IGFBP-2 alone did not have any significant ($P>0.49$) effect on DNA synthesis (Fig. 2a). However, when IGFBP-2 was added to cell culture medium containing IGF-I, IGFBP-2 abolished the bioactivity of IGF-I (Fig. 2b). Addition of IGFBP-3 to cell culture medium significantly inhibited the mitogenic effect of IGF-I and IGF-II but not des-(1-3) IGF-I (Fig. 3) in line with the reduced affinity of des-(1-3) IGF-I for IGF-binding proteins[14] In previous studies we found that IGFBP-3 alone also inhibited the mitogenic activity of serum and mammary tissue extracts from heifers at this stage of development[4].

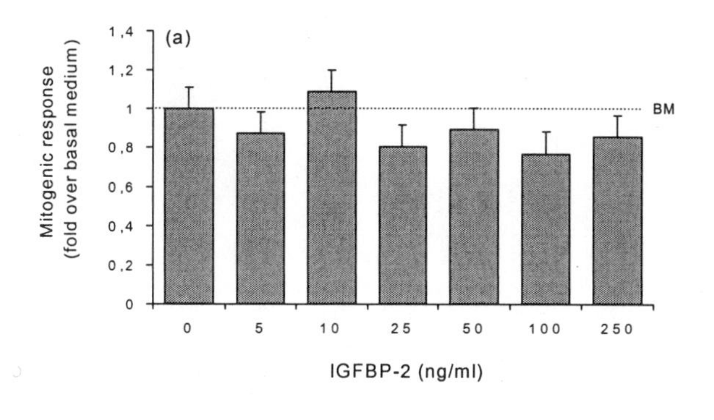


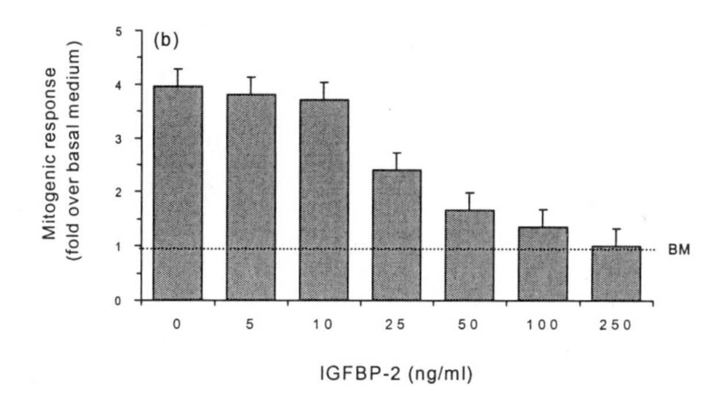


Fig. 2 Effect of IGFBP-2 on (a) DNA synthesis and (b) DNA synthesis stimulated by IGF-I (6.25 ng/ml), measured by incorporation of [methyl-3H]thymidine into mammary epithelial cells prepared from 8-9 mo old prepubertal heifers and grown in 3-dimensional collagen gels. Values are LSmeans ± SEM of values obtained from cultures with triplicate samples. Dotted line represent mitogenic response obtained with cells in basal medium.

The evidence therefore suggests that locally produced IGFBP-3 may therefore inhibit mammary cell growth, either by associating with IGF-I or IGF-II and modulating its activity, or by an IGF-independent mechanism.

# 4. THE EPIDERMAL GROWTH FACTOR (EGF) FAMILY

The epidermal growth factor family is comprised of at least ten proteins including epidermal growth factor (EGF), heparin binding EGF, transforming growth factor-α (TGF-α), amphiregulin and various heregulins.

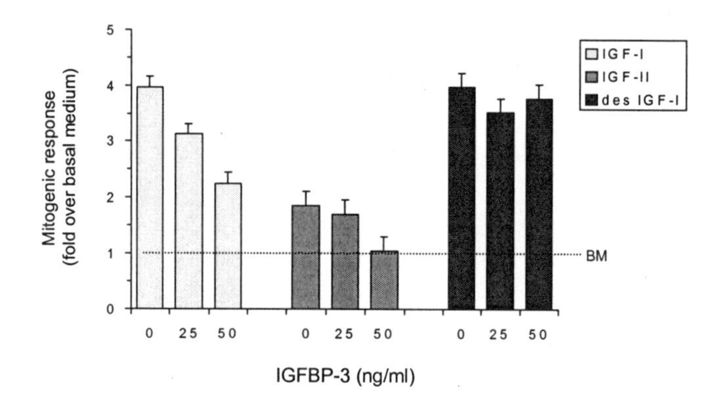


Fig. 3 Effect of IGFBP-3 on DNA synthesis stimulated by IGF-I, IGF-II or des-(1-3) IGF-I (all 6.25 ng/ml), measured by incorporation of [methyl-3H]thymidine into mammary epithelial cells prepared from 8-9 mo old prepubertal heifers and grown in 3-dimensional collagen gels. Values are LSmeans ± SEM of values obtained from cultures with triplicate samples. Dotted line represent mitogenic response obtained with cells in basal medium.

These proteins are a family of structurally related growth factors that act via the EGF (ErbB-1) receptor, which is a specific high-affinity receptor located on the plasma membrane. Currently, little is known concerning the mammogenic role of EGF and related growth factors in ruminants. Despite the existence of specific EGF receptors in mammary tissue from sheep and cows[15,16], expression of EGF messenger RNA is so far known only from a single study with ruminant tissue[17]. Expression of other proteins of the EGF family, however, occur in ruminants, as the bovine mammary gland expresses TGF-α[18,19]. Recently, expression of mRNA for amphiregulin was shown in the sheep mammary gland[20]. Amphiregulin, that was originally isolated from conditioned medium from MCF-7 human mammary carcinoma cell line[21,22], has to our knowledge never been investigated in bovine mammary tissue.

The mitogenic effect of growth factors of the EGF family has been examined in a limited number of studies with ruminant cells. EGF and TGF-α both stimulated DNA synthesis in mammary tissue explants from midpregnant heifers[23], in mammary epithelial cells from pregnant heifers and cows[9,24], and in alveolar cells prepared from pregnant sheep[25]. DNA synthesis was also stimulated by EGF in undifferentiated mammary epithelial cells from 4-6 mo old heifers[7], by TGF-α in alveolar cells from non-pregnant, pregnant and lactating sheep[16], and by amphiregulin in alveolar cells prepared from pregnant sheep[25]. Furthermore, EGF infused into the bovine mammary gland, caused an increase in DNA synthesis in the gland[26].

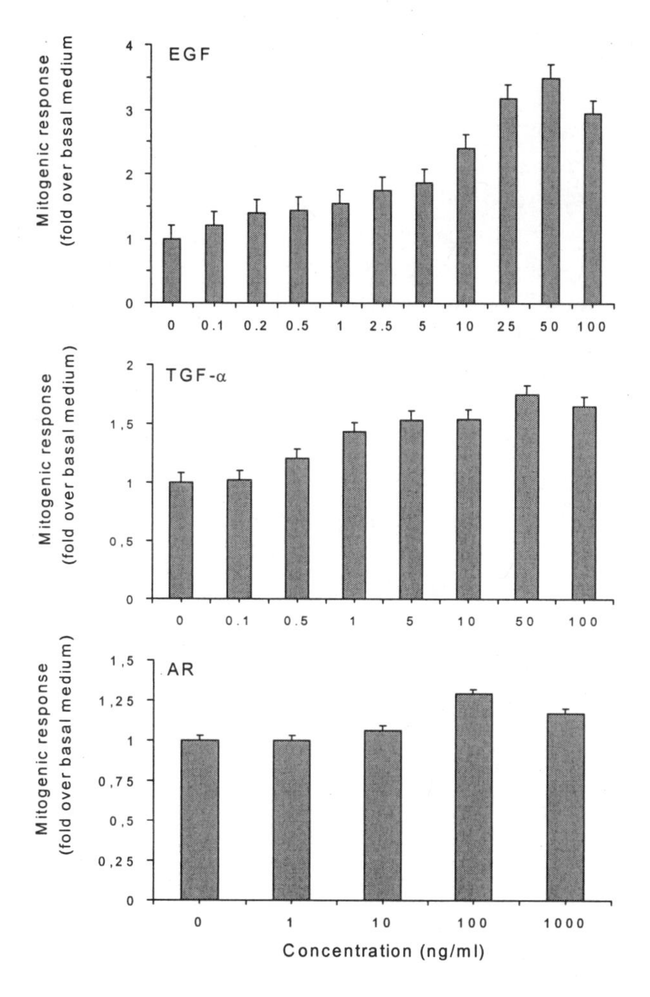


Fig. 4 Effect of rhEGF, rhTGF-α and rhAR on DNA synthesis measured by incorporation of [methyl-$^{3}$H]thymidine into mammary epithelial cells prepared from 8-9 mo prepubertal heifers and grown in 3-dimensional collagen gels. Values are LSmeans ± SEM of values obtained from cultures with triplicate samples.

To investigate the mitogen activity of EGF, TGF-α and amphiregulin in mammary tissue from prepubertal heifers, we tested their effect on DNA synthesis in undifferentiated mammary epithelial cells from 8-9 mo heifers. Proliferation of mammary epithelial cells was stimulated by EGF, TGF-α and amphiregulin but not as much as by growth factors from the IGF family (Fig. 4). Synthesis of DNA was increased about 3.5 and 1.75 fold over basal medium by EGF and TGF-α at a concentration of 50 ng/ml. The previous studies with alveolar cells from pregnant sheep[25] also showed significant effects of both EGF and TGF-α but the effect of EGF was smaller and not clearly dose-related compared with TGF-α. Likewise, Collier et al.[9] found that bovine mammary epithelial cells from pregnant heifers responded more to bovine TGF-α than to EGF. In the MAC-T bovine mammary epithelial cell line, EGF was reported to be without effect[27]. The mitogenic effect of amphiregulin was significantly ($P<0.01$) increased in our study with a maximal effect (~30% increase) at a concentration of 100 ng/ml (Fig. 4) but the effect was smaller than the effect of EGF and TGF-α. The presence of mRNA for EGF, TGF-α and amphiregulin in the ruminant mammary gland and the results presented here, suggest that these growth factors may be important in regulation of mammary growth and development in prepubertal heifers. However, more information is needed about the suggested presence of these factors in mammary tissue.

## 5. THE FIBROBLAST GROWTH FACTOR (FGF) FAMILY

The fibroblast growth factors (FGFs) are a family of heparin binding growth factors that consist of at least 15 structurally related polypeptide growth factors[28]. The FGFs are widely distributed throughout the developing and adult body and display diverse biological activities, most of which are mediated via FGF tyrosine kinase receptors (FGFRs), of which five have been described until now[29]. At least three of the FGFs are supposed to participate in the regulation of growth and function of mammary gland cells in ruminants. These are fibroblast growth factor-1 (FGF-1), also known as acidic FGF (aFGF), fibroblast growth factor-2 (FGF-2), also known as basic FGF (bFGF), and fibroblast growth factor-7 (FGF-7), also known as keratinocyte growth factor (KGF). Recently, it was shown that these FGFs and their receptors are expressed during development, lactation and involution in the bovine mammary gland[29]. The highest mRNA concentrations of FGF-1, -2 and -7 were detected in the glands of virgin heifers and primigravid heifers during involution. Tissue concentrations of FGF-1 and -2 was also shown to be highest in these periods.

The mitogenic effects of FGFs on mammary tissue has almost exclusively been studied in rodents. Using primary cell cultures, FGF-2 has been found to be growth-stimulatory for mouse mammary epithelial cells from all stages of mammary development[30,31]. FGF-1 and -7 also stimulated growth of mammary epithelial cells from virgin mice[32]. In ruminant mammary tissue, only one study exist in which the effect of FGF-2 has been investigated. FGF-2 (0.1- 50 ng/ml) showed dose-dependent mitogenic activity in undifferentiated mammary epithelial cells from 4-6 mo old heifers[33]. Using our primary mammary epithelial cell cultures, we have investigated the mitogenic effect of FGF-1 and -2. Fig. 5 shows that FGF-1 significantly increased DNA synthesis in concentrations above 0.5 ng/ml with maximum stimulation at the highest concentration (100 ng/ml) added to the cell culture medium. FGF-2, however, tended to increase DNA synthesis at concentrations below 10 ng/ml, but significantly decreased DNA synthesis when 50 and 100 ng/ml was added to culture medium. The stimulating effect of FGF-1 is in agreement with previous studies in mice, while the mitogenic response to FGF-2 contrast the dose-dependent mitogenic activity in mammary epithelial cells from 4-6 mo old heifers[33]. Anyhow, previous studies showing expression and tissue concentrations of FGFs and the present results suggest FGFs to be important in the local regulation of the bovine mammary gland.

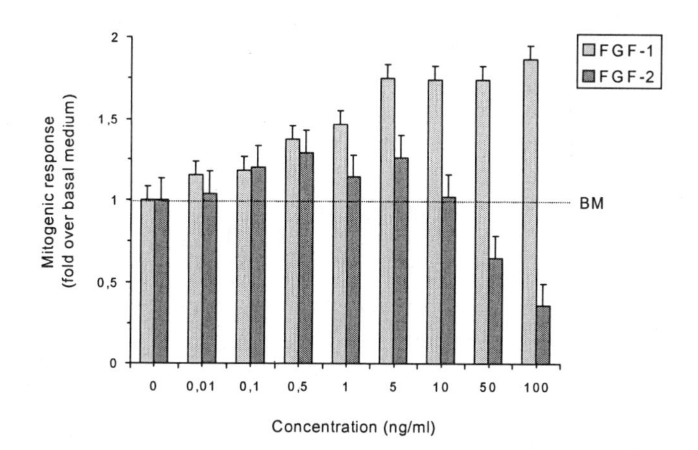


Fig. 5 Effect of bovine FGF-1 and FGF-2 on DNA synthesis measured by incorporation of [methyl-3H]thymidine into mammary epithelial cells prepared from 8-9 mo prepubertal heifers and grown in 3-dimensional collagen gels. Values are LSmeans ± SEM of values obtained from cultures with triplicate samples. Dotted line represent mitogenic response obtained with cells in basal medium.

# 6. THE TRANSFORMING GROWTH FACTOR-ß (TGF-ß) FAMILY

The TGF-ß family is composed of at least five related multifunctional proteins that can affect a variety of cellular functions such as extracellular matrix formation, differentiation and growth depending on the cell and tissue type[34,35]. TGF-ß is involved in a variety of physiological and pathological processes, including wound healing in which TGF-ß produced in blood platelets induces accelerated healing of incisional wounds[36]. Platelets represent the most concentrated natural source of TGF-ß1 (20 mg/kg[37]), but TGF-ß1 is believed to be secreted from nearly all cells[35]. Three isoforms of TGF-ß (TGF-ß1, TGF-ß2 and TGF-ß3) have been identified. The most well characterized member of the TGF-ß family is TGF-ß1, which in its biological active form is a 25 kDa disulphide-linked homodimer[38]. TGF-ß1 is secreted in an inactive form bound to a 75 kDa glycoprotein known as the latency associated peptide (LAP)[34]. This latent TGF-ß1 complex has to be activated to obtain its biological functions. This activation may be important in controlling the physiological functions of TGF-ß1 in vivo[34].

Very little is known about the role of TGF-ß in mammary development in ruminants. We have shown the existence of TGF-ß1 in prepubertal heifer serum in concentrations ranging from 7 to 30 ng/ml[39], and Plaut et al.[40] has shown the existence of specific receptor binding to bovine mammary membranes from prepubertal and pubertal heifers as well as from lactating and dry cows. In addition, Plath et al[19] showed expression of TGF-ß1 in the bovine mammary gland during mammogenesis, lactogenesis, galactopoiesis and involution. Both specific receptor binding and expression of TGF-ß1 mRNA was higher during the prepubertal and pubertal periods than during lactation. These data support a role of TGF-ß1 in regulation of pubertal mammary growth and development.

The actions of TGF-ß1 are mediated by binding to highly specific cell surface receptors of which most cells have three types, designated type I, II and III[35]. The type I and II receptors are supposed to be directly involved in signal transduction while the type III receptor has been postulated to enhance binding of TGF-ß1 to the signalling receptors. We have investigated the existence of the type I and II TGF-ß receptors in bovine mammary tissue from prepubertal heifers using immunohistochemistry. These studies showed extensive staining for both the type I and II receptor in the ductal epithelium making an effect of TGF-ß1, mediated through the receptor, possible in the heifer mammary glands (Purup et al., unpublished results).

In accordance with this, we have investigated the biological effect of TGF-ß1 in primary cultures of mammary epithelial cells. A biphasic effect of TGF-ß1 was observed (Fig. 6)[39]. Addition of 25 and 50 pg/ml of TGF-ß1 to cell culture medium stimulated DNA synthesis, while higher concentrations of TGF-ß1 (>100 pg/ml) inhibited DNA synthesis. Maximal inhibition occurred at 5 ng/ml, corresponding to a 66% inhibition of DNA synthesis. In addition, TGF-ß1 (5 ng/ml) inhibited 77 to 92% of the mitogenic effect of serum (5% added to cell culture medium) obtained from 24 prepubertal heifers, and adding increasing concentrations of TGF-ß antibody had a positive effect on mitogenic activity of serum[39].

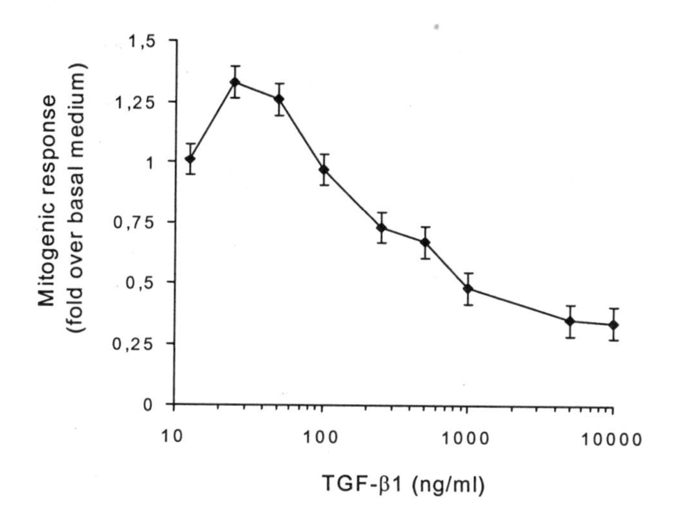


Fig. 6 Effect of TGF-ß1 on DNA synthesis in primary cultures of mammary epithelial cells prepared from 8-9 mo prepubertal heifers and grown in 3-dimensional collagen gels. TGF-ß1 (12.5, 25, 50, 100, 250, 1000, 5000, 10000 pg/ml) was added to basal medium and DNA synthesis measured by incorporation of [methyl-3H]thymidine in the last 24 h of the culture period. Values are LSmeans ± SEM. Data from Purup et al., 1999.

To study the potency and relative importance of TGF-ß1 in regulation of pubertal mammary growth, we investigated whether TGF-ß1 can influence the mitogenic effects of growth factors of the IGF and EGF families. TGF-ß1 in different concentrations (Fig. 7) was added to cell culture medium containing IGF-I (25 ng/ml), IGF-II (50 ng/ml), des(1-3)IGF-I (50 ng/ml), INS (100 ng/ml), EGF (25 ng/ml), TGF-α (5 ng/ml) and amphiregulin (100 ng/ml). The maximal effect of TGF-ß1 was obtained with concentrations of 500 pg/ml. This concentration is 30 times lower than the observed mean concentrations in serum from prepubertal heifers. The inhibition of the mitogenic effects induced by the different growth factors at this concentration of TGF-ß1 corresponded to approximately 50% of the proliferation obtained without TGF-ß1 in the medium. It is interesting that this relative growth inhibition was of the same size for growth factors from both the IGF and EGF family, while the inhibition measured as dpm was significant higher for the growth factors in the IGF family than in the EGF family. These results show that TGF-ß1 is a very potent regulator of pubertal mammary growth in heifers.

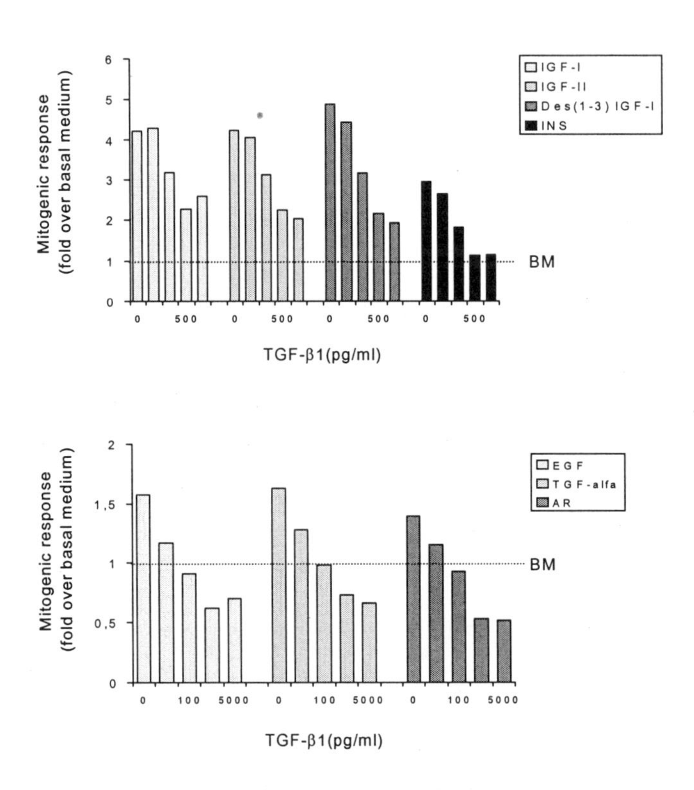


Fig. 7 Effect of TGF-ß1(0, 25, 100, 500, 5000 pg/ml) on DNA synthesis induced by IGF-I (25 ng/ml), IGF-II (50 ng/ml), Des(1-3) IGF-I (50 ng/ml), INS (100 ng/ml), EGF (25ng/ml), TGF-$\alpha$ (5 ng/ml) and amphiregulin (AR; 100 ng/ml). DNA synthesis was measured by incorporation of [methyl-$^3$H]thymidine into mammary epithelial cells prepared from 8-9 mo prepubertal heifers and grown in 3-dimensional collagen gels. Values are LSmeans ± SEM of values obtained from cultures with triplicate samples. Dotted line represent mitogenic response obtained with cells in basal medium.

## 7. OTHER GROWTH FACTORS

A number of other growth factors, both stimulatory and inhibitory, have been described as factors that might play a role in regulating growth and development of mammary gland. Among these are mammary-derived growth factor-1 (MDGF-1) and platelet-derived growth factor (PDGF). MDGF-1 has been detected in human and bovine milk as well as milk of other mammals[41]. Bovine MDGF-1 is predominant in mammary gland secretions from pregnant heifers and in colostrum and MDGF-1 from human milk was observed to be growth stimulatory in normal and immortalized human mammary epithelial cells and in MCF-7 cells[42]. PDGF is a potent mitogen in human serum which specifically stimulates the proliferation of mesenchymal cells. PDGF is secreted by epithelial cell lines derived from normal human tissue[43] and from a number of breast cancer cell lines[44]. Receptors for PDGF, however, do not seem to be present at either breast cancer or normal mammary epithelial cells, suggesting the effect of PDGF on epithelial cells to be indirect via the stroma. Mammary-derived growth inhibitor (MDGI), a member of a family of fatty acid-binding proteins, has been isolated from mammary tissue of pregnant and lactating ewes[45] as well as the lactating bovine mammary gland[46]. MDGI is expressed in lobulo-alveolar and ductal cells of pregnant and lactating animals but not in mammary epithelial cells of virgin animals[47]. The physiological role of MDGI remains speculative. The postulated growth inhibitory function of MDGI is based on studies with transformed cells but so far no inhibitory effect on normal mammary epithelial cells in vivo or in primary culture has been demonstrated. Its role in the mammary gland appears to be as a differentiation factor rather than a an anti-proliferative factor[47].

## 8. SUMMARY AND CONCLUSIONS

The regulation of mammary growth and development in heifers is accomplished by complex interactions of hormones and growth factors. Although estrogen and growth hormone are of primary importance for regulation of pubertal mammary growth in heifers, a large number of both systemic and mammary tissue specific growth factors are also essential. This conclusion is supported by results obtained from studies using primary mammary epithelial cells derived from prepubertal heifers and cultured in three-dimensional collagen gels as an in vitro model. Growth factors of the IGF and EGF growth factor families and FGF-1 are growth-stimulatory while FGF-2 and TGF-ß1 have biphasic effects, stimulating DNA synthesis at low concentrations and inhibiting DNA synthesis at higher concentrations.

Furthermore, the results show that the IGF binding proteins, IGFBP-2 and IGFBP-3, modulates IGF-I action on mammary cell growth. This is most likely due to association with IGF-I, also supported by the results showing that the mitogenic activity of des-(1-3) IGF-I was unaffected by addition of IGFBP-3 to culture medium.

In conclusion, the present results describing the biological effect of growth factors and binding proteins in vitro together with previous results of abundance of mRNA, protein and specific receptors, suggest that at least IGF-I, IGFBP-3, FGF-1, FGF-2 and TGF-ß1 are important factors for regulation of mammary development and growth in vivo in heifers. However, more information is needed concerning growth factors of the EGF family as well as other growth factors, especially about the suggested presence in mammary tissue, to conclude that these growth factors also are important in regulation of mammary growth in heifers.

**Acknowledgements**

The authors thank C.J. Juhl, A.K. Nielsen, and H.M. Purup for their skilled technical assistance.

## 9. REFERENCES

1. Sejrsen, K., and S. Purup. 1997. Influence of prepubertal feeding level on milk yield potential of dairy heifers: A review. J. Anim. Sci. 75, 828-835.
2. Forsyth IA. The mammary gland. In: Bailliere's Clinical Endocrinology and Metabolism 5, 809-832, 1991.
3. Oka T, Yoshimura M, Lavandero S, Wada K, Ohba Y. Control of growth and differentiation of the mammary gland by growth factors. J Dairy Sci 74, 2788-2800, 1991.
4. Weber MS, Purup S, Vestergaard M, Ellis SE, Sondergaard J, Akers RM, Sejrsen K. Contribution of insulin-like growth factor-I (IGF-I) and IGF-binding protein-3 (IGFBP-3) to mitogenic activity in bovine mammary extracts and serum. J Endocrinology 161, 365-373, 1999.
5. Jones JI, Clemmons DR. Insulin-like growth factors and their binding proteins: Biological actions. Endocrine Rev 16, 3-34, 1995.
6. Shamay, A., N. Cohen, M. Niwa, and A. Gertler. 1988. Effect of insulin-like growth factor I on deoxyribonucleic acid synthesis and galactopoiesis in bovine undifferentiated and lactating mammary tissue in vitro. Endocrinol 123, 804-809.
7. Peri I, Shamay A, McGrath MF, Collier RJ, Gertler A. Comparative mitogenic and galactopoietic effects of IGF-I, IGF-II and des-3-IGF-I in bovine mammary gland in vitro. Cell Biol International Reports 16, 359-368, 1992.
8. Forsyth IA. The insulin-like growth factor and epidermal growth factor families in mammary cell growth in ruminants: action and interaction with hormones. J Dairy Sci 79, 1085-1096, 1996.

9. Collier RJ, McGrath MF, Byatt JC, Zurfluh L. Regulation of bovine mammary growth by peptide hormones: involvement of receptors, growth factors and binding proteins. Livest Prod Sci 35, 21-33, 1993.
10. McGrath MF, Collier RJ, Clemmonds DR, Busby WH, Sweeny CA, Krivi GG. The direct in vitro effect of insulin-like growth factors (IGFs) on normal bovine mammary cell proliferation and production of IGF bindings proteins. Endocrinology 129, 671-678, 1991.
11. Baumrucker CR, Stemberger BM. Insulin and insulin-like growth factor-I stimulate DNA synthesis in bovine mammary tissue in vitro. J Anim Sci 67, 3503-3514, 1989.
12. Purup S, Sejrsen K, Akers RM. Effect of bovine GH and ovariectomy on mammary tissue sensitivity to IGF-I in prepubertal heifers. J Endocrinol 144, 153-158, 1995.
13. Weber MS. The role of insulin-like growth factor-I (IGF-I) and IGF-binding proteins in mammary gland development. Ph. D. Thesis, Virginia Polytechnic Institute and State University, 1998.
14. Forbes B, Szabo L, Baxter RC, Ballard FC, Wallace JC. Classification of the insulin-like growth factor binding proteins with three distinct categories according to their binding specification. Biochem Biophys Res Commun 157, 196-202, 1988.
15. Spitzer E, Grosse R. EGF receptors on plasma membranes purified from bovine mammary gland of lactating and pregnant animals. Biochem International 14, 581-588, 1987.
16. Moorby CD, Taylor JA, Forsyth IA. Transforming growth factor-β: receptor binding and action on DNA synthesis in the sheep mammary gland. J Endocrinology 144, 165-171, 1995.
17. Sheffield LG. Mastitis increases growth factor messenger ribonucleic acid in bovine mammary glands. J Dairy Sci 80, 2020-2024, 1997.
18. Koff MD, Plaut K. Expression of transforming growth factor-β-like messenger ribonucleic acid transcripts in the bovine mammary gland. J Dairy Sci 78, 1903-1908, 1995.
19. Plath A, Einspanier R, Peters F, Sinowatz F, Schams D. Expression of transforming growth factors alpha and beta-1 messenger RNA in the bovine mammary gland during different stages of development and lactation. J Endocrinol 155, 501-511, 1997.
20. Forsyth IA, Keable S, Taylor JA, Turvey A, Lennard S. Expression of amphiregulin in the sheep mammary gland. Mol Cell Endocrinology 126, 41-48, 1997.
21. Shouab M, McDonald VL, Bradley G, Todaro GJ. Amphiregulin: a bifunctional growth-modulating glycoprotein produced by the phorbol 12-myristate 13-acetate-treated human breast adenocarcinoma cell line MCF-7. Proc Natl Acad Sci USA 85, 6528-6532, 1988.
22. Shouab M, Plowman GD, McDonald VL, Bradley JG, Todaro GJ. Structure and function of human amphiregulin: a member of the epidermal growth factor family. Science 243, 1074-1076, 1989.
23. Sheffield LG. Hormonal regulation of epidermal growth factor content and signalling in bovine mammary tissue. Endocrinology 139, 4568-75, 1998.
24. Zurfluh LL, Bolten SL, Byatt JC, McGrath MF, Tou JS, Zupec ME, Krivi GG. Isolation of a genomic sequence encoding a biologically active bovine TGF-α protein. Growth Factors 3, 257-266, 1990.
25. Forsyth IA, Taylor JA, Moorby CD. DNA synthesis by ovine mammary alveolar epithelial cells: effects of heparin, epidermal growth factor-related peptides and interaction with stage of pregnancy. J Endocrinology 156, 283-290, 1998.
26. Collier RJ, McGrath MF. Effect of human epidermal growth factor (hEGF) on bovine mammary development in vivo. J Dairy Sci 71 (Suppl. 1) 228, 1988.

27. Woodward TL, Akers RM, Turner JD. Lack of mitogenic response to EGF, pituitary and ovarian hormones in bovine mammary epithelial cells. Endocrine 2, 529-535, 1994.
28. Szebenyi G, Fallon JF. Fibroblast growth factors as multifunctional signalling factors. International Rev Cytology 185, 45-106, 1999.
29. Plath A, Einspanier R, Gabler C, Peters F, Sinowatz F, Gospodarowicz, Schams D. Expression and localization of members of the fibroblast growth factor family in the bovine mammary gland. J Dairy Sci 81, 2604-2613, 1998.
30. Levay-Young BK, Imagawa W, Wallace DR, Nandi S. Basic fibroblast growth factor stimulates the growth and inhibits casein accumulation in mouse mammary epithelial cells in vitro. Mol Cell Endocrinology 62, 327-336, 1989.
31. Lavandero SA, Chappuzeau A, Sapag-Hagar M, Oka T. In vivo and in vitro evidence of basic fibroblast growth factor action in mouse mammary gland development. FEBS Lett 439, 351-356, 1988.
32. Imagawa W, Cunha GR, Young P, Nandi S. Keratinocyte growth factor and acidic fibroblast growth factor are mitogens for primary cultures of mammary epithelium. Biochem Biophys Res Commun 204, 1165-1169, 1994.
33. Sandowski Y, Peri I, Gertler A. Partial purification and characterization of putative paracrine/autocrine bovine mammary epithelium growth factors. Livest Prod Sci 35, 35-48, 1993.
34. Plaut K. Role of epidermal growth factor and transforming growth factors in mammary development and lactation. J Dairy Sci 76, 1526-1538, 1993.
35. Roberts AB, Sporn MB. The transforming growth factor beta. In: Peptide Growth Factors and Their receptors. MB Sporn and AB Roberts (Eds.). Springer-Verlag, Heidelberg, Germany, pp. 421-472, 1990.
36. Mustoe TA, Pierce GF, Thomason A, Gramates P, Sporn MB, Deuel TF. Transforming growth factor type beta induces accelerated healing of incisional wounds in rats. Science 237, 1333-1336, 1987.
37. Van den Eijnden-van Raaij AJM, Koornneef I, van Zoelen EJJ. A new method for high yield purification of type beta transforming growth factor from human platelets. Biochem Biophys Res Commun 157, 16-23, 1988.
38. Grainger DJ, Mosedale DE, Metcalfe JC, Weissberg PL, Kemp PR. Active and acid-activatable TGF-ß in human sera, platelets and plasma. Clin Chim Acta 235, 11-31, 1995.
39. Purup S, Vestergaard M, Weber MS, Plaut K, Akers RM, Sejrsen K . Local regulation of pubertal mammary growth in heifers. J Anim Sci, 1999 (In press).
40. Plaut K, Maple RL, Capuco AV, Bell AW. Validation of transforming growth factor-(1 binding assay for bovine mammary tissue. In: Intercellular Signalling in the Mammary Gland. CJ Wilde, M Peaker, and CH Knight (Eds.), Plenum Press Ltd., 91-93, 1995.
41. Bano M, Salomon DS, Kidwell WR. Purification of a mammary-derived growth factor from human milk and human mammary tumors. J Biol Chem 260, 5745-5752, 1985.
42. Bano M, Kidwell WR, Lippman ME, Dickson RB. Characterization of mammary-derived growth factor-1 receptors and response in human mammary epithelial cell lines. J Biol Chem 265, 1874-1880, 1990.
43. Bronzert DA, Bates SE, Sheridan JP. Transforming growth factor-( induces platelet-derived growth factor (PDGF) messenger RNA and PDGF secretion while inhibiting growth in normal human mammary epithelial cells. Mol Endocrinology 4, 981-989, 1990.
44. Rozengurt E, Sinnet-Smith J, Taylor-Papadimitriou J. Production of PDGF-like growth factor by breast cancer cell lines. International J Cancer 36, 247-252, 1985.

45. Politis I, Fantuz F, Baldi A. Identification of mammary-derived growth inhibitor in sheep mammary tissue. Small Ruminant Res 18, 151-155, 1995.
46. Böhmer FD, Kraft R, Otto A, Wernstedt C, Hellman U, Kurtz A, Muller T, Rohdi K, Etzold G, Lehmann W. Identification of a polypeptide growth inhibitor from bovine mammary gland. J Biol Chem 262, 15137-15143, 1987.
47. Kurtz A, Vogel F, Funa K, Heldin CH, Grosse R. Developmental regulation of mammary-derived growth inhibitor expression in bovine mammary tissue. J Cell Biol 110, 1779-1789, 1990.

5.

# Insulin-like Growth Factor Binding Protein-5 (IGFBP-5) Potentially Regulates Programmed Cell Death and Plasminogen Activation in the Mammary Gland

Elizabeth Tonner[1], Gordon Allan[1], Lulzim Shkreta[2], John Webster[3], C. Bruce A. Whitelaw[3], David J. Flint[1]

*[1]Hannah Research Institute, Ayr, UK; [2]University of Bologna, Italy; [3]Roslin Institute, Edinburgh, UK*

Key words: Insulin-like growth factor, apoptosis, plasminogen, casein

This study aims to investigate the mechanism by which prolactin and GH interact to maintain mammary epithelial cell function in the rat. IGF-I is an important survival factor for the mammary gland and we have demonstrated that the effects of GH and prolactin involve IGF-I. GH acts by increasing IGF-I whilst prolactin acts by inhibiting the expression of IGFBP-5 from the mammary epithelium. During mammary involution, when serum prolactin levels decline, IGFBP-5 expression is dramatically upregulated and it binds with high affinity to IGF-I preventing IGF-I interaction with the IGF-receptor and thus leading to epithelial cell apoptosis. We have identified a specific interaction of IGFBP-5 with $\alpha_{s2}$-casein. This milk protein has also been shown to bind plasminogen and its activator tissue-type plasminogen activator (tPA) leading to enhanced conversion of plasminogen to plasmin. Plasmin is an important initiator of re-modelling of the extracellular matrix during mammary involution. A potential interaction between the cell death and extracellular matrix remodelling is evident from the observation that IGFBP-5 binds to plasminogen activator inhibitor-I (PAI-1). We thus hypothesized that IGFBP-5 could activate cell death by sequestration of IGF-I and activate plasminogen cleavage by sequestering PAI-1. In support of this hypothesis we have shown that both prolactin and GH inhibit tPA activity and plasminogen activation in the involuting mammary gland. Our results suggest that GH and prolactin inhibit cell death and ECM remodelling via the IGF-axis and also indicate a novel role for the milk protein $\alpha_{s2}$-casein in this process. We have now established lines of transgenic mice expressing IGFBP-5 on the β-lactoglobulin promoter to explore its function in greater detail.

# 1. INTRODUCTION

The IGFBPs are a family of 6 proteins which bind IGF-1 and IGF-2 with high affinity. These proteins can inhibit or enhance IGF action and the same binding protein can have both inhibitory or stimulatory effects dependent upon its concentration, phosphorylation status and proteolysis. Direct effects of some of these IGFBPs (i.e. independent of their ability to bind IGFs) have also been described.

In this review we will focus upon the roles of these molecules in controlling cell death during mammary involution at the end of lactation and a possible IGF-independent effect of IGFBP-5 controlling the plasminogen system that is responsible for initiating the remodelling of the extracellular matrix (ECM).

## 1.1 IGFBPs in development

Few studies have examined IGFBP expression in the developing rodent mammary gland although both IGFBP-2 and -5 have been implicated in aspects of branching morphogenesis which occurs during pregnancy[33]. Similarities exist however with respect to the developing mammary gland and a variety of embryonic developmental processes and the mammary gland has been described as an ideal tissue to study these processes since its development occurs mainly in the adult. Based upon the observations of Wood et al[33] and our own studies in the involuting mammary gland we have also examined IGFBP-2 and -5 expression in the developing mouse, rat, and chick embryo.

Development in the embryonic limb bud shares many features with other systems since epithelial-mesenchymal interactions underlie morphogenesis, and the mammary gland is another example of this. Similarities between the limb bud and the mammary gland are apparent in relation to the IGF system. In the limb bud IGFBP-2 and –5 are expressed in the epithelial cells of the apical ectodermal ridge (AER), while IGF-1 is expressed in the neighbouring subridge mesoderm. This is similar to the mammary gland, where IGFBP-2 and –5 are expressed in mammary epithelial cells while IGF-1 is expressed by the stromal cells. In addition, several other factors shown to be important in limb bud morphogenesis have also been shown to be developmentally regulated in the foetal and postnatal mammary gland, including Msx-1, Msx-2, BMP-2 and BMP-4 [24].

Intriguingly, all of the human IGFBP genes are localized to the same chromosomal regions as different homeobox (Hox) gene families[1]; IGFBP-2 and –5 are linked to the Hox D gene family on human chromosome 2 in opposite transcriptional orientations. The functional significance of this linkage is unknown, but vertebrate Hox genes are well known for their important functions during embryonic development. Recently, it has been shown that loss of function of *Hoxa*9, *Hoxd*9 and *Hoxb*9 in mice leads to impairment of adult mammary gland development during and after pregnancy[4]. These studies suggest a direct role for these factors in directing differentiation of the mammary epithelium ductal system prior to lactation.

## 1.2 IGFBP expression in the mammary gland and milk during lactation

The milk of all species studied contains IGFs and IGFBPs which vary in a species-dependent manner[8]. The presence of IGFs in the milk is similar to that of other growth factors: higher in colostrum than in mature milk.

IGFBP-2 is the major IGFBP in human milk with concentrations 10-fold higher in milk than maternal serum suggesting that milk IGFBP-2 is made in the mammary gland[9].

Porcine milk contained IGFBP-1, 2, 3 and 4 and some of this IGFBP-3 was in association with ternary complexes[7,27]. IGFBP-2 and -3 mRNA are expressed in porcine mammary gland indicating that they are produced in the mammary gland rather than transported into the gland from blood[16].

Bovine colostrum[28] contains IGFBP-1,-2, -3, and -4 and has high IGF-1 levels although serum IGF-1 is low[32]. IGFBP-3 predominates in prepartum secretions and colostrum but is reduced post partum.

Rat milk contains IGFBP-3, IGFBP-2, and IGFBP-4 and lactating mammary gland expressed IGFBP-2 mRNA suggesting that it is synthesised in the gland[5,6,30]. IGFBP-4 expression has been described in rat mammary gland although its concentration in milk is low[5] and we were unable to detect IGFBP-4 expression until involution of the mammary gland[30]. Mammary gland IGFBP-3 mRNA is decreased at parturition and remains low in lactation suggesting that milk IGFBP-3 is serum derived[5,17]. A portion of the IGFBP-3 in milk was in association with ternary complexes[5]. We were unable to detect IGFBPs-1,-3 or –6 in rat mammary gland during lactation or involution[30]. Although IGFBP-5 and –6 have not been detected in the milk during lactation in any of the species studied, the mammary gland of the bitch expresses IGFBP-2 and IGFBP-5 and progestin administration decreases IGFBP-5 mRNA levels[20].

These studies have demonstrated the species-specific nature of the expression of IGFBPs in milk but, because of their descriptive nature they have failed to shed light on the roles played by the binding proteins at this time. Are they present to transport IGFs, to inhibit the actions of the IGFs or to inhibit the degradation of IGFs in the neonatal gut? It was our intention to ascribe functions to the IGFBPs present in the mammary gland and milk and we therefore examined changes in IGFBP expression induced by involution of the mammary gland as well as examining the hormonal control of IGFBP expression during lactation.

## 1.3 Changes in IGFBP expression during involution of the mammary gland

A dramatic increase in the concentration of IGFBP-5 occurs in rat milk within 24 h of mammary involution induced by removal of the suckling young[30]. Increased levels of IGFBP-5 mRNA, as well as more modest increases in IGFBP-2 and -4 (but not IGFBP-1, -3, and –6) were evident in the mammary gland and in situ hybridisation studies showed that expression occurred in the mammary epithelial cells.

IGF-1 has been implicated as a survival factor for many different cell types[22], including the mammary epithelium and since milk from involuting mammary glands contained large quantities IGFBP-5, we proposed that this served to inhibit IGF-1 actions in the gland. Our studies also showed that both prolactin and GH maintained mammary gland function by inhibiting programmed cell death or apoptosis[31]. We further demonstrated that only prolactin was able to influence IGFBP-5 synthesis[30] serving as a potent inhibitor of IGFBP-5 expression. These findings enabled us to propose a mechanism for the interactive effects of prolactin and GH. GH increases IGF-1 synthesis whilst prolactin represses IGFBP-5 synthesis thereby maximising the effect of IGF-1 on mammary epithelial cell survival. Several studies have indicated that IGFBP-5 interacts with a variety of components of the ECM. However we demonstrated that IGFBP-5 in milk associates with the casein micelle fraction. The affinity of IGFBP-5 for IGF-1 was approximately 5-8 x $10^{-10}$M, which is higher than that of the type-I IGF receptor[29] and the concentration of IGFBP-5 in milk was approximately 40 μg/ml. These findings support the proposal that IGFBP-5 acts as an inhibitor of IGF action in milk.

## 1.4 IGF-independent effects of IGFBPs

Although all of the IGFBPs bind the IGFs, these molecules also exhibit additional properties, which have led to suggestions that the IGFBPs have biological effects which are independent of the IGFs. Both IGFBP-1 and-2 contain an RGD sequence which is a common feature of molecules which interact with cell-surface integrins. Transfection of cells with IGFBP-1 led to increased rates of cell migration, an effect which was unaffected by the addition of IGF-1 [14]. IGFBP-3 has been shown to inhibit DNA synthesis in human breast cancer cells independently of its ability to bind IGF-1 [23].

Studies in which IGFBP-3 was shown to inhibit growth of a fibroblast cell line derived from a type-1 IGF receptor knock-out mouse were also compelling evidence for such IGF-independent effects.

We thus considered whether the high concentrations of IGFBP-5 in milk from involuting mammary glands was present solely to inhibit IGF actions or whether it was also involved in IGF-independent actions of this IGFBP.

## 1.5 Interactions of IGFBP-5 with the plasminogen system

IGFBP-5 was first identified in high concentrations in bone where it binds to hydroxyapatite, a crystalline form of calcium phosphate[19]. The casein micelle is composed of casein molecules stabilized by calcium phosphate nanoclusters. The caseins interact through phosphorylated residues, whereas non-phosphorylated proteins such as $\alpha$-lactalbumin are not incorporated into the micelles[12]. IGFBP-5 has a consensus recognition site for the true mammary casein kinases and it has been reported to be phosphorylated[15]. There is no evidence, however, that IGFBP-5 is bound to calcium phosphate but we have been able to show that IGFBP-5 is involved in a protein-protein interaction with $\alpha_{s2}$-casein and, in particular, the dimeric form of this milk protein. Dimeric $\alpha_{s2}$-casein has also been shown to bind to plasminogen and tissue-type plasminogen activator (t-PA) which results in enhanced activity of t-PA and the generation of the enzyme plasmin[11]. Plasmin cleaves a number of pro-enzymes, such as prostromelysins and procollagenases, and thereby initiates the degradation of the ECM which occurs at the end of lactation[18]. A possible functional interaction of IGFBP-5 and components of the plasminogen system on the casein micelle became a distinct possibility with the observation that IGFBP-5 binds to plasminogen activator inhibitor-I (PAI-1) [21]. PAI-1 binds t-PA and inhibits its actions and thus it is conceivable that IGFBP-5 plays a dual role of binding IGF-1 (to inhibit its actions and induce apoptosis) whilst also binding PAI-1 thereby activating t-PA and consequent ECM remodelling (see figure). Such a mechanism would link cell death and ECM degradation, to ensure a highly coordinated tissue-remodelling process. If this were the case we would anticipate that prolactin, by inhibiting IGFBP-5 synthesis, would increase the bioavailability of PAI-1 and suppress plasmin production. We have recently shown that this is indeed the case, since if lactating rats have their litters removed but are treated with prolactin, both t-PA activity and plasmin production are inhibited. GH also inhibited t-PA activity and this could involve IGF-1 since IGF-1 has been shown to increase PAI-1 mRNA expression[10]. We are currently examining these possibilities with in vitro assays for tPA and plasmin activity as well as producing transgenic mice expressing IGFBP-5 on the β-lactoglobulin promoter and developing various IGFBP-5 mutants to examine the effects of exogenous IGFBPs in vitro and in vivo.

Does IGFBP-5 co-ordinate activation of cell death and remodelling of the extracellular matrix?

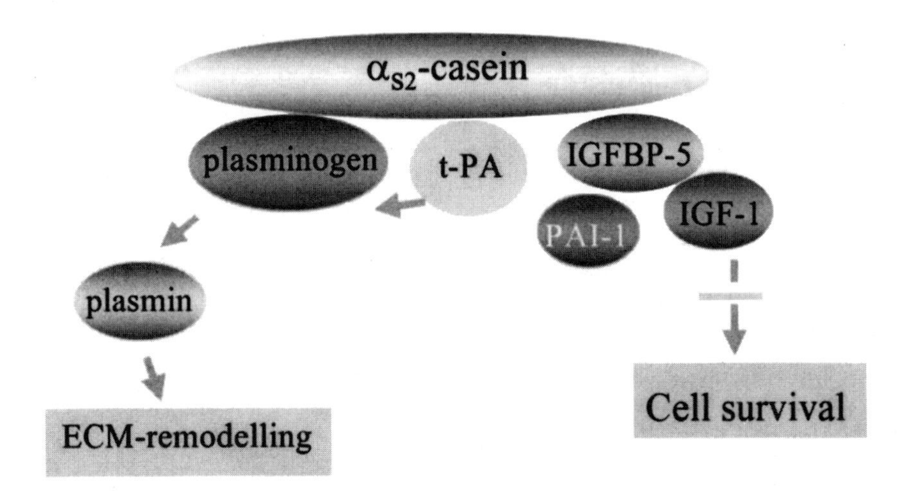


## 2. CONCLUSIONS

Perhaps one of the most intriguing aspects of IGFBP biology is the complexity of the IGFBPs produced and the way in which different species appear to adopt different IGFBPs to perform the equivalent function. For example IGFBP-3 appears to play a more important role than IGFBP-5 in apoptosis of human mammary cells. These two molecules are most closely related in an evolutionary sense and both possess other interesting properties including cell surface receptors[2,3,23,34] and share nuclear localisation signals[13,25,26] which may indicate that these molecules also have intracellular actions.

In this article we have developed a hypothesis which provides a mechanism to explain the interactive effects of GH and prolactin in maintaining mammary epithelial cell survival and also a possible mechanism which links epithelial cell death and extracellular matrix degradation. Our model proposes that GH increases IGF-1 production with IGF-1 acting as a survival factor. Prolactin acts by repressing expression of IGFBP-5 which allows IGF-1 to be maximally effective. The demonstration that IGFBP-5 binds to $\alpha_{s2}$-casein, as do plasminogen and tPA suggest that this milk protein assumes a novel role in the involuting mammary gland by serving as a matrix for the assembly of multi-protein complexes involved in tissue reorganization. The observation that IGFBP-5 binds to PAI-1 suggests that IGFBP-5 may possess IGF-independent effects by sequestering PAI-1 and thereby activating tPA activity and plasmin production. This offers a mechanism whereby apoptosis and ECM degradation can be highly coordinated. This proposal is supported by the observation the GH and prolactin can both suppress plasmin activation as well as inhibiting apoptosis.

These hypotheses will shortly be tested using transgenic mice expressing IGFBP-5 in the mammary gland as well as examining various IGFBP mutants in terms of their IGF-binding capabilities, or their ability to bind to $\alpha_{s2}$-casein and influence remodelling of the extracellular matrix.

**Acknowledgements**

We thank Miss Margaret Gardner for skilled technical assistance. This work was funded by the Scottish Executive Rural Affairs Department.

# 3. REFERENCES

1. Allander SV, Ehrenborg E, Luthman H, Powell DR. Conservation of IGFBP structure during evolution: cloning of chicken insulin-like growth factor binding protein-5. Progress in Growth Factor Research 6, 159-165, 1995.
2. Andress DL. Heparin modulates the binding of insulin-like growth factor (IGF) binding protein-5 to a membrane protein in osteoblastic cells. J Biol Chem 270, 28289-28296, 1995.
3. Andress DL. Insulin-like growth factor-binding protein-5 (IGFBP-5) stimulates phosphorylation of the IGFBP-5 receptor. Am J Physiol 274, E744-750, 1998.
4. Chen F, Capecchi MR. Paralogous mouse *Hox* genes, *Hoxa9*, *Hoxb9*, and *Hoxd9*, function together to control development of the mammary gland in response to pregnancy. Proceedings of the National Academy of Sciences USA 96, 541-546, 1998.
5. Donovan SM, Hintz RL, Rosenfeld RG. Investigation into the potential physiological sources of rat milk IGF-1 and IGF-binding proteins. J Endocrinol 145, 569-578, 1995.
6. Donovan SM, Hintz RL, Wilson DM, Rosenfeld RG. Insulin-like growth factors 1 and 2 and their binding proteins in rat milk. Pediatr Res 29, 50-55, 1991.
7. Donovan SM, McNeil LK, Jiminez-Flores R, Odle J. Insulin-like growth factors and Insulin-like growth factor binding proteins in porcine serum and milk throughout lactation. Pediatr Research 36, 159-168, 1994.
8. Donovan SM, Odle J. Growth factors in milk as mediators of infant development. Annual Review of Nutrition 14, 147-67, 1994.

9. Eriksson U, Duc G, Froesch ER, Zapf J. Insulin-like growth factors (IGF) 1 and 2 and IGF binding proteins (IGFBPs) in human colostrum/transitory milk during the first week postpartum: comparison with neonatal and maternal serum. Biochemical and Biophysical Research Communications 196, 267-273, 1993.
10. Fattal PG, Schneider DJ, Sobel BE, Billadello JJ. Post-transcriptional regulation of expression of plasminogen activator inhibitor type-1 mRNA by insulin and insulin-like growth factor 1. J Biol Chem 267, 12412-12415, 1992.
11. Heegaard CW, Andreasen PA, Petersen TE, Rasmussen LK. Binding of plasminogen and tissue-type plasminogen activator to dimeric $\alpha_{s2}$-casein accelerates plasmin generation. Fibrinolysis and Proteolysis 11, 29-36, 1997.
12. Holt C. The milk salts and their interactions with casein. In Advanced Dairy Chemistry, Vol 3 Editor PF Fox. pp 233-254, Chapman and Hall, London, 1997.
13. Jaques G, Noll K, Wegmann B, WittenS , Kogan E, Radulescu RT, Havemann K. Nuclear localization of insulin-like growth factor binding protein 3 in a lung cancer cell line. Endocrinology 138:1767-1770, 1997.
14. Jones JI, Gockerman A, Busby WH Jr, Wright G, Clemmons DR. Insulin-like growth factor binding protein –1 stimulates cell migration and binds to the $\alpha5\beta1$ integrin by means of its Arg-Gly-Asp sequence. Proc Natl Acad Sci USA 90, 10553-10557, 1993.
15. Jones JI, Gockerman A, Clemmons DR. Insulin-like growth factor binding protein-5 (IGFBP-5) binds to extracellular matrix and is phosphorylated. 74$^{th}$ Annual Meeting of the Endocrine Society, San Antonio TX, (A1284), 1992.
16. Lee CY, Bazer FW, Simmen FA. Expression of components of the insulin-like growth factor system in pig mammary glands and serum during pregnancy and pseudopregnancy: effects of oestrogen. J Endocrinol 137, 473-483, 1993.
17. Marcotty C, Frankenne F, Meuris S, Hennen G. Immunolocalisation and expression of insulin-like growth factor I (IGF-1) in the mammary gland during rat gestation and lactation. Mol Cell Endocrinol 99, 237-243, 1994.
18. Matrisian LM. Metalloproteinases and their inhibitors in matrix remodelling. Trends Genet 6, 121-125, 1990.
19. Mohan S, Bautista CM, Wergedal J, Baylink DJ. Isolation of an inhibitory insulin-like growth factor (IGF) binding protein from bone cell conditioned medium: a potential local regulator of IGF action. Proc Natl Acad Sci USA 86, 8338-8342, 1989.
20. Mol JA, Selman PJ, Sprang EP, van Neck JW, Oosterlaken-Dijksterhuis MA. The role of progestins, insulin-like growth factor (IGF) and IGF-binding proteins in the normal and neoplastic mammary gland of the bitch: a review. J Reprod Fertil Suppl 51, 339-344, 1997.
21. Nam TJ, W Busby W Jr, Clemmons DR. Insulin-like growth factor binding protein-5 binds to plasminogen activator inhibitor-1. Endocrinology 138, 2972-2978, 1997.
22. O'Connor R. Survival factors and apoptosis. Adv Biochem Eng Biotechnol 62, 137-166, 1998.
23. Oh Y, Muller HL, Lamson G, Rosenfeld RG. Insulin-like growth factor (IGF)-independent action of IGF-binding protein-3 in Hs578T human breast cancer cells. Cell surface binding and growth inhibition. J Biol Chem 268, 14964-14971, 1993.
24. Phippard DJ, Weber-Hall SJ, Sharpe PT, Naylor MS, Jayatalake H, Maas R, Woo ID, Roberts-Clark, Francis-West PH, Liu Y-H, Maxson R, Hill RE, Dale TC. Regulation of *Msx-1*, *Msx-2*, *Bmp-2* and *Bmp-4* during foetal and postnatal mammary gland development. Development 122, 2729-2737, 1996.
25. Radulescu RT. Nuclear localization signal in insulin-like growth factor-binding protein type 3. Trends BiochemSci 19:278, 1994.

26. Schedlich LJ, Young TF, Firth SM, Baxter RC. Insulin-like growth factor-binding protein (IGFBP)-3 and IGFBP-5 share a common nuclear transport pathway in T47D human breast carcinoma cells. J Biol Chem 273, 18347-52, 1998.
27. Simmen FA, Simmen RC, Reinhart G. Maternal and neonatal somatomedin C/insulin-like growth factor-I (IGF-1) and IGF binding proteins during early lactation in the pig. Developmental Biology 130, 16-27, 1998.
28. Skaar TC, Vega JR, Pyke SN, Baumrucker CR. Changes in insulin-like growth factor-binding proteins in bovine mammary secretions associated with pregnancy and parturition. J Endocrinol 131, 127-133, 1991.
29. Steele Perkins G, Turner J, Edman JC, Hari J, Pierce SB, Stover C, Rutter WJ, Roth RA. Expression and characterisation of a functional human insulin-like growth factor receptor. J Biol Chem 263, 11486-11492.
30. Tonner E, Barber MC, Travers MT, Logan A, Flint DJ. Hormonal control of insulin-like growth factor-binding protein–5 production in the involuting mammary gland of the rat. Endocrinology 138, 5101-5107, 1997.
31. Travers MT, Barber MC, Tonner E, Quarrie L, Wilde CJ, Flint DJ. The role of prolactin and growth hormone in the regulation of casein gene expression and mammary cell survival: relationships to milk synthesis and secretion. Endocrinology 137, 1530-1539, 1996.
32. Vega JR, Gibson CA, Skaar TC, Hadsell DL, Baumrucker CR. Insulin-like growth factor (IGF)-1 and -2 and IGF binding proteins in serum and mammary secretions during the dry period and early lactation in dairy cows. J Animal Science. 69, 2538-2547, 1991.
33. Wood TL. The IGFs and IGFBPs in ductal and alveolar development in rodent mammary glands. Journal of Mammary Gland Biol Neoplasia, in press.
34. Yamanaka Y, Fowlkes JL, Wilson EM, Rosenfeld RG, Oh Y. Characterization of insulin-like growth factor binding protein-3 (IGFBP-3) binding to human breast cancer cells: kinetics of IGFBP-3 binding and identification of receptor binding domain on the IGFBP-3 molecule. Endocrinology 140, 1319-1328, 1999.

6.

# Somatostatin and Opioid Receptors in Mammary Tissue.

*Role in cancer cell growth*

[1]Anastassia Hatzoglou, [1]Efstathia Bakogeorgou, [1]Marillena Kampa, [3]Simone Panagiotou, [3]Pierre-Marie Martin, [2]Spyros Loukas, [1]Elias Castanas

*[1]Lab. of Experimental Endocrinology, School of Medicine, Heraklion, [2]Institute of Biology, NCSR Democritos, Athens, Greece, [3]Experimental Cancerology, School of Medicine, Marseille France.*

Key words: Opioid, Somatostatin, Peptide receptors, cell proliferation

Abstract: Somatostatin and opioid systems, are the two main inhibitory systems in mammals. Both classes of substances have been identified in normal and malignant mammary gland, as well as their cognitive receptors. They have been implied in the inhibition of cell growth of cancer cells and cell lines, in a dose-dependant and reversible manner. Somatostatin acts through homologous receptors (SSTRs), belonging to five distinct classes (SSTR1-5). We, and others have identified SSTR2 and 3 as been the only SSTRs present in the breast. Furthermore, opioids act through the three classes of opioid receptors ($\mu$, $\delta$,$\kappa$). In the breast, kappa opioid receptor subtypes ($\kappa_1$-$\kappa_3$) are the most widely expressed. We further have shown that opioids, in addition to their binding to opioid receptors, compete for binding to SSTRs. This functional interaction, together with other identified modes of opioid action in the breast (modulation of steroid receptors, proteases' secretion, interaction with cytoskeletal elements), will be discussed, taking into consideration also the possible local production of casomorphins (casein-derived opioids), which are very potent antiproliferative agents.

## 1. INTRODUCTION

Different neuropeptides and hormones have been identified in breast cancer tissue, and breast cancer cell lines. They include growth hormone, prolactin, vasoactive intestinal polypeptide, GnRH, GHRH, IGF-1, and finally somatostatin and opioids[4,5,9,10,27,44,47,51]. These two peptides are the major negative regulators in mammals, implicated in a variety of processes, from hormone secretion to the modulation of cell proliferation. In the present review, we will focus on the role and possible interactions of these two systems in the human breast.

# 2. ROLE OF SOMATOSTATIN AND OPIOID SYSTEM IN THE BREAST.

## 2.1 Somatostatin system

Somatostatin, a naturally occurring deka-tetrapeptide, is produced mainly in the hypo-thalamus and the pancreas, but it has also been identified in a variety of normal and cancer tissues[38]. Two main forms of somatostatin are present in biological fluids: Somatostatin 14 and somatostatin 28. [35] The biological action of somatostatin is mediated through specific receptors, belonging to the seven-loop transmembrane receptor superfamily[38]. Five different receptors (named SSTR1-5) have been cloned and pharmacologically characterized in a variety of tissues[38].

In normal breast tissue, somatostatin immunoreactivity was detected in the stroma only, while epithelial cells show somatostatin production only after malignant transformation[5]. In addition somatostatin receptors have been identified in breast cancer tissue[41], and human breast cancer cell lines[15,45]. The existence of somatostatin receptors has been correlated in a great number of studies with other prognostic factors. The results are contradictory, depending on the number of tumors studied and the method of analysis, but, the general conclusion is that the assay of somatostatin receptors does not have any predictive significance on the development of the disease or the response to treatment[9,20,21]. Identification of receptor subtypes has also been assayed in malignant breast tissue. SSTR2 was the most abundantly found subtype in the breast, followed by SSTR3[42,43]. In other studies, and depending on the technique used, SSRT1 and SSTR4 were equally identified, but in a much lesser extend than the two other subtypes[8]. Finally, peritumoral blood veins, seem also to possess somatostatin receptors[40], and this finding indicates a possible regulatory mechanism for tumor growth[7].

The action of somatostatin in the breast might be direct or indirect. Direct action includes the inhibitory effect of the agent on cell proliferation[15,45]. This effect is dose dependant, occurring at concentrations similar to the affinity of somatostatin for its receptors. It was verified in a number of cell lines, both hormone-sensitive and independent, as well as in transplantable tumors. Indirect action implies the effect of somatostatin on growth hormone secretion and IGF-1 concentration[9,25]. This later factor was indeed found to have a stimulatory effect on cell growth. Nevertheless, clinical trials in small or larger series, with the use of somatostatin alone or in combination with other factors do not have shown, until now any positive result. A final implication of the somatostatinergic system in diagnosis is the scintigraphy with somatostatin analogs. It has been used during the last years, and although it was reported that it could distinguish breast cancer tumors, their metastasis and finally metastatic tumors in which the primary focus was not identified, its discriminative value was not judged satisfactory, unless with the somatostatin analogs used until now, with the possible exception of intraoperative detection[6].

## 2.2 Opioid system

Endogenous opioids derive from three different precursor proteins, namely proenkephalin A and B (prodynorphin) and proopiomelanocortin (POMC) [16]. These three proteins give rise to a number of opioid peptides with different affinities towards opioid receptors. These later receptors belong equally to the seven-loop membrane receptor superfamily[33,37], and are distinguished both pharmacologically and biochemically to three main categories: delta, mu and kappa. In addition, pharmacological evidences exist about a further subdivision of delta receptors to delta 1 and 2, of mu receptors to mu1 and mu2 and finally of kappa receptors to kappa1-3 [1,2].

In addition to endogenous opioid peptides, a number of food derives opioids have been identified, from different proteins including hemoglobin, gluten, and caseins. Both alpha and beta caseins include peptides with a potent opioid activity (see Kampa et al[18], for a discussion). These later have been tested and found to decrease the proliferation of breast and prostate cells. In addition, we have identified a very potent opioid pentapeptide, derived from human alpha$_{S1}$-casein[18], with remarkable antiproliferative activity in different systems, including breast[18], prostate[17], bladder (in preparation), kidney[14,31,32], and different cells of hemopoietic origin (in preparation). It was named alpha$_{S1}$-casomorphin, and its structure is Tyr-Val-Pro-Phe-Pro-$NH_2$ [18].

Opioids were found in a great diversity of primary human tumors. Between them breast tumors were found to possess endogenous opioid peptides by different techniques[4,44,51]. Furthermore, opioid receptors were equally found in primary tumors, as well as in different breast cancer cell lines. Using opioid alkaloids, endogenous opioid peptide analogs and casomorphin peptides, we[12,13], as well as other groups[24] have shown that a dose-dependent and mostly reversible inhibition of cell proliferation can be obtained. Nevertheless, either the reversibility of this effect was not complete, either opioid agonists were not inhibited by selective antagonists. In addition, as was the case for morphine, no specific receptors were identified[12]. These results indicate a possible mediation of the antiproliferative effect by another membrane receptor system.

In general seven loops membrane receptors act through adenylate cyclase increased, or decreased activity. Somatostatin and opioid receptors decrease intracellular cAMP, through an inhibition of adenylate cyclase[39]. In addition, in breast cancer cell lines, as well as in different other organs (kidney, prostate), we have identified further mechanisms of opioid action: Interaction with the cellular cytoskeletal elements[28,29], both after an acute and a prolonged application. Furthermore, a decrease of steroid receptor levels[30], and a decrease of the steroid regulated secretion of enzymes were also identified[29]. These different mechanisms, dependent and/or independent from steroid hormones and cAMP, indicate that the action of these drugs might be much more extended from previously described. It is further interesting to note that $\alpha_{S1}$-casomorphin is a very potent substance for all of the above actions, indicating that, probably, this peptide, might be a promising agent in cancer chemotherapy.

## 3. INTERACTION BETWEEN OPIOID AND SOMATOSTATIN SYSTEMS

The seven loop membrane receptor superfamily possesses a number of homologies between its members (see Kieffer et al[19], for a description). This result, in addition with the possible coexpression of multiple receptors on the

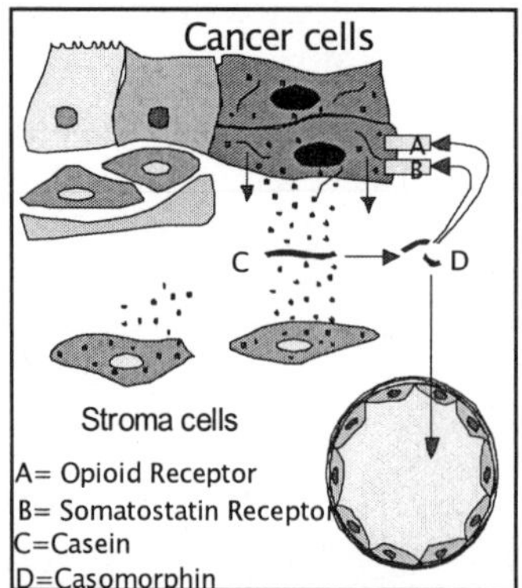


**Fig. 1.** Hypothetical model for the action of casomorphin peptides in breast cancer

same cell, makes possible an interaction either at the ligand, or at a post-receptor level. Indeed, such interactions have been reported for the somatostatin and the EGF[48], VIP[49], or the opioid receptor[26,50], and the opioid with the adrenergic[22,23,34,46], or the dopamine receptor[3].

The coexpression of opioids and somatostatin receptors in breast cancer cell lines, the inhibitory role of both systems in cell proliferation, and a possible common mode of action of these two families of receptors, via an inhibition of adenylate cyclase[37,38], together with the partial inhibition of opioid agonists by selective antagonists, makes possible such an interaction. We have indeed reported that some opioid alkaloids inhibit somatostatin binding in the human T47D breast cancer cell line[15]. In addition, different casomorphins interact equally with somatostatin receptors[13], indicating that this cross talking, at the receptor level might be of a biological importance. This interaction is not restricted in the breast, but was also found in other organs as the prostate[17] and the kidney[14]. Therefore, this effect might be a general possible cross-talking between the two main inhibitory mammalian systems. Finally, mathematical simulations have shown that, although several systems could share a common signalling pathway, they might have compartmentated intracellular access to second messengers, such as cyclase or GTPasas molecules[11].

## 4. CONCLUSION

In view of the above results, a possible scheme emerges for the role of opioids and somatostatin in breast cancer. It is presented in Fig. 1. In this hypothetical action, cancer cells become depolarized by their malignant transformation. In addition they produce a disruption of the basal membrane. Therefore, their secretion, which normally occurs towards the acinar lumen occurs equally towards the stroma. Stromal cells, and transformed epithelial cancer cells, produce high amounts of different proteases, which in turn could attack and partially degrade casein molecules, giving rise to casomorphin peptides. These peptides could bind to opioid receptors of malignant cells, inhibiting their proliferation, by the different mechanisms involved above, and by lowering the availability of the cells to steroids, which are responsible, at an early stage of carcinogenesis for growth. In addition, opioids, by binding to somatostatin receptors of the same malignant cells, and peritumoral vessels, they potentiate their inhibitory action. This scheme of biological action provides a possible control of tumor progression and spread, and could be of value at early stages of breast cancer.

In addition to their production by casein degradation, opioids and somatostatin could arrive to tumor cell environment by several other sources. Indeed, the general circulation has a somatostatin and opioid content, provided by the secretion of these neuropeptides by other endocrine cells such as the hypothalamus, the pituitary, the pancreas or the adrenals. Local production, especially by neuroendocrine cells, production by infiltrating lymphocytes, is another potential source of these neuropeptides. It is therefore plausible to propose that these two systems (opioids and somatostatin) may play a crucial role for the development of tumor control, especially at initial stages, and that possible pharmaceutical substances, acting via these two types of receptors might be a valuable addition for cancer control.

## 5. REFERENCES


1. Castanas, E., Bourhim, N., Giraud, P., Boudouresque, F., Cantau, P., Oliver, C., Interaction of opiates with opioid binding sites in the bovine adrenal medulla: I Interaction with delta and mu sites., J. Neurochem. 45, 677, 1985a.
2. Castanas, E., Bourhim, N., Giraud, P., Boudouresque, F., Cantau, P., Oliver, C., Interaction of opiates with opioid binding sites in the bovine adrenal medulla: II Interaction with kappa sites., J. Neurochem. 45, 688, 1985b.
3. Castanas, E., P. Jaquet, G. Gunz, P. Cantau and P. Giraud, Direct action of opiates on bromocriptine-inhibited prolactin release by human prolactinoma cells in primary culture, J Clin Endocrinol Metab 61, 963, 1985.
4. Chatikhine, V.A., A. Chevrier, C. Chauzy, C. Duval, J. d'Anjou, N. Girard and B. Delpech, Expression of opioid peptides in cells and stroma of human breast cancer and adenofibromas, Cancer Lett 77, 51, 1994.
5. Ciocca, D.R., L.A. Puy, L.C. Fasoli, O. Tello, J.C. Aznar, F.E. Gago, S.I. Papa and R. Sonego, Corticotropin-releasing hormone, luteinizing hormone-releasing hormone, growth hormone-releasing hormone, and somatostatin-like immunoreactivities in biopsies from breast cancer patients, Breast Cancer Res Treat 15, 175, 1990.
6. Cuntz, M.C., E.A. Levine, T.M. O'Dorisio, J.C. Watson, D.A. Wray, G.D. Espenan, C. McKnight, J.R. Meier, L.J. Weber, R. Mera, M.S. O'Dorisio and E.A. Woltering, Intraoperative gamma detection of 125I-lanreotide in women with primary breast cancer [In Process Citation], Ann Surg Oncol 6, 367, 1999.
7. Denzler, B. and J.C. Reubi, Expression of somatostatin receptors in peritumoral veins of human tumors, Cancer 85, 188, 1999.
8. Evans, A.A., T. Crook, S.A. Laws, A.C. Gough, G.T. Royle and J.N. Primrose, Analysis of somatostatin receptor subtype mRNA expression in human breast cancer, Br J Cancer 75, 798, 1997.
9. Foekens, J.A., H. Portengen, W.L. van Putten, A.M. Trapman, J.C. Reubi, J. Alexieva-Figusch and J.G. Klijn, Prognostic value of receptors for insulin-like growth factor 1, somatostatin, and epidermal growth factor in human breast cancer, Cancer Res 49, 7002, 1989.
10. Gespach, C., W. Bawab, P. de Cremoux and F. Calvo, Pharmacology, molecular identification and functional characteristics of vasoactive intestinal peptide receptors in human breast cancer cells, Cancer Res 48, 5079, 1988.
11. Graeser, D. and R.R. Neubig, Compartmentation of receptors and guanine nucleotide-binding proteins in NG108-15 cells: lack of cross-talk in agonist binding among the alpha 2-adrenergic, muscarinic, and opiate receptors, Mol Pharmacol 43, 434, 1993.

12. Hatzoglou, A., E. Bakogeorgou and E. Castanas, The antiproliferative effect of opioid receptor agonists on the T47D human breast cancer cell line, is partially mediated through opioid receptors, Eur J Pharmacol 296, 199, 1996a.
13. Hatzoglou, A., E. Bakogeorgou, C. Hatzoglou, P.M. Martin and E. Castanas, Antiproliferative and receptor binding properties of alpha- and beta- casomorphins in the T47D human breast cancer cell line, Eur J Pharmacol 310, 217, 1996b.
14. Hatzoglou, A., E. Bakogeorgou, E. Papakonstanti, C. Stournaras, D.S. Emmanouel and E. Castanas, Identification and characterization of opioid and somatostatin binding sites in the opossum kidney (OK) cell line and their effect on growth, J Cell Biochem 63, 410, 1996c.
15. Hatzoglou, A., L. Ouafik, E. Bakogeorgou, K. Thermos and E. Castanas, Morphine cross-reacts with somatostatin receptor SSTR2 in the T47D human breast cancer cell line and decreases cell growth, Cancer Res 55, 5632, 1995.
16. Hollt, V., Opioid peptide processing and receptor selectivity, Annu Rev Pharmacol Toxicol 26, 59, 1986.
17. Kampa, M., E. Bakogeorgou, A. Hatzoglou, A. Damianaki, P.M. Martin and E. Castanas, Opioid alkaloids and casomorphin peptides decrease the proliferation of prostatic cancer cell lines (LNCaP, PC3 and DU145) through a partial interaction with opioid receptors, Eur J Pharmacol 335, 255, 1997.
18. Kampa, M., S. Loukas, A. Hatzoglou, P. Martin, P.M. Martin and E. Castanas, Identification of a novel opioid peptide (Tyr-Val-Pro-Phe-Pro) derived from human alpha S1 casein (alpha S1-casomorphin, and alpha S1- casomorphin amide), Biochem J 319, 903, 1996.
19. Kieffer, B.L., K. Befort, C. Gaveriaux-Ruff and C.G. Hirth, The delta-opioid receptor: Isolation of a cDNA by expression cloning and pharmacological characterization., Proc natl Acad Sci (USA) 89, 12048, 1992.
20. Klijn, J.G., B. Setyono-Han, M. Bontenbal, C. Seynaeve and J. Foekens, Novel endocrine therapies in breast cancer, Acta Oncol 35, 30, 1996.
21. Lamberts, S.W., W.W. de Herder, P.M. van Koetsveld, J.W. Koper, A.J. van der Lely, H.A. Visser-Wisselaar and L.J. Hofland, Somatostatin receptors: clinical implications for endocrinology and oncology, Ciba Found Symp 190, 222, 1995.
22. Lameh, J., S. Eiger and W. Sadee, Interaction among mu-opioid receptors and alpha 2-adrenoceptors on SH- SY5Y human neuroblastoma cells, Eur J Pharmacol 227, 19, 1992.
23. Lee, S., C.R. Rosenberg and J.M. Musacchio, Cross-dependence to opioid and alpha 2-adrenergic receptor agonists in NG108-15 cells, Faseb J 2, 52, 1988.
24. Maneckjee, R., R. Biswas and B.K. Vonderhaar, Binding of opioids to human MCF-7 breast cancer cells and their effects on growth, Cancer Res 50, 2234, 1990.
25. Manni, A., Somatostatin and growth hormone regulation in cancer, Biotherapy 4, 31, 1992.
26. Maurer, R., B.H. Gaehwiler, H.H. Buescher and R.C.R. Hill, D., Opiate antagonistic properties of an octapeptide somatostatin analog., Proc. Natl. Acad. Sci. (USA) 79, 4815, 1982.
27. Nass, S.J. and N.E. Davidson, The biology of breast cancer [In Process Citation], Hematol Oncol Clin North Am 13, 311, 1999.
28. Panagiotou, S., E. Bakogeorgou, E. Papakonstanti, A. Hatzoglou, F. Wallet, C. Dussert, C. Stournaras, P.M. Martin and E. Castanas, Opioid agonists modify breast cancer cell proliferation by blocking cells to the G2/M phase of the cycle: involvement of cytoskeletal elements, J Cell Biochem 73, 204, 1999.

29. Panagiotou, S., A. Hatzoglou, F. Calvo, P.M. Martin and E. Castanas, Modulation of the estrogen-regulated proteins cathepsin D and pS2 by opioid agonists in hormone-sensitive breast cancer cell lines (MCF7 and T47D): evidence for an interaction between the two systems, J Cell Biochem 71, 416, 1998.
30. Panagiotou, S., P.-M. Martin and E. Castanas, Opioid agonists decrease the concentration of steroid receptors in hormone-sensitive breast cancer cell lines., in: 2emes Journees of the IFR Jean RocheMarseille, France), 1997.
31. Papakonstanti, E.A., E. Bakogeorgou, E. Castanas, D.S. Emmanouel, R. Hartig and C. Stournaras, Early alterations of actin cytoskeleton in OK cells by opioids., J. Cell. Biochem. , (accepted), 1998.
32. Papakostanti, E.A., D.S. Emmanouel, A. Gravanis and C. Stournaras, Na+/Pi cotransport alters rapidly cytoskeletal protein polymerization dynamics in Opossum Kidney cells., Biochem. J. 315, 1996.
33. Raynor, K., H. Kong, S. Law, J. Heerding, M. Tallent, F. Livingston, J. Hines and T. Reisine, Molecular biology of opioid receptors, NIDA Res Monogr 161, 83, 1996.
34. Reggiani, A., A. Carenzi and D.D. Bella, Opioids and beta-receptors interaction: further studies in cultured cells, NIDA Res Monogr 75, 347, 1986.
35. Reichlin, S., Somatostatin, in: Brain Peptides, eds. D.T. Krieger, M.J. Brownstein and J.B. Martin (John Wiley and sons, Inc., New York) p. 711, 1983.
36. Reisine, T., Somatostatin, Cell Mol Neurobiol 15, 597, 1995.
37. Reisine, T. and G.I. Bell, Molecular biology of opioid receptors [see comments], Trends Neurosci 16, 506, 1993.
38. Reisine, T. and G.I. Bell, Molecular properties of somatostatin receptors, Neuroscience 67, 777, 1995.
39. Reisine, T., S.F. Law, A. Blake and M. Tallent, Molecular mechanisms of opiate receptor coupling to G proteins and effector systems, Ann N Y Acad Sci 780, 168, 1996.
40. Reubi, J.C., U. Horisberger and J. Laissue, High density of somatostatin receptors in veins surrounding human cancer tissue: role in tumor-host interaction?, Int J Cancer 56, 681, 1994a.
41. Reubi, J.C., E. Krenning, S.W. Lamberts and L. Kvols, Somatostatin receptors in malignant tissues, J Steroid Biochem Mol Biol 37, 1073, 1990.
42. Reubi, J.C., J.C. Schaer, B. Waser and G. Mengod, Expression and localization of somatostatin receptor SSTR1, SSTR2, and SSTR3 messenger RNAs in primary human tumors using in situ hybridization, Cancer Res 54, 3455, 1994b.
43. Schulz, S., J. Schmitt, D. Wiborny, H. Schmidt, S. Olbricht, W. Weise, A. Roessner, C. Gramsch and V. Hollt, Immunocytochemical detection of somatostatin receptors sst1, sst2A, sst2B, and sst3 in paraffin-embedded breast cancer tissue using subtype- specific antibodies, Clin Cancer Res 4, 2047, 1998.
44. Scopsi, L., E. Balslev, N. Brunner, H.S. Poulsen, J. Andersen, F. Rank and L.I. Larsson, Immunoreactive opioid peptides in human breast cancer, Am J Pathol 134, 473, 1989.
45. Setyono-Han, B., M.S. Henkelman, J.A. Foekens and G.M. Klijn, Direct inhibitory effects of somatostatin (analogues) on the growth of human breast cancer cells, Cancer Res 47, 1566, 1987.
46. Simantov, R., H. Nadler and R. Levy, A genetic approach to reveal the action of the opiate receptor in selected neuroblastoma-glioma cells. Interaction with alpha-adrenoceptors, calmodulin and Ca2+-ATPase, Eur J Biochem 128, 461, 1982.
47. Spring-Mills, E.J., S.B. Stearns, P.J. Numann and P.H. Smith, Immunocytochemical localization of insulin- and somatostatin-like material in human breast tumors, Life Sci 35, 185, 1984.

48. Vidal, C., I. Rauly, M. Zeggari, N. Delesque, J.P. Esteve, N. Saint-Laurent, N. Vaysse and C. Susini, Up-regulation of somatostatin receptors by epidermal growth factor and gastrin in pancreatic cancer cells, Mol Pharmacol 46, 97, 1994.
49. Virgolini, I., Q. Yang, S. Li, P. Angelberger, N. Neuhold, B. Niederle, W. Scheithauer and P. Valent, Cross-competition between vasoactive intestinal peptide and somatostatin for binding to tumor cell membrane receptors, Cancer Res 54, 690, 1994.
50. Walker, J.M., W.D. Bowen, S.T. Atkins, M.K. Hemstreet and D.H. Coy, μ- Opiate binding and morphine antagonism by octapeptide analogs of somatostatin., Peptides 8, 869, 1987.
51. Zagon, I.S., P.J. McLaughlin, S.R. Goodman and R.E. Rhodes, Opioid receptors and endogenous opioids in diverse human and animal cancers, J Natl Cancer Inst 79, 1059, 1987.

7.

# Heparan Sulphate

## *Regulation of growth factors in the mammary gland*

Kirsty L. Bateman, Maryse Delehedde, Nicolas Sergeant, Isabelle Wartelle, Rishma Vidyasagar, and David G. Fernig
*School of Biological Sciences, University of Liverpool, UK.*

Key words Heparan sulphate, glycosaminoglycan

Abstract: Heparan sulphate (HS) is a linear glycosaminoglycan found ubiquitously on the surface and in the pericellular matrix of metazoan cells that is covalently attached to core proteins to form HS proteoglycans. HS interacts specifically with a large number (>100) of extracellular regulatory proteins, many of which are key players in mammary development and function. HS thus acts a receptor, in which capacity it regulates the bioavailability, localisation and activity of these regulatory proteins.

## 1. HS: STRUCTURAL BASIS OF FUNCTION

The initial product of HS biosynthesis is a linear polysaccharide of the disaccharide glucuronic acid β1-4 N-acetyl glucosamine. This starting structure, usually of several hundred saccharides in length, is the substrate for a series of modification reactions, the first of which results in the replacement of the N-acetyl group on glucosamine by a sulphate. The N-sulphate acts as a signal for the majority of the subsequent modifications, which include 6-O and 3-O sulphation of the glucosamine, epimerization of the glucuronic acid to iduronic acid and the 2-O sulphation of the latter saccharide. The sequence complexity of HS arises from two factors. Firstly, only a fraction of all potential modification reactions take place. Secondly, the modifications are clustered so that the HS chain consists of alternating unmodified domains and highly modified S-domains, enriched in iduronate residues and sulphate groups. The potential repertoire of S-domain structures is much larger than the number of proteins encoded by the human genome. Since proteins bind to S-domains there is sufficient structural diversity in HS to ensure that each protein-HS interaction is highly specific.

*Table 1.* Examples of HS-binding regulatory proteins

| | |
|---|---|
| Fibroblast growth family[1-3] | IGF-1 binding proteins (BP3, BP5) |
| Transforming growth factor β1 (TGFβ1)[4] | HARP/HB-GAM |
| Wnt family[5, 6] | Platelet-derived growth factor |
| Amphiregulin, betacellulin heparin-binding EGF (HB-EGF), heregulins/neuregulins[7-9] | Hepatocyte growth factor /scatter factor (HGF/SF) 10 |
| Cytokines, e.g., interleukins interferon γ [11, 12] | Collagens, fibronectin, laminin. [13, 14] |
| Lipoprotein lipase | |

The majority of growth factors and morphogens implicated in the regulation of mammary gland development bind to HS (Table 1) and it appears that each of these proteins recognises a different structure in HS. In addition, key regulatory binding proteins such as IGFBP3 and IGFBP5 also bind to HS. The three main components of the extracellular matrix and the basement membrane, collagens, fibronectin and laminin, also interact with specific S-domains in HS, though it would appear that the collagens collectively recognise saccharide sequences of similar, if not identical structure. Finally, lipoprotein lipase, a key enzyme in lipid metabolism, which is an important feature of lactation is anchored to its target cells through HS.

## 2. REGULATORY ACTIVITIES OF HS

HS critically controls two key features of extracellular regulatory proteins, their localisation and sequestration and their generation of intracellular signals.

The sequestration of a growth factor on extracellular HS was first proposed 12 years ago[15]. This can prevent the diffusion of the growth factor within a tissue compartment or between tissue compartments, as well as allowing a local store of the growth factor to act on a restricted number of cells. In the mammary gland there is direct evidence for HS performing this function with regard to FGF-2 and FGF-3, and indirect evidence for many other HS-binding proteins. A regulatory role for HS is supported by the observation of changes in the structure of the HS chains which accompany the growth and development of many epithelia, including the mammary gland[16]. Moreover, the association between the growth state of the cells (quiescent duct versus growing terminal end bud) and the number of HS receptors for FGF-2 associated with these structures in vivo suggests that HS dynamically regulates the activity of FGF-2 in the developing mammary gland[17]. Other results support a similar role for HS in the regulation of the activity of HGF/SF and Wnt family members and it is likely from studies in other systems that HS has an analogous function towards all the growth factors and cytokines in Table 1.

Many of the HS-binding proteins possess a dual receptor system, which consists of a conventional signalling receptor, e.g., receptor tyrosine kinase or integrin and an HS receptor. Upon ligand binding the conventional signalling receptor activates specific intracellular signalling pathways. The dual receptor systems provide a means to increase the combinatorial outcome of cellular signals generated by a single effector.

FGF-2 is the classic growth factor with a dual receptor system[1]. It elicits a cellular response through its FGFR receptor tyrosine kinase. In the presence of the activating HS receptor, a specific temporal pattern of activation of signalling pathways is observed. For example, following the addition of FGF-2 to $G_0$ synchronised rat mammary (Rama) 27 fibroblasts, the phosphorylation of mitogen-activated protein kinase (MAPK) peaks after 10 min, and then declines to half maximal levels, which are maintained well into S-phase, 18 h later. The early peak of MAPK phosphorylation is probably related to the activation of the c-fos promoter. In the same cells treated with chlorate, which prevents the biosynthesis of the HS receptor, the signalling pathways are activated in a transient manner, which is sufficient for the stimulation of c-fos transcription, but not for entry into the cell cycle.

HGF/SF, a product of the mammary stroma, has been thought to target exclusively the epithelium. Indeed HGF/SF does not stimulate the growth of Rama 27 fibroblasts and these cells possess very few HS receptors for HGF/SF[18]. However, Rama 27 fibroblasts do produce c-met, the tyrosine kinase receptor for HGF/SF. In the absence of its HS receptor, HGF/SF stimulates the phosphorylation of MAPK in Rama 27 fibroblasts biphasically, a response that is neither sustained nor sufficient for cell proliferation.

There is also evidence for a contribution of the proteoglycan core protein to dual receptor systems. Human mammary (Huma) myoepithelial-like 109 cells possess both the HS18 and the c-met receptors for HGF. When these cells are grown on a plastic substratum, they express the proteoglycans perlecan and syndecan-2 on their cell surfaces and, surprisingly, HGF/SF has no effect on the proliferation of these cells. However, under these conditions HGF/SF does cause an early and transient stimulation of the phosphorylation of MAPK. In contrast, HGF/SF induces a sustained activation of MAPK and stimulates the proliferation of Huma 109 cells grown on a fibronectin or collagen substratum. Compared to cells grown on a plastic substratum, Huma 109 myoepithelial-like cells grown on fibronectin or collagen substrata show a marked reduction in the expression of cell surface perlecan, but not syndecan-2.

There are, as yet, no studies in transgenic mice in which a perturbation to HS biosynthesis in the mammary gland has been introduced. A number of studies have examined the effects of chlorate, an inhibitor of the sulphation of HS on the differentiation of mammary epithelial cells in collagen gels. The results show that both the basal morphogenic potential of the cells and the ability of FGFs and HGF/SF to stimulate ductal morphogenesis in this system are dependent on HS.

In conclusion HS is a ubiquitous component of the cell surface and pericellular matrix that binds to and regulates the activities of many of the key proteins which control the development and function of the mammary gland. Moreover, HS chains are carried on the core proteins of proteoglycans and each chain will possess several S-domains. Thus HS, strategically placed at the cell surface and in the pericellular matrix, has the potential to integrate signals arriving from disparate effectors.

### Acknowledgments

We apologise for not citing the key papers of numerous colleagues, due to the lack of space. We thank the AICR, the Cancer and Polio Research Fund, the Mizutani Foundation for Glycoscience, the Medical Research Council, The Royal Society and the NWCRF for financial support.

## 3. REFERENCES

1. Fernig, D. G., and Gallagher, J. T., 1994, Fibroblast growth factors: an information network controlling tissue growth, morphogenesis and repair. *Prog. Growth Factor Res.* **5**: 353-377.
2. Rahmoune, H., Gallagher, J. T., Rudland, P. S., and Fernig, D. G., 1998, Interaction of heparan sulphate from mammary cells with extracellular regulatory proteins. Acidic and fibroblast growth factor: regulation of the activity of bFGF by high and low affinity binding sites in heparan sulphate. *J. Biol. Chem.* **273**: 7303-7310.
3. Turnbull, J. E., Fernig, D. G., Ke, Y. Q., Wilkinson, M. C., and Gallagher, J. T., 1992, Identification of the basic fibroblast growth factor binding sequence in fibroblast heparan sulfate. *J. Biol. Chem.* **267**: 10337-10341.
4. Lyon, M., Rushton, G., and Gallagher, J. T., 1997, The interaction of the transforming growth factor-betas with heparin/heparan sulfate is isoform-specific. *J. Biol. Chem.* **272**: 18000-6.
5. Lin, X. H., and Perrimon, N., 1999, Dally cooperates with Drosophila Frizzled 2 to transduce Wingless signalling. *Nature.* **400**: 281-284.
6. Tsuda, M., Kamimura, K., Nakato, H., Archer, M., Staatz, W., Fox, B., Humphrey, M., et al., 1999, The cell-surface proteoglycan Dally regulates Wingless signalling in Drosophila. *Nature.* **400**: 276-280.

7. Aviezer, D., and Yayon, A., 1994, Heparin-dependent binding and autophosphorylation of epidermal growth factor (EGF) receptor by heparin-binding EGF-like growth factor but not by EGF. *Proc Natl Acad Sci U S A*. **91**: 12173-7.
8. Higashiyama, S., Abraham, J. A., and Klagsbrun, M., 1993, Heparin-binding EGF-like growth factor stimulation of smooth muscle cell migration: dependence on interactions with cell surface heparan sulfate. *J Cell Biol*. **122**: 933-40.
9. Luetteke, N. C., Qui, T. H., Fenton, S. E., Troyer, K. L., Riedel, R. F., Chang, A., and Lee, D. C., 1999, Targeted inactivation of the EGF and amphiregulin genes reveals distinct roles for EGF receptor ligands in mouse mammary gland development. *Development*. **126**: 2739-2750.
10. Lyon, M., Deakin, J. A., Mizuno, K., Nakamura, T., and Gallagher, J. T., 1994, Interaction of hepatocyte growth factor with heparan sulfate. Elucidation of the major heparan sulfate structural determinants. *J Biol Chem*. **269**: 11216-23.
11. Hasan, M., Najjam, S., Gordon, M. Y., Gibbs, R. V., and Rider, C. C., 1999, IL-12 is a heparin-binding cytokine. *J Immunol*. **162**: 1064-70.
12. Lortat-Jacob, H., and Grimaud, J. A., 1991, Interferon-gamma binds to heparan sulfate by a cluster of amino acids located in the C-terminal part of the molecule. *FEBS Lett*. **280**: 152-4.
13. LeBaron, R. G., Hook, A., Esko, J. D., Gay, S., and Hook, M., 1989, Binding of heparan sulfate to type V collagen. A mechanism of cell- substrate adhesion. *J Biol Chem*. **264**: 7950-6.
14. Walker, A., and Gallagher, J. T., 1996, Structural domains of heparan sulphate for specific recognition of the c-terminal heparin-binding domain of human plasma fibronectin (HEPII). *Biochem. J.* **317**: 871-877.
15. Vlodavsky, I., Folkman, J., Sullivan, R., Fridman, R., Ishai-Michaeli, R., Sasse, J., and Klagsbrun, M., 1987, Endothelial cell-derived basic fibroblast growth factor: synthesis and deposition into subendothelial extracellular matrix. *Proc. Natl Acad. Sci. USA*. **84**: 2292-2296.
16. Rudland, P. S., Fernig, D. G., and Leinster, S. (ed.), 1998, *Mammary Development and Cancer*, vol. 63. Portland Press, Colchester, UK.
17. Rudland, P. S., Barraclough, R., Fernig, D. G., and Smith, J. A. 1997. Mammary stem cells in normal development and cancer, p. 147-232. *In* C. Potton (ed.), Stem cells and cancer. Churchill Livingstone.
18. Rahmoune, H., Gallagher, J. T., Rudland, P. S., and Fernig, D. G., 1998, Interaction of heparan sulphate from mammary cells with hepatocyte growth factor/scatter factor. *Biochemistry (USA)*. **37**: 6003-6008.

8.

# Progestin-induced Mammary Growth Hormone (GH) Production

Jan A. Mol, Irma Lantinga-van Leeuwen, [2]Evert van Garderen, Ad Rijnberk
*Dept. of Clinical Sciences of Companion Animals and Pathology[2], Faculty of Veterinary Medicine, Utrecht University, The Netherlands*

Key words: growth hormone, dog, progestins

Abstract: Toxicity studies using beagle dogs revealed in the 1980s that synthetic progestins may induce a syndrome of growth hormone (GH) excess, known as acromegaly, and the development of predominantly benign mammary hyperplasia. In the early 1990s is was discovered that progestin-induced GH excess in the dog originates within the mammary gland. This mammary-derived GH may have endocrine, para/autocrine as well as exocrine effects. The expression of GH mRNA is also found in cats and humans indicating that mammary GH expression is not unique for the dog. The mammary gene is identical to the pituitary-expressed gene and uses the same promoter. Nevertheless a striking difference exists in the mammary gland. Pit-1, which is a prerequisite factor for pituitary GH mRNA expression, is likely not involved in the mammary gene expression. These studies shed new light on the mechanism of progesterone-induced mammary hyperplasia and urges for further research on potential adverse effects of synthetic progestins.

## 1. INTRODUCTION

Steroid hormones play an essential role in the proliferation and differentiation of the mammary gland. As a consequence steroid hormones make the mammary epithelium prone to oncogenic derailment. In addition they may stimulate the promotion and progression of mammary tumors[1]. Studies using progesterone receptor knockout mice (PRKO) have revealed the importance of progesterone in mammary development. Progesterone is essential for the lobulo-alveolar outgrowth of mammary epithelium[2]. The highest DNA labeling indices are found in the progesterone-dominated phase of the sexual cycle[3,4]. Progesterone exerts its effect through activation of type A- or type B-progesterone receptors, which bind to specific DNA response elements after dimerization[4,5]. In addition specific co-activator or co-repressor proteins are essential for interaction with the general transcription complex and modulation of the transcription process[6]. When wild-type mammary epithelium is transplanted into mammary tissue of PR-/- mice, progesterone stimulates the normal outgrowth of the neighbouring PR-/- epithelium[7]. This indicates that progesterone stimulates epithelial proliferation in part by a paracrine factor.

## 2. MAMMARY GROWTH HORMONE

Synthetic progestins as well as endogenous progesterone may induce excessive plasma GH concentrations in the dog resulting in acromegaly and insulin resistance[8-10]. The elevated plasma GH concentrations were characterized by the absence of a pulse pattern and insensitivity to stimulation and inhibition tests,[11,12] with the exception of an inhibition by the progesterone antagonist RUU 486[11]. These autonomous characteristics, and the absence of a decrease of plasma GH concentrations after complete hypophysectomy was the impetus for a search for an extra-pituitary source. This resulted in the finding that the mammary gland was the source of progestin-induced plasma GH concentrations[13]. This was confirmed by an arterio-venous gradient over the mammary gland, an immediate decline of plasma GH concentrations after complete mammectomy and the immunohistochemical staining of GH positive cells in foci of hyperplastic mammary tissue[13,14].

The expression of the gene encoding GH was also proven by RT-PCR, and Northern blot in the canine mammary gland[15,16]. The stimulation of GH mRNA expression after progestin treatment in the cat[15] and the presence of GH mRNA in the human mammary gland[17] did prove that mammary GH expression is not a dog specific phenomenon[18].

## 3. REGULATION OF MAMMARY GH GENE EXPRESSION

Improved insights in the regulation of mammary GH synthesis may lead to specific treatments aimed to interfere with mammary GH production.

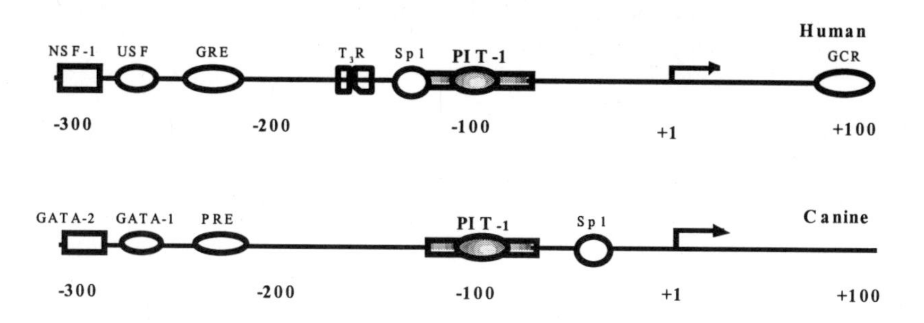


Fig. 1 Comparison of the human and canine GH gene promoter region[19,20].

Analysis of the mammary GH mRNA revealed that the mammary transcript is identical to that of the pituitary, including the 5'UTR. This suggests that also the same promoter is used. An essential element in the GH gene promoter is a Pit-1 response element. Mutations in Pit-1 (also called POU1F1) result into dwarfism due to disturbances in the development of GH (and prolactin and TSH) producing cells in the pituitary. Comparison of the canine GH gene promoter with that of the human GH gene showed the presence of a Pit-1 and a progesterone receptor response element[20] (Fig. 1). However, analysis of Pit-1 mRNA within the mammary gland revealed that mammary GH expression occurred in the absence of Pit-1 expression[16]. In addition the finding that German shepherd dwarfs, which have a pituitary anomaly resulting in GH deficiency, produce GH after progesterone stimulation further strengthen the differences in GH gene expression between mammary and pituitary tissue[21].

Immunohistochemically all canine GH producing mammary cells stain for the PR but not all PR containing cells express GH. In malignant tumors GH mRNA can be found in the absence of detectable PR by ligand binding assays. These findings indicate that additional transcription factors are required for mammary GH expression[14,15].

## 4. BIOLOGICAL FUNCTIONS OF MAMMARY GH

The release of mammary produced GH into the circulation results in the dog into clinical features of GH excess. However, mammary GH may have also local autocrine or paracrine effects on proliferation and differentiation of mammary epithelium (Fig. 2). Apart form direct effects on recruitment of stem cells[22], GH may also stimulate the local expression of IGF-I in stromal cells of the mammary fat pad[23].

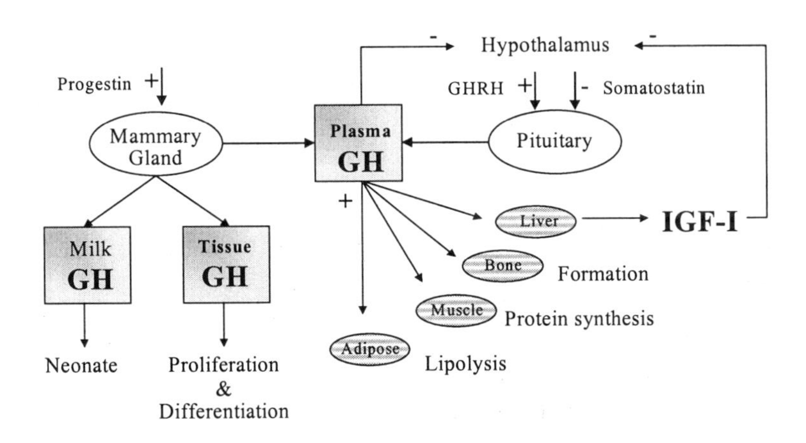


Fig. 2 Schematic representation of the contribution of mammary GH to the endocrine effects of pituitary produced GH, local para/autocrine tissue effects, and exocrine effects via milk.

It has been shown that especially in colostrum supra-physiological GH concentrations higher than 1000 μg/l can be found[24]. The effects on immune function or postnatal development of the gastro-intestinal system of the neonate remain to be elucidated.

## 5. MAMMARY GH AND BREAST CANCER

The local production of GH has implications for the insights in mammary gland tumor formation and progression, and treatment of mammary cancer. Expression of mammary GH is associated with local expression of IGFs and their binding proteins (IGFBPs), thus creating a proliferative environment for the glandular epithelium[18]. Also treatment with GH may induce mammary gland hyperplasia as has been shown in aging primates[25]. The stimulated cell proliferation may increase the possibilities for tumor initiation by mutagenic compounds, or stimulate tumor promotion and progression.

Against the background of these considerations treatment of mammary cancer with progesterone agonists may need to be re-evaluated for potential adverse effects. Whether progestins have beneficial or adverse effects may be dependent upon the specificity of the progestin. Frequently used compounds such as MPA and megestrol acetate have also profound glucocorticoid and androgen agonistic effects, which may in part be responsible for the observed beneficial effects in patients with mammary cancer[27,28].

**Acknowledgement**

This work was supported in part by the Dutch Cancer Society.

# 6. REFERENCES

1. Pike MC, Spicer DV, Dahmoush L, Press MF. Estrogens, progestogens, normal breast cell proliferation, and breast cancer risk. Epidemiol Rev 15, 17-35, 1993.
2. Humphreys RC, Lydon J, O'Malley BW, Rosen JM. Mammary gland development is mediated by both stromal and epithelial progesterone receptors. Mol Endocrinol 11, 801-811, 1997.
3. Anderson TJ, Battersby S, King RJB. Oral contraceptive use influences resting breast proliferation. Hum. Pathol. 20, 1139-1144, 1989.
4. Clarke CL, Sutherland RL. Progestin regulation of cellular proliferation. Endocr. Rev. 11, 266-301, 1990.
5. Graham JD, Clarke CL. Physiological action of progesterone in target tissues. Endocr Rev 18, 502-519, 1997.
6. Torchia J, Glass C., Rosenfeld MG. Co-activators and co-repressors in the integration of transcriptional responses. Current Opinion Cell Biol. 10: 373-383, 1998.
7. Brisken C, Park S, Vass T, Lydon JP, O'Malley BW, Weinberg RA. A paracrine role for the epithelial progesterone receptor in mammary gland development. Proc Natl Acad Sci USA 28: 5076-5081, 1998.
8. Concannon P, Altszuler N, Hampshire J, Butler WR, Hansel W. Growth hormone, prolactin, and cortisol in dogs developing mammary nodules and an acromegaly-like appearance during treatment with medroxyprogesterone acetate. Endocrinology 106, 1173-1177, 1980.
9. Rijnberk A, Eigenmann JE, Belshaw BE, Hampshire J, Altszuler Acromegaly associated with transient overproduction of growth hormone in a dog. N J Am Vet Med Assoc 177, 534-537, 1980.
10. Eigenmann JE, Eigenmann RY, Rijnberk A, van der Gaag I, Zapf J, Froesch ER. Progesterone-controlled growth hormone overproduction and naturally occurring canine diabetes and acromegaly. Acta Endocrinol (Copenh) 104, 167-76, 1983.
11. Watson ADJ, Rutteman GR, Rijnberk A, Mol JA. Effect of somatostatin analogue SMS 201-995 and antiprogestin agent RU 486 in canine acromegaly. Front Horm Res 17, 193-198, 1987.

12. Selman PJ, Mol JA, Rutteman GR, and Rijnberk A. Progestin treatment in the dog. I Effects on growth hormone, insulin-like growth factor I and glucose homeostasis. Eur J Endocrinol 131, 413-421, 1994.
13. Selman PJ, Mol JA, Rutteman GR, van Garderen E, Rijnberk A. Progestin-induced growth hormone excess in the dog originates in the mammary gland. Endocrinology 134, 287-292, 1994.
14. Van Garderen E, de Wit M, Voorhout WF, Rutteman GR, Mol JA, Nederbragt H, Misdorp W: Expression of growth hormone in canine mammary tissue and mammary tumors. Am J Pathol 150, 1037-1047, 1997.
15. Mol JA, van Garderen E, Selman PJ, Wolfswinkel J, Rijnberk A, Rutteman GR. Growth hormone mRNA in mammary gland tumors of dogs and cats. J Clin Invest 95, 2028-2034, 1995.
16. Lantinga-van Leeuwen IS, Oudshoorn M, Mol JA. Canine mammary growth hormone gene transcription initiates at the pituitary-specific start site in the absence of Pit-1. Mol Cell Endocrinol 150, 121-128, 1999.
17. Mol JA, Henzen-Logmans SC, Hageman Ph, Misdorp W, Blankenstein MR, Rijnberk A. Expression of the gene encoding GH in the human mammary gland. J Clin Endocr Metab 80, 3094-3096,1995.
18. Mol JA, van Garderen E, Rutteman GR, Rijnberk A. New insights in the molecular mechanism of progestin-induced proliferation of mammary epithelium; induction of the local biosynthesis of growth hormone (GH) in the mammary gland of dogs, cats and humans. J Steroid Biochem Mol Biol 57, 67-71, 1996.
19. Tuggle CK, Trenkle A. Control of growth hormone synthesis. Dom Animal Endocrinol 13, 1-33, 1996.
20. Lantinga-van Leeuwen IS, Mol JA. Cloning and characterisation of the 5'flanking region of the canine growth hormone gene. Submitted.
21. Kooistra HS, Voorhout G, Selman PJ, Rijnberk A: Progestin-induced growth hormone (GH) production in the treatment of dogs with congenital GH deficiency. Domest Anim Endocrinol 15, 93-102, 1998
22. Green H, Morikawa M, Nixon T. A dual effector theory of growth-hormone action. Differentiation 29, 195-198, 1985
23. Walden PD, Ruan W, Feldman M, Kleinberg DL. Evidence that the mammary fat pad mediates the action of growth hormone in mammary gland development. Endocrinol. 139: 659-662, 1998
24. Schoenmakers I, Kooistra HS, Okkens AC, Hazewinkel HAW, Bevers MM, Mol JA. Growth hormone concentrations in mammary secretions and plasma of the periparturient bitch and in plasma of the neonate. J Reprod Fert Suppl. 51: 363-367, 1997.
25. Ng ST, Zhou J, Adesanya OO, Wang J, LeRoith D, Bondy CA. Growth hormone treatment induces mammary gland hyperplasia in aging primates. Nature Medicine 10, 1141-1144, 1997.
26. Feldman M, Ruan W, Cunningham BC, Wells JA, Kleinberg DL. Evidence that the growth hormone receptor mediates differentiation and development of the mammary gland. Endocrinology 133, 1602-1608, 1993.
27. Birrell SN, Hall RE, Tilley WD. Role of the androgen receptor in human breast cancer. J Mammary Gland Biol Neoplasia 3, 95103, 1998.
28. Bentel JM, Birrell SN, Pickering MA, Holds DJ, Horsfall DJ, Tilley WD. Androgen receptor agonist activity of the synthetic progestin, medroxyprogesterone acetate, in human breast cancer cells. Mol Cell Endocrinol 154: 11-20, 1999.

# 9.

# Translocation and Action of Polypeptide Hormones within the Nucleus

## *Relevance to Lactogenic Transduction*

Charles V. Clevenger and Michael A. Rycyzyn
*Department of Pathology & Laboratory Medicine, University of Pennsylvania Medical Center, Philadelphia, USA*

Key words: Prolactin, Growth Hormone, Retrotransport, Peptide hormones, Nuclear.

Abstract: The action of polypeptide hormones at the cell surface as mediated by transmembrane receptors is well recognised. However, a growing body of evidence also indicates that such hormones are also translocated and act directly within the cell nucleus. This chapter will overview what is known of the action of one such example, namely prolactin (PRL), from its classic action at the cell surface, to its novel function within the nucleus.

## 1. FUNCTION OF PRL IN RESPONSIVE TISSUES

Prolactin (PRL) was originally identified as a neuroendocrine hormone of pituitary origin[1]. While the primary function of this hormone was initially thought to lie solely within the breast, the functional pleiotropism of this peptide with regards to reproduction, osmoregulation, and behavior was subsequently recognised[2]. Several lines of evidence have now also demonstrated an immunoregulatory role for this peptide[3]. Structural analysis of PRL has revealed it to be related to members of the cytokine/hemato-poietin family such as growth hormone (GH), erythropoietin, granulocyte-macrophage colony stimulating factor (GM-CSF) and the interleukins 2-7[4]. Synthesis of PRL is not limited to the hypophysis, as numerous extra-pituitary sites of PRL expression including the decidua, breast, and T-lymphocytes have been detected[5-7]. The receptor for prolactin (PRLr) is present on numerous tissues including mammary epithelia, T and B lymphocytes, and macrophages[8,9]. Acting through its receptor, PRL signalling stimulates cell proliferation, survival, and cellular differentiation in a tissue- and microenvironment-dependent manner. With respect to the mammary and immune systems, these data indicate that PRL acts at the endocrine, paracrine, and autocrine levels in regulating T-lymphocyte proliferation and survival[10-13] and the terminal maturation of mammary tissues[14,15]. Several lines of evidence have also indicated that PRL may act as both an endocrine and autocrine/paracrine progression factor for mammary carcinoma in both rodents and humans[8,16-19].

# 2. TRANSDUCTION OF THE PRL SIGNAL

## 2.1 CELL SURFACE RECEPTOR-ASSOCIATED SIGNALS

The effects of PRL on responsive tissues are mediated by the interaction of ligand with its receptor, the PRLr. As the initial event in PRL-induced signalling, binding of PRL induces PRLr dimerization[20,21]. Receptor dimerization mediates the juxtaposition of the intracytoplasmic domains of the PRLr. The intracellular (signalling) domain of the PRLr contains a region of membrane-proximal homology to other cytokine receptors, i.e. the Box 1/Variable Box/Box 2/X box, as well as a unique C-terminal tail. The box 1 and 2 motifs have been implicated in Jak2 binding, and respectively consist of hydrophobic/proline and hydrophobic/acidic residues. The box 1 motif is essential for PRLr function; its deletion abrogates PRLr function[22]. The tyrosine residues present within the C-terminus of the rat PRLr may also contribute to the engagement of Stat 5 and the activation of Jak2[23]. PRLr dimerization results in the rapid phosphorylation of the PRLr signalling domains[24] and the activation of PRLr-associated kinases such as Jak2[25,26] and Fyn[27], Shc-Grb2-Sos[28,29], Vav[30], and Bag-1/Bcl-2[31]. These events induce several signalling cascades, contributing to the transactivation of PRL-responsive gene loci involved in proliferation (i.e. IRF-1, cyclin B, histone H3) and the differentiated mammary phenotype (i.e. milk proteins such as β-casein)[32,33,26].

## *2.2* NUCLEAR TRANSLOCATION OF PRL

The internalisation of PRL occurs within 30 min of its addition to cells expressing the PRLr. Di-leucine motifs within the intracellular domain of the PRLr are thought to mediate the internalisation of both ligand and receptor into an endosomal/multivesicular body/lysosomal pathway[9,34,35]. While some of this internalised hormone is degraded, an appreciable quantity can be stored by the murine T-cells for up to one week of culture. By itself (i.e. in the absence of other mitogenic hormones), PRL is weakly mitogenic to breast cancer cell cultures and non-mitogenic to cultures of murine T lymphocytes[9,32]. Recent data however indicate that in both lymphocytes and mammary epithelium PRL acts as a potent survival factor in the absence of other growth factors[31].

PRL, however, does act as a potent co-mitogen with both IL2 (on T-cells) and epidermal growth factor (EGF; on human breast cancer cells). These data indicate that PRL is necessary (but not sufficient) for cell cycle transit into S-phase during IL2-driven T-cell proliferation; in its absence, the expression of genes necessary for S-phase entry (i.e. histone, cyclins) does not occur[32]. When both IL2-stimulated T-cells and EGF-stimulated breast cancer cells are cultured in media containing PRL, appreciable quantities of PRL (up to 10-20% of total intracellular PRL) can be detected within the cell nucleus by biochemical, immunofluorescent, and immunogold electron microscopy approaches[9,35,36]. The internalisation of proteins from the extracellular medium through a *trans*-Golgi/ER pathway into the cytosol or nucleus has been previously observed with several bacterial toxins and viral proteins, and is a process known as retrotransport[37-41]. In the absence of co-mitogenic stimulation (such as supplied by IL2 or EGF), the nuclear retrotranslocation of PRL does not occur, an observation noted by other laboratories[42]. While the nuclear retrotransport of PRL could represent an epiphenomenon of cell proliferation, we hypothesised that this event was necessary for cell cycle progression. To confirm this hypothesis, three eukaryotic expression constructs of PRL were synthesised: 1) wild-type PRL, bearing its N-terminal ER leader sequence (termed "PRL/WT"), 2) a deletion construct of PRL lacking its leader sequence ("PRL/ER-"), and 3) a chimeric construct of PRL which replaced its leader sequence with the SV40 large T nuclear translocation signal sequence ("PRL/NT+"). When these constructs were transfected into the IL2- and PRL-responsive T-cell line Nb2, the expressed proteins were found within the extracellular medium, cytoplasm, and the nucleus, respectively, and were bioactive and of the appropriate size[43]. Only the transfectant that secreted PRL into the extracellular medium (PRL/WT) was capable of proliferation in the absence of any mitogenic stimulation. In the presence of IL2, both the PRL/WT and PRL/NT+ transfectants demonstrated markedly increased proliferation (5-10 fold increased over either the parental of PRL/ER- lines). In the presence of exogenous neutralising anti-PRL antiserum (which blocked the action of extracellular PRL), however, significant proliferation and survival of only the PRL/NT+ transfectant was noted[43]. These data demonstrated that nuclear PRL contributed to IL2-stimulated proliferation by providing a necessary, but not sufficient, function within the cell nucleus.

The nuclear retrotranslocation of PRL has been observed in several PRL-responsive tissues including the breast, T-lymphocytes, liver[44,45], ovary[46], and adrenal[47]. Other peptide hormones such as EGF, NGF, and PDGF[48,49], insulin[50-52], FGF[53], and IL5[54,55] have also been observed within the nucleus after their introduction into the extracellular medium. These data would indicate that the nuclear retrotranslocation of peptide hormones is a widespread phenomenon that could regulate numerous physiologic processes.

## 3. MECHANISMS OF PRL RETROTRANSPORT TO THE NUCLEUS, THE ROLE OF CYCLOPHILIN B

Given that PRL lacks intrinsic localisation motifs or enzymatic activity, it was reasoned that its nuclear retrotransport and action was mediated by PRL-associated chaperones. To identify these binding partners, yeast two-hybrid analysis was employed with PRL as "bait". This analysis has revealed that the peptidyl-prolyl isomerase (PPI) cyclophilin B (CypB) interacts with PRL[56]. The cyclophilins are a family of peptidyl-prolyl isomerases (PPI) that serve as protein chaperones and mediate the immunosuppressive effects of cyclosporine (CsA). Structural motifs within N- and C-termini of CypB mediate its ER/extracellular/nuclear localisation. Indeed, CypB can be found in serum and breast milk at concentrations of 150 ng/ml. Our data demonstrate that CypB directly interacted with both PRL and GH *in vitro* and *in vivo* through the use of recombinant CypB, PRL, and GH and antibodies targeted against these proteins. This interaction was significantly enhanced by the inclusion of cyclosporine A. The exogenous addition of physiologic concentrations of CypB into the defined medium of responsive cell lines potentiated PRL- and GH-driven proliferation ten- and forty-fold, respectively. CypB by itself was non-mitogenic, nor did it potentiate the action of either interleukin-2 or -3. CypB did not alter the affinity of the PRLr for its ligand, or increase the activation of PRLr-associated Jak2. The potentiation of PRL-action by CypB, however, was accompanied by a dramatic increase in the nuclear retrotranslocation of PRL. A CypB mutant, termed CypB-NT, was generated that lacked the wild-type N-terminal nuclear localisation sequence. Although CypB-NT demonstrated levels of PRL binding and PPI activity equivalent to wild type CypB, it was incapable of mediating the nuclear retrotransport of PRL or enhancing PRL-driven proliferation. These data reveal that CypB serves as a reverse chaperone for PRL that potentiates the action of this hormone. Given that cyclophilins associate with and/or modulate the activity of both the ER transporter Sec61 and known transcription factors, the interaction between CypB and PRL may provide a direct mechanism for somatolactogenic action within the nucleus.

## 4. CONCLUSION

While classically viewed as signalling only from the cell surface, the action of PRL and other related peptide hormones is also directly mediated within the nucleus. Thus, both peptide and steroid hormones demonstrate analogous signalling mechanisms, i.e. immediate-early signalling from the surface, and delayed-sustained signalling from within the nucleus. Given the fundamental nature of the process of hormone retrotransport, novel strategies aimed at interrupting this signalling pathway may be of significant biologic and clinical utility.

**Acknowlegdments**

This work was supported by grants from the National Institutes of Health, R01 CA69294 and DK50771 (to C.V.C.). M.A.R. is supported by the NIH training grant T32 CA09140.

## 5. REFERENCES

1. Riddle, O. and Braucher, P. F., 1931, Studies on the physiology of reproduction in birds. Control of the special secretion of the crop-gland in pigeons by an anterior pituitary hormone. *Am J Physiol* **97**: 617-625.
2. Nicoll, C. S., 1974, Physiological actions of prolactin. In *Handbook of Physiology,* Section VII, (Greep, R. O. and Astwood, E. B., eds.), American Physiological Society, Washington, DC, pp. 253-292.
3. Clevenger, C. V., Freier, D. O., and Kline, J. B., 1998, Prolactin receptor signal transduction in cells of the immune system. *Journal of Endocrinology* **157**: 187-197.
4. Bazan, J. F., 1990, Haematopoietic receptors and helical cytokines. *Immunology Today* **11**: 350-354.
5. Clevenger, C. V. and Plank, T. L., 1997, Prolactin as an autocrine/paracrine growth factor in breast tissue. *Journal of Mammary Gland Biology and Neoplasia* **2**: 59-68.
6. DiMattia, G. E., Gellersen, B., Bohnet, H. G., and Friesen, H. G., 1986, A human B-lymphoblastoid cell line produces prolactin. *Endocrinology* **122**: 2508-2517.
7. Neal, K. D., Montgomery, D. W., Truong, T. M., and Yu-Lee, L.-Y., 1992, Prolactin gene expression in human thymocytes. *Molecular and Cellular Endocrinology* **87**: R19-R23.
8. Clevenger, C. V., Chang, W.-P., Ngo, W., Pasha, T. L. M., Montone, K. T., and Tomaszewski, J. E., 1995, Expression of prolactin and prolactin receptor in human breast carcinoma: Evidence for an autocrine/paracrine loop. *Am J Pathol* **146**: 1-11.

9. Clevenger, C. V., Russell, D. H., Appasamy, P. M., and Prystowsky, M. B., 1990, Regulation of IL2-driven T-lymphocyte proliferation by prolactin. *Proceedings of the National Academy of Science USA* **87**: 6460-6464.
10. Gala, R. R., 1991, Prolactin and Growth Hormone in the Regulation of the Immune System. *PSEBM* **198**: 513-527.
11. Yu-Lee, L. Y., 1997, Molecular Actions of Prolactin in the Immune System. *Proceedings of the Society for Experimental Biology and Medicine* **215**: 35-52.
12. Kooijman, R., Hooghe-Peters, E. L., and Hooghe, R., 1996, Prolactin, Growth Hormone, and Insulin-like Growth Factor-1 in the Immune System. *Advances in Immunology* **63**: 377-454.
13. Prystowsky, M. B. and Clevenger, C. V., 1994, Prolactin as a second messenger for interleukin 2. *Immunomethods* **5**: 49-55.
14. Horseman, N. D., Zhao, W., Montecino-Rodriguiez, E., Tanaka, M., Nakashima, K., Engle, S. J., Smith, F., Markoff, E., and Dorshkind, K., 1997, Defective mammopoiesis, but normal hematopoiesis, in mice with a targeted disruption of the prolactin gene. *EMBO J* **16**: 6926-6935.
15. Shiu, R. P. C., Murphy, L. C., Tsuyuki, D., Myal, Y., Lee-Wing, M., and Iwasiow, B., 1987, Biological actions of prolactin in human breast carcinoma. *Recent Progress in Hormone Research* **43**: 277-299.
16. Reynolds, C., Montone, K. T., Powell, C. M., Tomaszewski, J. E., and Clevenger, C. V., 1997, Distribution of prolactin and its receptor in human breast carcinoma. *Endocrinology* **138**: 5555-5560.
17. Fields, K., Kulig, E., and Lloyd, R. V., 1993, Detection of prolactin messenger RNA in mammary and other normal and neoplastic tissues by polymerase chain reaction. *Lab Invest* **68**: 354-360.
18. Ormandy, C. J., Hall, R. E., Manning, D. L., Robertson, J. F. R., Blamey, R. W., Kelly, P. A., Nicholson, R. I., and Sutherland, R. L., 1997, Coexpression and cross-regulation of the prolactin receptor and sex steroid hormone receptors in breast cancer. *Journal of Clinical Endocrinology and Metabolism* **82**: 3692-3699.
19. Mertani, H., Garcia-Cabellero, T., Lambert, A., Gerard, F., Palayer, C., Boutin, J. M., Vonderhaar, B. K., Waters, M. J., Lobie, P. E., and Morel, G., 1998, Cellular expression of growth hormone and prolactin receptors in human breast disorders. *Int J Cancer* **79**: 202-211.
20. Gertler, A., Grosclaude, J., Strasburger, C. J., Nir, S., and Djiane, J., 1996, Real-time measurements of the interactions between lactogenic hormones and prolactin-receptor extracellular domains from several species support the model hormone-induced transient receptor dimerization. *Journal of Biologic Chemistry* **271**: 24482-24491.
21. Sommers, W., Ultsch, M., DeVos, A. M., and Kossiakoff, A. A., 1994, The X-ray structure of a growth hormone-prolactin receptor complex. *Nature* **372**: 478-481.
22. Lebrun, J.-J., Ali, S., Ullrich, A., and Kelly, P. A., 1995, Proline-rich sequence-mediated Jak2 association to the prolactin receptor is required but not sufficient for signal transduction. *Journal of Biologic Chemistry* **270**: 10664-10670.
23. Lebrun, J.-J., Ali, S., Goffin, V., Ullrich, A., and Kelly, P. A., 1995, A single phosphotyrosine residue of the prolactin receptor is responsible for activation of gene transcription. *Proceedings of the National Academy of Science USA* **92**: 4031-4035.
24. Chang, W.-P., Ye, Y., and Clevenger, C. V., 1998, Stoichiometric structure/function analysis of the prolactin receptor signaling domains by receptor chimeras. *Molecular and Cellular Biology* **18**: 896-905.

25. Campbell, G. S., Argetsinger, L. S., Ihle, J. N., Kelly, P. A., Rillema, J. A., and Carter-Su, C., 1994, Activation of JAK2 tyrosine kinase by prolactin receptors in Nb2 cells and mouse mammary gland explants. *Proceedings of the National Academy of Science USA* **91**: 5232-5236.
26. Rui, H., Lebrun, J.-J., Kirken, R. A., Kelly, P. A., and Farrar, W. L., 1994, JAK2 activation and cell proliferation induced by antibody-mediated prolactin receptor dimerization. *Endocrinology* **135**: 1299-1306.
27. Clevenger, C. V. and Medaglia, M. V., 1994, The protein tyrosine kinase $p59^{fyn}$ is associated with prolactin receptor and is activated by prolactin stimulation of T-lymphocytes. *Molecular Endocrinology* **8**: 674-681.
28. Clevenger, C. V., Torigoe, T., and Reed, J. C., 1994, Prolactin induces rapid phosphorylation and activation of prolactin receptor associated Raf-1 kinase in a T-cell line. *Journal of Biologic Chemistry* **269**: 5559-5565.
29. Erwin, R. A., Kirken, R. A., Malbarba, M. G., Farrar, W. L., and Rui, H., 1995, Prolactin activates Ras via signaling proteins SHC, growth factor receptor bound 2, and son of sevenless. *Endocrinology* **136**: 3512-3518.
30. Clevenger, C. V., Ngo, W., Luger, S. M., and Gewirtz, A. M., 1995, Vav is necessary for prolactin-stimulated proliferation and is translocated into the nucleus of a T-cell line. *J Biol Chem* **270**: 13246-13253.
31. Clevenger, C. V., Thickman, K., Ngo, W., Chang, W.-P., Takayama, S., and Reed, J. C., 1997, Role of Bag-1 in the survival and proliferation of the cytokine-dependent lymphocyte lines, Ba/F3 and Nb2. *Mol Endocrinol* **11**: 608-618.
32. Clevenger, C. V., Sillman, A. L., Hanley-Hyde, J., and Prystowsky, M. B., 1992, Requirement for prolactin during cell cycle regulated gene expression in cloned T-lymphocytes. *Endocrinology* **130**: 3216-3222.
33. Yu-Lee, L. Y., 1990, Prolactin stimulates transcription of growth-related genes in Nb2 T lymphoma cells. *Molecular and Cellular Endocrinology* **68**: 21-28.
34. Vincent, V., Goffin, V., Rozakis-Adcock, M., Mornon, J.-P., and Kelly, P. A., 1997, Identification of Cytoplasmic motifs required for short prolactin receptor internalization. *Journal of Biologic Chemistry* **272**: 7062-7068.
35. Clevenger, C. V., Sillman, A. L., and Prystowsky, M. B., 1990, Interleukin-2 driven nuclear translocation of prolactin in cloned T-lymphocytes. *Endocrinology* **127**: 3151-3159.
36. Rao, Y.-P., Buckley, D. J., Olson, M. D., and Buckley, A. R., 1995, Nuclear translocation of prolactin: Collaboration of tyrosine kinase and protein kinase C activation in rat Nb2 node lymphoma cells. *J Cell Physiol* **163**: 266-276.
37. De Virgilio, M., Weninger, H., and Ivessar, N. I., 1998, Ubiquitination is required for the retrotranslocation of a short-lived luminal endoplasmic reticulum glycoprotein to the cytosol for degradation by the proteasome. *Journal of Biologic Chemistry* **273**: 9734-9743.
38. Hazes, B. and Read, R. J., 1997, Accumulating evidence suggests that several AB toxins subvert the endoplasmic reticulum-associated protein degradation pathway to enter target cells. *Biochemistry* **36**: 11051-11054.
39. Johannes, L. and Goud, B., 1998, Surfing on a retrograde wave: How does Shiga toxin reach the endoplasmic reticulum. *Trends in Cell Biology* **8**: 158-162.
40. Wiertz, E. J. H. J., Tortorella, D., Bogyo, M., Yu, J., Mothes, W., Jones, T. R., Rapaport, T. A., and Ploegh, H. L., 1997, Sec61-mediated transfer of a membrane protein from the endoplasmic reticulum to the proteasome for destruction. *Nature* **384**: 432-438.

41. Wiedlocha, A., Falnes, P. O., Rapak, A., Muraguchi, A., Klingenberg, O., and Olsnes, S., 1996, Stimulation of proliferation of a human osteosarcoma cell line by exogenous acidic fibroblast growth factor requires both activation of receptor tyrosine kinase and growth factor internalization. *Molecular and Cellular Biology* **16**: 270-280.
42. Perrot-Applanat, M., Gualillo, O., Buteau, H., Edery, M., and Kelly, P. A., 1997, Internalization of prolactin receptor and prolactin in transfected cells does not involve nuclear translocation. *J Cell Sci* **110**: 1123-1132.
43. Clevenger, C. V., Altmann, S. W., and Prystowsky, M. B., 1991, Requirement of nuclear prolactin for interleukin-2-stimulated proliferation of T lymphocytes. *Science* **253**: 77-79.
44. Josefsberg, Z., Posner, B. I., Patel, B., and Bergeron, J. J. M., 1979, The uptake of prolactin into female rat liver. *Journal of Biologic Chemistry* **254**: 209-215.
45. Buckley, A. R., Crowe, P. D., and Russell, D. H., 1988, Rapid activation of protein kinase C in isolated rat liver nuclei by prolactin, a known hepatic mitogen. *Proceedings of the National Academy of Science USA* **85**: 8649-8653.
46. Nolin, J. M., 1978, Intracellular prolactin in rat corpus luteum and adrenal cortex. *Endocrinology* **102**: 402-406.
47. Nolin, J. M., 1980, Incoporation of regulatory peptide hormones by individual cells of the adrenal cortex: Prolactin-adrenocorticotropin differences. *Peptides* **1**: 249-255.
48. Maher, D. W., Lee, B. A., and Donoghue, D. J., 1989, The alternatively spliced exon of the platelet-derived growth factor A chain encodes a nuclear targeting signal. *Molecular and Cellular Biology* **9**: 2251-2253.
49. Rakowicz-Szulczynska, E. M., Rodeck, U., Herlyn, M., and Koprowski, H., 1986, Chromatin binding of epidermal growth factor, nerve growth factor, and platelet-derived growth factor in cells bearing the appropriate surface receptors. *Proceedings of the National Academy of Science USA* **83**: 3728-3732.
50. Smith, R. M., Goldberg, R. I., and Jarett, L., 1988, Preparation and characterization of colloidal gold-insulin complex with binding and biological activities identical to native insulin. *Journal of Histochemistry and Cytochemistry* **36**: 359-365.
51. Smith, R. M. and Jarett, L., 1987, Ultrastructural evidence for the accumulation of insulin in nuclei of intact 3T3-L1 adipocytes by an insulin-receptor mediated process. *Proceedings of the National Academy of Science USA* **84**: 459-463.
52. Soler, A. P., Thompson, K. A., Smith, R. M., and Jarett, L., 1989, Immunological demonstration of the accumulation of insulin, but not insulin receptors, in nuclei of insulin-treated cells. *Proceedings of the National Academy of Science USA* **86**: 6640-6644.
53. Baldin, V., Roman, A. M., Bosc-Bierne, I., Amalric, F., and Bouche, G., 1990, Translocation of bFGF to the nucleus is G1 phase cell cycle specific in bovine aortic endothelial cells. *EMBO Journal* **9**: 1511-1517.
54. Jans, D. A., Briggs, L. J., Gustin, S. E., Jans, P., Ford, S., and Young, I. G., 1997, A functional bipartite nuclear localization signal in the cytokine interleukin-5. *FEBS Letters* **406**: 315-320.
55. Jans, D. A., Briggs, L. J., Gustin, S. E., Jans, P., Ford, S., and Young, I. G., 1997, The cytokine interleukin-5 (IL-5) effects cotransport of its receptor subunits to the nucleus in vitro. *FEBS Letters* **410**: 368-372.
56. Rycyzyn, M. A., Reilly, S. C., O'Malley, K., and Clevenger, C. V., 1999, Role of cyclophilin B in somatolactogenic transduction and nuclear retrotranslocation. Submitted.

10.

# Mammary Gland Development and the Prolactin Receptor

Nadine Binart[1], Christopher J. Ormandy[2] and Paul A. Kelly[1]
*[1]INSERM Unité 344 - Endocrinologie Moléculaire, Faculté de Médecine Necker, Paris Cedex, France. [2]Cancer Research Program, Garvan Institute of Medical Research, Sydney, Australia*

Key words: prolactin, prolactin receptor

Abstract: Prolactin (PRL), synthesized by the anterior pituitary and to a lesser extent by numerous extrapituitary tissues, affects more physiological processes than all other pituitary hormones combined. This hormone is involved in >300 separate effects in various vertebrate species where its role has been well documented. The initial step in its action is the binding to a specific membrane receptor which belongs to the superfamily of class 1 cytokine receptors. The function of this receptor is mediated, at least in part, by two families of signaling molecules: Janus kinases and signal transducers and activators of transcription. PRL-binding sites have been identified in a number of cells and tissues of adult animals. Disruption of the gene for the PRL receptor has provided a new animal model with which to better understand the actions of PRL on mammary morphogenesis and mammary gland gene expression. The recent availability of genetic mouse models provides new insights into mammary developmental biology and how the action of a hormone at specific stages of development can have effects later in life on processes such as mammary development and breast cancer initiation and progression.

## 1. INTRODUCTION

In the normal mammary gland, development, growth and differentiation are under the control of a variety of growth factors and hormones. Prolactin is considered a major player[1]. Functional differentiation of the gland as measured by induction of milk protein synthesis both *in vivo* and *in vitro* is dependent on prolactin. To summarize our understanding of PRL actions in the development of mammary gland, we will discuss PRL secretion and mammary gland morphogenesis, and will describe recent findings regarding this hormone in the context of mammary development.

## 2. PITUITARY AND EXTRAPITUITARY PRL SECRETION

Three decades after PRL was identified, the amino acid sequence of sheep PRL (also referred to as lactogenic hormone or luteotropic hormone depending on its biological properties) was determined and shown to be a protein of 199 amino acids[2]. At the end of the 1970's, the nucleotide sequence of PRL cDNAs from several species was identified[3]. As anticipated from earlier structural studies, the primary structure of PRL appeared closely related to that of two other hormones, growth hormone (GH), also of pituitary origin, and placental lactogen (PL), secreted by mammalian placenta[4,5]. Today, genetic[6,7], structural[3,8], binding[8] and structure- functional[3,8] studies of these three hormones, as well as the more recently identified somatolactin and prolactin-related proteins, have clearly demonstrated that they all belong to a unique family of proteins.

In addition to being synthesized and secreted by lactotrophic cells of the anterior pituitary gland, PRL is also produced by numerous other cells and tissues[4]. The subject of extrapituitary PRL has recently been reviewed[4] and thus will only be briefly discussed. In addition to the anterior pituitary gland, PRL gene expression has been confirmed in various regions of the brain, decidua, myometrium, lacrimal gland, thymus, spleen, circulating lymphocytes and lymphoid cells of bone marrow, and tumors, skin fibroblasts, and sweat glands (reviewed in[4]). PRL can thus be found in several fluid compartments in addition to serum, such as cerebrospinal fluid, amniotic fluid, tears, follicular fluid, and sweat.

In addition, the mammary epithelial cell is an important site of PRL synthesis and secretion. PRL is present in significant concentrations in milk and milk PRL is absorbed by the neonatal gut and is thought to cause changes in the maturation of the hypothalamic neuroendocrine system[9]. Interestingly, hypophysectomized rats retain ~ 20% of biologically active PRL in the circulation, which increases to ~ 50% of normal levels with time. Neutralization of circulating PRL with anti-PRL antibodies results in immune dysfunction and death[10], suggesting that extrapituitary PRL is important, and under some circumstances, can compensate for pituitary PRL.

Pituitary PRL acts via a classical endocrine pathway, meaning it is secreted by a gland, transported by the circulatory system and acts on peripheral target cells via specific receptors located on the plasma membrane. The PRL that is produced by different non pituitary/peripheral cell types can act in a more direct fashion, that is as a growth factor, neurotransmitter, or immunomodulator, in an autocrine or paracrine manner. Thus, locally produced PRL can act on adjacent cells (paracrine) or on the PRL-secreting cell itself (autocrine). Using paracrine or autocrine mechanisms, it would thus be possible to activate many of the actions associated with PRL without ever affecting the circulating concentration of the hormone.

## 3. MAMMARY GLAND ORGANOGENESIS

At birth, mice show rudiments of indistinguishable mammary ductal architecture. Pubertal mammary gland development is controlled by pituitary and gonadal hormones that act directly and indirectly. After growth until the onset of puberty, terminal end buds (TEBs) form and ductal epithelium begins. Proliferation at this stage occurs mainly in the body cells of the terminal end bud. Later during puberty, under the influence of PRL and progesterone, lobular buds branch off from the ductal system[11]. Organogenesis of the mammary gland is completed when the ductal system has grown to the full extent of the fat pad and lobule buds have sprouted at regular intervals along the ducts. Up to midpregnancy, ductal elongation, branching, and the number of lobules increased. Under the influence of PRL, placental lactogens, progesterone, and local growth factors, the lobuloalveolar epithelium undergoes extensive proliferation. At parturition, the lobuloalveolar epithelium is converted to a secretory phenotype and the full complement of milk proteins and lactogenic enzymes are synthesized. Involution of the lobuloalveolar system occurs at the end of the lactation in response to milk stasis, and decrease of systemic lactogens. Thus, PRL and placental lactogens, which bind to the PRL receptor, act during these different stages: lobule budding, lobuloalveolar expansion during pregnancy, and lactational differentiation and maintenance of milk secretion.

## 4. STUDIES OF MICE WITH TARGETED GENE DISRUPTIONS

Studies of mice with targeted disruptions of PRL and the PRL receptor (PRLR) genes have demonstrated that this hormone is an obligate regulator of mammary organogenesis, lobuloalveolar growth, and functional differentiation[12,13]. In nulliparous mice ( PRL$^{-/-}$) the mammary glands consists of a primary and secondary branched ductal system with numerous terminal end buds along the ducts. Terminal end buds are absent in the mature glands of the normal mice. Thus the complete absence of PRL results in the arrest of mammary organogenesis at an immature pubertal state. Females with one functional allele of the PRL receptor (PRLR$^{+/-}$) showed almost complete failure of lactation after their first pregnancy. The severity of this phenotype was reduced when the females were mated at 20 weeks and was absent following a second pregnancy. Histological and whole mount analysis of virgin mammary glands are reported elsewhere[14]. In PRLR$^{-/-}$ mice mammary development is reduced, but relatively normal up to puberty. A partial rescue of pregnancy by administration of progesterone allowed analysis of mammary development at day 5.5. Progesterone is required for ductal branching, and the addition of this steroid to maintain the pregnancy is also able to rescue ductal sidebranching in PRLR$^{-/-}$ females, as ductal bifurcation appeared to be normal (Fig 1). To distinguish between the developmental defects intrinsic to the epithelium and those resulting from systemic endocrine alterations, mammary epithelium from PRLR$^{-/-}$ mice was transplanted into mammary fat pads of wild-type mice. In virgin mice, the PRLR$^{-/-}$ epithelial transplants developed normally at puberty; indicating an indirect effect of prolactin on ductal development. During pregnancy, these transplants showed normal side branching and formation of alveolar buds, however without any lobuloalveolar development[15]. These experiments indicate that during pregnancy PRL acts directly on the mammary epithelium to produce lobuloalveolar development.

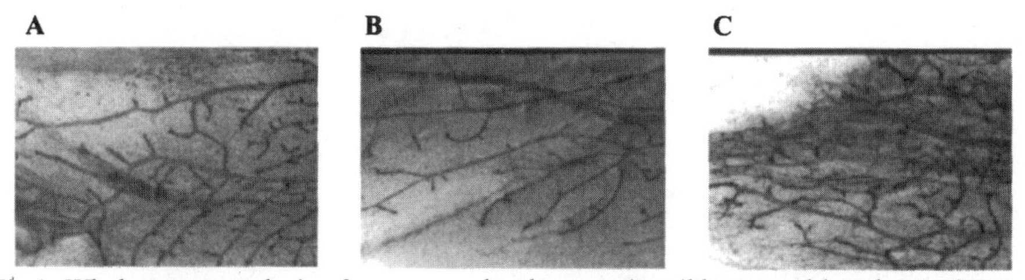


Fig.1. Whole mount analysis of mammary development in wild-type and knockout mice at day 5.5 of pregnancy. Whole mounts of mammary glands from mice at 12 weeks of age were prepared from wild-type (A) or knockout (B and C ) animals. Administration of progesterone leads to the maintenance of pregnancy in knockout mice (C). The fourth inguinal mammary gland is shown.

The complete absence of the progesterone receptor gene results in a gland lacking terminal end buds but displaying only some branched ducts[16]. Estradiol receptor $\alpha$ knockout females are infertile, and a lack of ductal growth and differentiation has been reported[17]. In mice deficient in Stat5a, a primary mediator of PRL action, mammary lobuloalveolar outgrowth during pregnancy is curtailed and females are unable to lactate after parturition because of a failure of terminal differentiation[18]. Similarly, mammary gland development is also impaired in Stat5b$^{-/-}$ females and, although milk protein genes are expressed, there is insufficient milk to feed pups[19]. Interestingly, Stat5b, but not Stat5a deficient females exhibit severely compromised fertility. Mice homozygous for a germline mutation in A-*myb*, a nuclear protein regulator of transcription, show a marked underdevelopment of the breast epithelial compartment following pregnancy demonstrating a critical role of A-*myb* in mammary gland development[20]. Mice lacking cyclin D1 gene also exhibit a dramatic impairment of mammary gland development leading to inability to lactate their litters[21].

In conclusion, the phenotypes of animals lacking functional genes encoding PRLR, PRL, Stat5a or Stat5b confirm the essential role of the lactogenic receptor and its signaling pathways in mammary gland development, whereas estradiol and progesterone receptors are also important but perhaps do not play as central a function as prolactin.

## 5. MECHANISMS

The role of locally derived growth factors in the mediation of PRL-induced mammary gland development remains unknown. Several growth factors are known to play specific and nonredundant paracrine roles at different stages of mammary development. These include epidermal growth factor (EGF), neuregulin (NRG), Wnt gene products, keratinocyte growth factor (KGF), hepatocyte growth factor (HGF) and insulin-like growth factor-1 (IGF-1). The interplay between endocrine hormones and epithelial and stromal factors is necessary for normal mammary development. Neuregulin stimulates alveolar development and secretory activity[22,23]. This member of EGF family is expressed during pregnancy in the stroma[24] so it may be involved in mediating effects which are important during pregnancy. Cyclin D1 knockout mice are devoid of PRL-dependent lobuloalveolar structures in the mature gland[25], reminiscent of those in progesterone receptor knockout mice. The schematic role of prolactin actions is summarized in Fig 2.

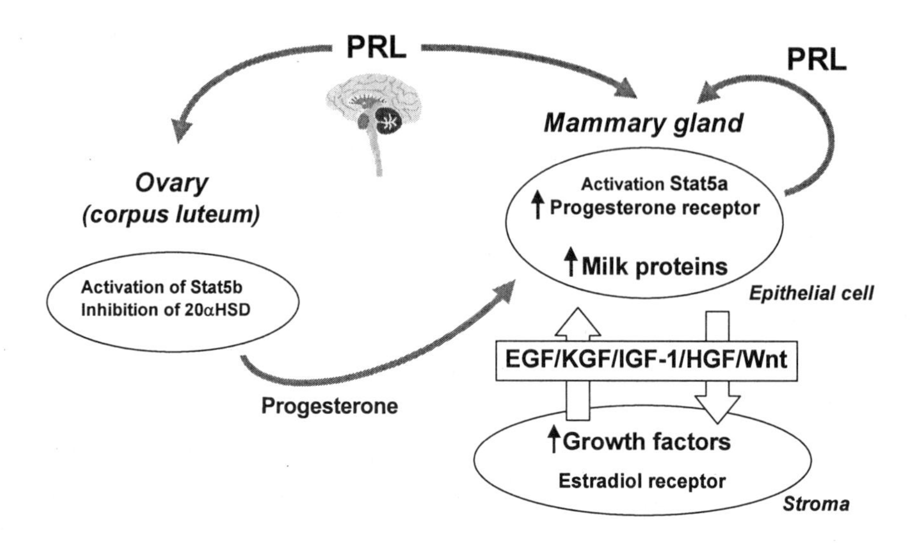


Fig. 2. Schematic action of PRL on mammary epithelial cells. PRL acts directly on the mammary gland epithelium, and indirectly through the corpus luteum via the signal transducer and activator of transcription Stat5b. In the mammary gland, the major mediator is Stat5a. Growth factor signalling between the mammary epithelium and stroma/adipose compartment induces growth and morphogenesis of the epithelium. Thus, PRL acts through endocrine and autocrine/paracrine pathways.

## 6. CONCLUSIONS

The technique of gene targeting in mice has been used to develop experimental models where the effects of the complete absence of any lactogen or PRL-mediated effects can be studied. It is clear that there are multiple actions associated with PRL. It will be important to correlate known effects with local production of PRL in some cases in order to distinguish classical endocrine from autocrine/paracrine effects in the mammary gland. The fact that extrapituitary PRL can compensate for pituitary PRL raises the interesting possibility that there may be other effects of PRL than those originally observed in hypophysectomized rats, including a potential role in human breast cancer. The PRL receptor knockout mouse model remains an interesting system to look for effects only activated by PRL or other lactogenic hormones.

In general these model systems will be in the future crucial to facilitate and identify the PRL regulated genes important in the development of the mammary gland.

## 7. REFERENCES


1. Neville, M.C. and Daniel, C.W., 1987, The Mammary Gland: Development regulation and function (New York: Plenum Press), pp 383-438.
2. Li, C.H., Dixon, J.S., Lo, T.B., Pankov, Y.M., and Schmidt, K.D., 1969, Amino acid sequence of ovine lactogenic hormone. *Nature* **224**: 695-696.
3. Nicoll, C.S., Mayer, G.L., and Russell, S.M., 1986, Structural features of prolactins and growth hormones that can be related to their biological properties. *Endocr. Rev.* **7**: 169-203.
4. Ben-Jonathan, N., Mershon, J.L., Allen, D.L., and Steinmetz, R.W., 1996, Extrapituitary prolactin: distribution, regulation, functions, and clinical aspects. *Endocr. Rev.* **17**: 639-669.
5. Kelly, M.A., Rubinstein, M., Asa, S.L., Zhang, G., Saez, C., Bunzow, J.R., Allen, R.G., Hnasko, R., Ben-Jonathan, N., Grandy, D.K., and Low, M.J., 1997, Pituitary lactotroph hyperplasia and chronic hyperprolactinemia in dopamine D2 receptor-deficient mice. *Neuron* **19**: 103-113.
6. Niall, H.D., Hogan, M.L., Sauer, R., Rosenblum, I.Y., and Greenwood, F.C., 1971, Sequences of pituitary and placental lactogenic and growth hormones: evolution from a primordial peptide by gene duplication. *Proc. Natl. Acad. Sci. USA* **68**: 866-870.
7. Miller, W.L. and Eberhardt, N.L., 1983, Structure and evolution of the growth hormone gene family. *Endocr. Rev.* **4**: 97-130.
8. Goffin, V., Shiverick, K.T., Kelly, P.A., and Martial, J.A., 1996, Sequence-function relationships within the expanding family of prolactin, growth hormone, placental lactogen and related proteins in mammals. *Endocr. Rev.* **17**: 385-410.
9. Kacsoh, B., Veress, Z., Toth, B.E., Avery, L.M., and Grosvenor, C.E., 1993, Bioactive and immunoreactive variants of prolactin in milk and serum of lactating rats and their pups. *J. Endocrinol.* **138**: 243-257.
10. Nagy, E. and Berczi, I., 1991, Hypophysectomized rats depend on residual prolactin for survival. *Endocrinology* **128**: 2776-2784.
11. Imagawa, W., Yang, J., Guzman, R., and Nandi, S., 1994, Control of mammary development. In The Physiology of Reproduction. E. Knobil, J.D. Neil, L.L. Ewing, G.S. Greenwald, C.L. Markert, and D.W. Pfaff, eds. (New York: Raven Press), pp. 1033-1063.
12. Horseman, N.D., Zhao, W., Montecino-Rodriguez, E., Tanaka, M., Nakashima, K., Engle, S.J., Smith, F., Markoff, E., and Dorshkind, K., 1997, Defective mammopoiesis, but normal hematopoiesis, in mice with a targeted disruption of the prolactin gene. *EMBO J.* **16**: 6926-6935.
13. Ormandy, C.J., Camus, A., Barra, J., Damotte, D., Lucas, B.K., Buteau, H., Edery, M., Brousse, N., Babinet, C., Binart, N., and Kelly, P.A., 1997, Null mutation of the prolactin receptor gene produces multiple reproductive defects in the mouse. *Genes Dev.* **11**: 167-178.
14. Ormandy, C.J., Binart, N., and Kelly, P.A., 1997, Mammary gland development in prolactin receptor knockout mice. *J. Mammary Gland. Biol. Neopl.* **2**: 355-364.

15. Brisken, C., Kaur, S., Chavarria, T.E., Binart, N., Sutherland, R.L., Weinberg, R.A., Kelly, P.A., and Ormandy, C.J., 1999, Prolactin controls mammary gland development via direct and indirect mechanisms. *Dev. Biol.* **210**: 96-106.
16. Lydon, J.P., DeMayo, F.J., Funk, C.R., Mani, S.K., Hughes, C.A., Montgomery, C.A., Shyamala, G., Conneely, O.M., and O'Malley, B.W., 1995, Mice lacking progesterone receptor exhibit pleiotropic reproductive abnormalities. *Genes Dev.* **9**: 2266-2278.
17. Korach, K.S., 1994, Insights from the study of animals lacking functional estrogen receptor. *Science* **266**: 1524-1527.
18. Liu, X., Robinson, G.W., Wagner, K.U., Garrett, L., Wynshaw-Boris, A., and Hennighausen, L., 1997, Stat5a is mandatory for adult mammary gland development and lactogenesis. *Genes Dev.* **11**: 179-186.
19. Udy, G.B., Towers, R.P., Snell, R.G., Wilkins, R.J., Park, S.H., Ram, P.A., Waxman, D.J., and Davey, H.W., 1997, Requirement of Stat5b for sexual dimorphism of body growth rates and liver gene expression. *Proc. Natl. Acad. Sci. USA* **94**: 7239-7244.
20. Toscani, A., Mettus, R.V., Coupland, R., Simpkins, H., Litvin, J., Orth, J., Hatton, K.S., and Reddy, E.P., 1997, Arrest of spermatogenesis and defective breast development in mice lacking A-myb. *Nature* **386**: 713-717.
21. Fantl, V., Stamp, G., Andrews, A., Rosewell, I., and Dickson, C., 1995, Mice lacking cyclin D1 are small and show defects in eye and mammary gland development. *Genes Dev.* **9**: 2364-2372.
22. Jones, F.E., Jerry, D.J., Guarino, B.C., Andrews, G.C., and Stern, D.F., 1996, Heregulin induces in vivo proliferation and differentiation of mammary epithelium into secretory lobuloalveoli. *Cell Growth Differ.* **7**: 1031-1038.
23. Krane, I.M. and Leder, P., 1996, NDF/heregulin induces persistence of terminal end buds and adenocarcinomas in the mammary glands of transgenic mice. *Oncogene* **12**: 1781-1788.
24. Yang, Y., Spitzer, E., Meyer, D., Sachs, M., Niemann, C., Hartmann, G., Weidner, K.M., Birchmeier, C., and Birchmeier, W., 1995, Sequential requirement of hepatocyte growth factor and neuregulin in the morphogenesis and differentiation of the mammary gland. *J.* Cell Biol. 131: 215-226.
25. Sicinski, P., Donaher, J.L., Parker, S.B., Li, T., Fazeli, A., Gardner, H., Haslam, S.Z., Bronson, R.T., Elledge, S.J., and Weinberg, R.A., 1995, Cyclin D1 provides a link between development and oncogenesis in the retina and breast. Cell 82: 621-630.

11.

# Paracrine Mechanisms of Mouse Mammary Ductal Growth

[1]G. R. Cunha, [1]J. F. Wiesen, [1]Z. Werb, [1]P. Young, [1]Y. K. Hom, [2]P. S. Cooke, and [3]D. B. Lubahn
*[1]Anatomy Department University of California, San Francisco, California, USA; [2]Department of Veterinary Biosciences, University of Illinois, Urbana, IL, USA; [3]Departments of Biochemistry, University of Missouri, Columbia, MO, USA*

Abstract: Ductal growth during puberty is stimulated by estrogens, which elicit their effects via specific estrogen receptors, ERα and ERβ. Analysis of mice with targeted disruption of ERα or ERβ has emphasized the importance of ERα in mammary gland development. In the mouse mammary gland, ERα are expressed in both epithelial and stromal cells (Kurita and Cunha, unpublished), which raises the possibility that the growth and morphogenetic effects of estrogen could be mediated via either epithelial or stromal ER. The aim of this paper is to review the role of epithelial versus stromal ER in mammary ductal-alveolar growth to assess the importance of paracrine mechanisms.

## 1. INTRODUCTION

For many years it has been tacitly assumed that actions of estradiol and on epithelium are mediated directly through ER in epithelial cells themselves. Mammary ductal growth gland is induced by estradiol. The presence of ERα in mammary epithelium combined with the ability of estradiol to induce proliferation in mammary epithelium certainly suggests a causal link; i.e., that effects of estradiol on mammary epithelial proliferation are mediated directly through ER in these cells. However, such a causal link between the presence of mammary epithelial ER and the effects of estradiol on proliferation of mammary epithelium has never been established. In fact, a rapidly emerging body of evidence suggests that this is clearly not true and that stromal ER mediates many effects of estradiol on estrogen target epithelium.[1]

Methodologies for producing αERKO mice, in which the ERα gene has been rendered non-functional by gene targeting,[2] has allowed examination of phenotypic and functional consequences of an absence of ERα.[3]. We have recently developed a new experimental system, which utilizes tissues from ERKO mice to study the mechanism of estradiol action.[1] This system involves separating and recombining epithelial and stromal tissue from αERKO mice with that of a wild-type mice, which express ERα. This tissue separation/recombination technique provides a unique method for experimentally controlling the ERα status of both stroma and epithelium. Thus, tissue recombinations can be prepared, which lack ERα in both stromal and epithelial compartments, express ERα in either epithelium or stroma, or express ERα in both epithelium and stroma. Tissue recombinants are then transplanted into host animals. By analyzing effects of a lack of stromal and/or epithelial ERα on estradiol response such as ductal growth, the role of ERα in each tissue compartment can be definitively determined.

## 2. STROMAL ERα IS REQUIRED FOR DUCTAL GROWTH IN THE MOUSE MAMMARY GLAND

Ductal mammary growth is profoundly impaired in αERKO mice. By the end of puberty, the mammary fat pad (FP) was almost entirely filled with ducts in wild-type mice, while in αERKO mice most of the FP was devoid of ducts[4]. Analysis of tissue recombinants were prepared with mammary epithelium (MGE) and mammary fat pad (FP) of wild-type (wt) and αERKO mice demonstrated that tissue recombinants composed of wt-FP+wt-MGE developed an extensively branched ductal network, which completely filled the fat pad. In contrast, ductal growth was minimal in αERKO-FP+αERKO-MGE tissue recombinants. When αERKO mammary epithelium was grown in association with wild-type FP (wt-FP+αERKO-MGE), αERKO mammary epithelium underwent extensive ductal growth. Although there was some variability in the amount of ductal growth in individual tissue recombinants, ductal growth was always more extensive in wt-FP+wt-MGE and wt-FP+αERKO-MGE versus αERKO-FP+αERKO-MGE tissue recombinants. Surprisingly, ductal tissue was never recognized in wholemounts of αERKO-FP+wt-MGE tissue recombinants. However, serial sections of αERKO-FP+wt-MGE tissue recombinants revealed that all contained small foci of mammary ducts.[4] These findings demonstrate that stromal fat pad cells are critical estrogen targets, and that estrogen elicits mammary ductal growth through stromal ERα. Epithelial ERα is neither necessary nor sufficient for ductal growth. These findings in mammary gland are in complete agreement with comparable tissue recombinant studies in the uterus and vagina.[1]

Stromal epidermal growth factor receptor plays a critical role in mammary epithelial growth. Growth factors such as epidermal growth factor are known to play a role in stromal-epithelial interactions which are critical in determining patterns of mammary growth, development, and ductal morphogenesis. To determine the role of signaling through the epidermal growth factor receptor (EGFR) in mammary ductal growth and branching, mice with a targeted null mutation in the EGFR were used. Such EGFR-KO mice die perinatally, and thus transplantation methods were used to study growth and development of EGFR-KO mammary glands.[5] When transplanted under renal capsules of athymic female mice, neonatal mammary glands of wild-type mice underwent extensive ductal growth with complete filling of the FP. Conversely, neonatal mammary glands of EGFR-KO mice exhibited impaired ductal growth with incomplete filling of the FP. These findings demonstrated that EGFR is essential for mammary ductal growth. To determine whether impaired ductal growth was due to an absence of EGFR in epithelium (E), FP or both, EGFR-KO or wild-type MGE was transplanted into wild-type gland-free FPs. Surprisingly, transplants of either EGFR-KO or wild-type MGE into wild-type mammary FPs exhibited extensive ductal growth with comparable complete filling of the FP in transplants into intact virgin female hosts. Similarly, lobulo-alveolar development was equivalent in transplants of EGFR-KO or wild-type MGE into wild-type mammary FPs, when the hosts also received a pituitary isograft as a source of prolactin.[5] These findings suggested that the absence of EGFR-signaling in the epithelium is not required for mammary ductal growth and lobulo-alveolar development. Instead, these findings suggested that the impaired ductal growth exhibited by EGFR-KO mammary glands was due to an absence of EGFR signaling in the mammary FP.

To determine the irrespective roles of stromal versus epithelial EGFR in mammary ductal growth, neonatal EGFR-KO and wild type (wt) mammary glands were surgically separated into FP and main epithelial duct (E) and then recombined as follows: wt-FP+wt-E, wt-FP+EGFR-KO-E, EGFR-KO-FP+wt-E, and EGFR-KO-FP+EGFR-KO-E+. When tissue recombinants contained wild type stroma, ductal development proceeded regardless of the epithelial source (wt or EGFR). This outcome corroborates results of transplantation of epithelium into cleared FPs. However, when tissue recombinants contained EGFR-KO stroma, ductal growth was meager regardless of the epithelial source. These data indicate that signaling through EGFR must occur in the stroma surrounding the epithelial ducts to induce normal ductal proliferation and morphogenesis in the mammary gland.[5]

## 3. CONCLUSION

Mammary ductal growth is estrogen-dependent and is profoundly impaired in αERKO mice.[6] Tissue recombinant studies using FPs and mammary epithelia from αERKO and wt mice demonstrate that estrogen stimulates mammary ductal growth via a paracrine mechanism, acting through stromal ERα. Epithelial ERα is neither necessary nor sufficient for ductal growth.[4] EGFR signaling is thought to be a downstream effector of estrogen action in several estrogen target organs.[7] Mammary epithelium expresses EGFR, and epidermal growth factor and TGFα are mitogens for mammary epithelial cells.[8] Despite this, our data clearly show that ductal growth does not require signaling through epithelial EGFR in the mammary gland in vivo. Instead, EGFR is absolutely necessary in the stromal FP to induce estrogen-dependent ductal growth. Our results suggest that under stimulatory estrogenic conditions the stroma responds to estrogen action through an EGFR-mediated signaling event that is required for stimulation of epithelial growth and development. In contrast, epithelial EGFR and ERα are neither necessary nor sufficient for ductal growth. What are the gene targets of estrogen action in mammary epithelial cells? One target of estrogen action in mammary epithelium is upregulation of the progesterone receptor (PR), which is only detectable in mammary epithelium[9]. The importance of epithelial PR in the mammary gland is demonstrated as impaired lobulo-alveolar development in PR null mice.[10] The targets of EGFR signaling remain to be determined.

**Acknowledgments**

The work from our laboratory was supported by the following grants: CA 58207, CA44768, CA57621, and Department of Defense grant DAMDH17-97-7324 to JFW.

## 4. REFERENCES

1. Cooke PS, Buchanan DL, Lubahn DB, Cunha GR 1998 Mechanism of estrogen action: lessons from the ERKO mouse. Biology of Reproduction 59:470-475
2. Lubahn DB, Moyer JS, Golding TS, Couse JF, Korach KS, Smithies O 1993 Alteration of reproductive function but not prenatal sexual development after insertional disruption of the mouse estrogen receptor gene. Proc Natl Acad Sci USA 90:11162-11166
3. Korach KS, Couse JF, Curtis SW, Washburn TF, Lindzey J, Kimbro KS, Eddy EM, Migliaccio S, Snedeker SM, Lubahn DB, Schomberg DW, Smith EP 1996 Estrogen receptor gene disruption: molecular characterization and experimental and clinical phenotypes. Recent Prog Horm Res 51:159-186

4. Cunha GR, Young P, Hom YK, Cooke PS, Taylor JA, Lubahn DB 1997 Elucidation of a role of stromal steroid hormone receptors in mammary gland growth and development by tissue recombination experiments. J Mammary Gland Biology and Neoplasia 2:393-402
5. Wiesen JF, Young P, Werb Z, Cunha GR 1999 Signaling through the stromal epidermal growth factor receptor is necessary for mammary ductal development. Development 126:335-44
6. Bocchinfuso WP, Hively WP, Couse JF, Varmus HE, Korach KS 1999 A mouse mammary tumor virus-Wnt-1 transgene induces mammary gland hyperplasia and tumorigenesis in mice lacking estrogen receptor-alpha. Cancer Res 59:1869-76
7. Nelson KG, Takahashi T, Bossert NL, Walmer DK, McLachlan JA 1991 Epidermal growth factor replaces estrogen in the stimulation of female genital-tract growth and differentiation. Proc Natl Acad Sci USA 88:21-25
8. Imagawa W, Bandyopadhyay GK, Nandi S 1990 Regulation of mammary epithelial cell growth in mice and rats. Endocrine Reviews 11:494-523
9. Shyamala G, Barcellos-Hoff MH, Toft D, Yang X 1997 In situ localization of progesterone receptors in normal mouse mammary glands: absence of receptors in the connective and adipose stroma and a heterogeneous distribution in the epithelium. J Steroid Biochem Mol Biol 63:251-259
10. Lydon JP, DeMayo FJ, Funk CR, Mani SK, Hughes AR, Montgomery CA, Shyamala G, Conneely OM, O'Malley BW 1995 Mice lacking progesterone receptor exhibit pleiotropic reproductive abnormalities. Genes and Development 9:2266-2278

12.

# Targeting of PKA in Mammary Epithelial Cells

*Mechanisms and functional consequences*

[1]Roger A. Clegg, [1]Rachel A. Gardner, [1]Rushika N. Sumathipala, [2]Françoise Lavialle, [2]Raphaël Boisgard and [2]Michèle Ollivier-Bousquet

*[1]Hannah Research Institute, Ayr, Scotland UK: [2]Unité de Biologie des Transports Cellulaires, Laboratoire de Biologie Cellulaire et Moléculaire, INRA, Jouy-en-Josas CEDEX, France*

Key words: PKA, mammary cell, AKAP, secretion, targeting.

Abstract: Targeting of protein kinases, promoting association with specific partner-molecules and localisation to particular sites within the cell, has come to be recognised as a key mechanism for attributing specificity to these enzymes. In mammary epithelial cells, the repertoire of acute regulatory roles played by cyclic AMP-dependent protein kinase (PKA) differs from that in other lipogenic cell-types. Furthermore, PKA is implicated in the regulation of mammary-specific function, mediating a tonic stimulation of the flux of newly-synthesised casein through its basal secretory pathway. Both these observations imply mammary-specific properties of either PKA targeting systems or of PKA itself. Evidence for the latter is currently lacking. Pulse-chase labelling experiments in the presence and absence of selective effectors of PKA have enabled the site(s) of action of this protein kinase on casein secretion to be localised to the early stages of the secretory pathway. Possible mechanisms are considered for the physical targeting of PKA to the membrane-enclosed components of the secretory pathway and evidence for their occurrence in mammary epithelial cells is presented.

## 1. INTRODUCTION

Targeting of protein kinases is a mechanism of imposing functional selectivity on their action, additional to that which their own sequence-recognition specificity gives them[1,2]. Furthermore, the characteristics of an individual targeting mechanism may be such as to enable a single enzyme to manifest diverse specificities both between cells and, within a single cell, at different points in time or during a programme of growth and differentiation.

## 1.1 PKA isozyme diversity

Cyclic AMP-dependent protein kinase (PKA) is one of a relatively small number of generic protein kinases acting pleiotropically as primary transducers of regulatory signals within cells. It is catalytically activated in proportion to the prevailing concentration of cAMP in its immediate environment. In its inactive state, PKA exists as a heterotetramer consisting of two catalytic (C-) and two regulatory (R-) subunits. Multiple isozymes of both R- and C- subunit are encoded by different genes (RIα,β; RIIα,β; Cα,β,γ[3,4,5]. Splice variants of C-subunit also exist, most strikingly in *C. elegans* but also in mammalian species[6-9]. The precise function of all this diversity is not known with certainty either in *C. elegans* or in mammals, but evidence is beginning to emerge to support the suggestions that determinants of targeting[8] and/or stability and turnover of the C-subunit protein molecule[10] may reside in these alternative terminal sequences.

## 1.2 Basal activity and activation of PKA

The presence of R-subunit sequences occluding the substrate cleft of C-subunit causes the latter to remain catalytically inactive in the holoenzyme. The cAMP signal is transduced via its high-affinity binding to two sites on each R-subunit, causing them to dissociate as an R-subunit dimer, concomitantly liberating two catalytically active C-subunits. Phosphorylation of proteins by C-subunit can lead to a variety of biological effects depending on the identity of the phospho-acceptor substrate protein. Examples of these include: enzymes catalysing biosynthesis, degradation and metabolic interconversion; transcriptional regulators; ion channels; transmembrane receptors and other signalling mediator proteins[11]. The PKA activation cycle is reversed by the action of cAMP phosphodiesterases. When making experimental measurements of PKA activity, a "snapshot" of the dissociation status at the time of sampling can be made by assaying in the absence of added cAMP to determine the "basal" or "expressed" activity. The "total" activity (measured in the presence of a saturating excess of cAMP) measures the maximum potential of the tissue for C-subunit catalysed protein kinase activity in the event of complete dissociation of all the cellular holoenzyme. In real situations within living cells, activation of PKA is not an "all-or-none" phenomenon but a proportional response to the prevailing cAMP level which may itself vary from place to place within the cell[12]. Thus the tonic or basal level of PKA activation is an important determinant of cellular activity (see e.g. Thorens et al.[13]). This function of PKA has attracted much less attention than its complementary role in the mediation of cellular responses of cells to acute modulators of cAMP levels, such as β-adrenergic agents or glucagon.

### 1.3 PKA targeting

The biological effects of PKA can be modulated by restricting its distribution within the cell. One way in which this can be achieved is by the expression of AKAPs (A-kinase anchoring proteins) which bind to R-subunits and to a secondary binding partner, thereby tethering the holoenzyme to that secondary partner[1,2]. Cells express a number of AKAPs, and a temporal dimension can be added to the spatial regulation of PKA if various AKAPs (having specificity for different secondary binding partners) are differentially expressed with respect to time, for instance, throughout the cell cycle or during differentiation[14]. In addition to holoenzyme targeting via AKAPs, it is possible that spatial localisation of PKA catalytic activity can be achieved by intrinsic targeting of C-subunit. The myristoylated amino terminus of C-subunit is a candidate motif for involvement in intrinsic targeting[15]. Although the majority of free C-subunit molecules are uniformly distributed throughout the cytosol of most cells, several examples of the specific interaction of C-subunit with partner proteins (in addition to R-subunit and the specific heat-stable inhibitor protein isoforms[16]) have recently been described[17,18], and atypical sub-cellular distribution of C-subunit has been reported in ovine spermatozoa which express a non-myristoylated splice-variant of the kinase[8].

## 2. RESULTS AND DISCUSSION

In mammary epithelial cells, three separate lines of evidence point towards sequestration/targeting of PKA. The first of these concerns the regulation of acetyl-CoA carboxylase (ACC). The phosphorylation cascade controlling its activity is normally initiated by PKA activation[19]; however, this is abrogated in intact mammary epithelial cells although it operates efficiently with intrinsic mammary components in cell-free systems[20]. Secondly, there is evidence for the membrane-localisation of a small fraction of the total mammary cellular PKA[21]. Thirdly, the tonic level of PKA activation in mammary epithelial cells has been shown to modulate the rate of the constitutive pathway of casein secretion and this effect has been localised to early steps (ER→Golgi; Golgi→TGN) in the pathway[22]. The effect of inhibition of basal PKA activity on casein secretion is illustrated in Fig. 1.

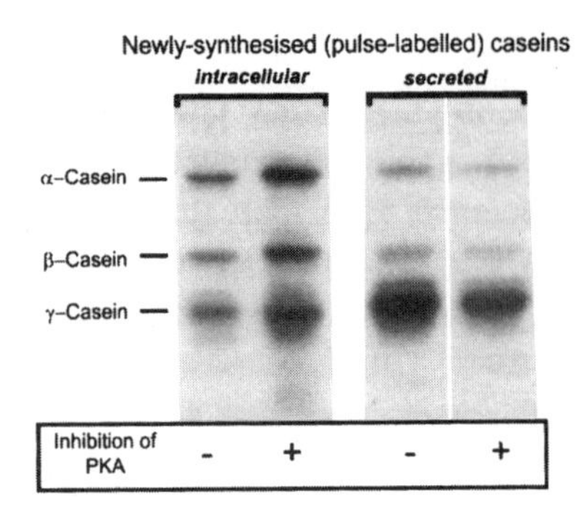


Fig. 1. *Suppression of constitutive secretion of newly-synthesised caseins in explants of rat mammary gland by inhibition of PKA. SDS PAGE analysis of pulse-labelled proteins.*

## 2.1 Association of PKA and AKAPs with membranes of secretory pathway elements

The occurrence of PKA in unfractionated (microsomal) membranes from unstimulated lactating rat mammary tissue has been confirmed by three independent methods: (1) Western blotting with anti-C-subunit antibodies; (2) photoaffinity-labelling of R-subunits with 8-azido-[$^{32}$P]cAMP; (3) enzymic assay of catalytic activity. Following stimulation of adenylate cyclase in intact cells with forskolin or isoprenaline, or direct activation of PKA with cpt cAMP, the following were found to be increased in microsomal membranes: PKA activity (total and expressed components); concentration of total C-subunit; concentration of RII-subunit (Fig. 2). Facultative targeting via conventional AKAPs is implied by this selective sequestration of RII subunit.

The AKAP content of microsomes from lactating rat mammary tissue was investigated by overlay (ligand blotting) with $^{32}$P-labelled recombinant RII∀ subunit. Major AKAPs with apparent molecular weights of 129000, 106000, 75000 were detected along with endogenous RII subunit at 54000. As previously described[23], expression of the 129000 and 75000 AKAPs was tightly developmentally regulated: they were expressed during lactation but not during pregnancy nor beyond 24h of involution following induction of milk stasis by litter-removal. AKAP distribution between cytosol and membranes was unaffected by PKA activation in intact cells (Fig. 2).

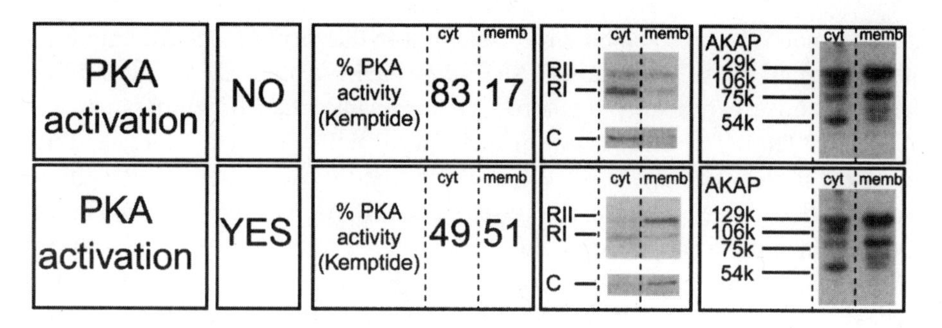


Fig. 2. Membrane-associated components of the PKA system in mammary tissue from lactating rats: effects of PKA activation before tissue fractionation.

Microsomal membranes were further fractionated on the basis of their buoyant density in sucrose gradients. Scanning down these gradients, AKAPs of $M_r$ 106000, 75000 and intrinsic RII predominated in light membrane fragments enriched in plasma membrane markers. RII was not found in denser membrane fragments, whereas 106000 and 75000 components persisted into fragments of intermediate density, enriched in Golgi membrane markers, but not into denser regions of the gradient. AKAP 129000 began to appear in Golgi/secretory vesicle-enriched fragments and became the predominant AKAP in the denser ER-derived membrane fragments. ER membrane was also, on the basis of protein content, the most abundant component in the gradient. Assay of PKA catalytic activity intrinsic to these membrane fractions, using Kemptide (a synthetic peptide phospho-acceptor substrate) showed firstly that "expressed" activity was virtually absent from all fractions and secondly that "total" activity distributed in a peak coincident with ER membranes. Potential phospho-acceptor substrate proteins of PKA resident in individual membrane fractions were identified by in vitro incubation of membranes (20μg protein) with [(-$^{32}$P]ATP in the absence of cAMP. Multiple phospho-proteins were found, and although only their $M_r$ and not their identity was revealed in these experiments, it is noteworthy that all phosphorylations were suppressed by the addition of PKI (a peptide inhibitor selective for PKA) and were only modestly enhanced by the addition to each incubation of 5μg of pure recombinant C-subunit.

## 3. CONCLUSIONS AND PERSPECTIVES

These results indicate the presence of two populations of C-subunit molecule in mammary microsomal membranes. In one, C-subunit is stoichiometrically associated with R-subunit in the form of heterotetrameric holoenzymes; these are totally cAMP-dependent in their Kemptide kinase activity and are probably anchored predominantly to ER membranes via AKAP 129000. The association of this AKAP with ER membranes, and its apparent molecular size, suggest that it may be the ER-targeted (N1 splice-variant) derivative of the carboxyl-extended D-AKAP1 splice isoform 1c (known in rodents as AKAP 121), hitherto thought to be expressed only in liver[24]. The other population consists of free C-subunit, occluded from access to Kemptide but able to catalyze the cAMP-independent phosphorylation of membrane-resident proteins that are themselves, at best, poorly accessible to exogenous free C-subunit. These endogenous free C-subunit molecules are tethered to the membrane via an intrinsic targeting mechanism relying only on the ability of C-subunit itself to interact with membrane component(s) – protein or lipid. The hypothesis that terminal domains of C-subunit are involved in its intrinsic targeting has been considered above. Future characterisation of this membrane-associated sub-population will further test this hypothesis and elucidate the mechanism of intrinsic targeting of C-subunit.

**Acknowledgements**

Aspects of this work have been funded by BBSRC and the British Council. Collaboration has been fostered by EU Action COST 825; the UK and French research teams are grateful to SERAD and INRA respectively for continuing support.

# 4. REFERENCES

1. Pawson, T., and Scott, J.D., 1997, Signalling through scaffold, anchoring, and adaptor proteins. *Science* **278**: 2075-2080.
2. Faux, M.C., and Scott, J.D., 1996, More on target with protein phosphorylation: conferring specificity by location. *TIBS* **21**: 312-315.
3. Beebe, S.J., 1994, The cAMP-dependent protein kinases and cAMP signal transduction. *Sem. Cancer Biol.* **5**: 285-294.
4. Uhler, M.D., Chrivia, J.C., McKnight, G.S., 1986, Evidence for a second isoform of the catalytic subunit of cAMP-dependent protein kinase. *J. Biol. Chem.* **261**: 15360-15363.
5. Beebe, S.J., Salomonsky, P., Jahnsen, T., and Li, Y., 1992, The Cγ subunit is a unique isoenzyme of the cAMP-dependent protein kinase. *J. Biol. Chem.* **267**: 25505-25512.
6. Tabish, M., Clegg, R.A., Rees, H.H., and Fisher, M.J., 1999, Organization and alternative splicing of the *Caenorhabditis elegans* cAMP-dependent protein kinase catalytic subunit gene (*kin-1*). *Biochem. J.* **339**: 209-216.
7. Guthrie, C.R., Skålhegg, B.S., and McKnight, G.S., 1997, Two novel brain-specific splice variants of the murine Cβ gene of cAMP-dependent protein kinase. *J. Biol. Chem.* **272**: 29560-29565.

8. San Augustin, J.T., Leszyk, J.D., Nuwaysir, L.M. and Witman, G.B., 1998, The catalytic subunit of the cyclic AMP-dependent protein kinase of ovine sperm flagella has a unique amino-terminal sequence. *J. Biol. Chem.* **273**: 24874-24883.
9. Thomis, D.C., Floyd-Smith, G., and Samuel, C.E., 1992, Mechanism of interferon action – cDNA structure and regulation of a novel splice-site variant of the catalytic subunit of human protein kinase A from interferon-treated cells. *J. Biol. Chem.* **267**: 10723-10728.
10. Chestukhin, A., Litovchick, L., Muradov, K., Batkin, M., and Shaltiel, S., 1997, Unveiling the substrate specificity of meprin β on the basis of the site in protein kinase A cleaved by the kinase-splitting membranal protease. *J. Biol. Chem.* **272**: 3153-3160.
11. Meinkoth, J.L., Alberts, A.S., Went, W., Fantozzi D., Taylor, S.S., Hagiwara, M., Montminy, M., Feramisco, J.R., 1993, Signal-transduction through the cAMP-dependent protein kinase. *Mol. Cell. Biochem.* **128**: 179-186.
12. Houslay, M.D., and Milligan, G., 1997, Tailoring cAMP-signalling responses through isoform multiplicity. *TIBS* **22**: 217-224.
13. Thorens, B., Dériaz, N., Bosco, D., DeVos, A., Pipeleers, D., Schuit, F., Medea, P., and Porret, A.,1996, Protein kinase A-dependent phosphorylation of GLUT2 in pancreatic β cells. *J. Biol. Chem.* **271**: 8075-8081.
14. Feliciello, A., Rubin, C.S., Avvedimento, E.V., and Gottesman, M.E., 1998, Expression of A kinase anchor protein 121 is regulated by hormones in thyroid and testicular germ cells. *J. Biol. Chem.* **273**: 23361-23366.
15. Boutin, J.A., 1997, Myristoylation. *Cell Signal.* **9**: 15-35.
16. Wiley, J.C., Wailes, L.A., Idzerda, R.L. and McKnight, G.S., 1999, Role of regulatory subunits and protein kinase inhibitor (PKI) in determining nuclear localization and activity of the catalytic subunit of protein kinase A. *J. Biol. Chem.* **274**: 6381-6387.
17. Wang, J., Zhou, Y., Wen, H., and Levitan, I.B., 1999, Simultaneous binding of two protein kinases to a calcium-dependent potassium channel. *J. Neurosci.* **19**: RC4 (1-7).
18. Zhong, H., SuYang, H., Erdjument-Bromage, H., Tempst, P., and Ghosh S., 1997, The transcriptional activity of NF-kB is regulated by the IkB-associated PKAc subunit through a cyclic AMP-independent mechanism. *Cell* **89**: 413-424.
19. Hardie, D.G., 1989, Regulation of fatty acid synthesis via phosphorylation of acetyl-CoA carboxylase. *Progr. Lipid Res.* **28**: 117-146.
20. Clegg, R.A., West, D.W., and Aitchison, R.E.D., 1987, Protein phosphorylation in rat mammary acini and in cytosol preparations in vitro: phosphorylation of acetyl-CoA carboxylase is unaffected by cyclic AMP. *Biochem. J.* **241**: 447-454.
21. Clegg, R.A., and Connor, K., 1991, Cyclic AMP-dependent protein kinase in mammary epithelial cells: activity and subcellular distribution are acutely modulated by isoprenaline. *Cell Signal.* **3**: 201-208.
22. Clegg, R.A., Gardner, R.A., Lavialle, F., Boisgard, R., and Ollivier-Bousquet, M., 1998, Casein secretion in mammary tissue: tonic regulation of basal secretion by protein kinase A. *Mol. Cell. Endocrinol.* **141**: 163-177.
23. Clegg, R.A., Gordge, P.C., and Miller, W.R., 1999, Expression of enzymes of covalent protein modification during regulated and dysregulated proliferation of mammary epithelial cells: PKA, PKC and NMT. *Adv. Enzyme Regul.* **39**: 175-203.
24. Huang, L.J-s., Wang, L., Ma, Y., Durick, K., Perkins, G., Deerinck, T.J., Ellisman, M.H., and Taylor, S.S., 1999, $NH_2$-terminal targeting motifs direct dual specificity A-kinase – anchoring protein 1 (D-AKAP1) to either mitochondria or endoplasmic reticulum. *J. Cell Biol.* **145**: 951-959.

13.

# The PEA3 Group of ETS-related Transcription Factors

*Role in breast cancer metastasis*

[1,2]Yvan de Launoit, [1,3]Anne Chotteau-Lelievre, [1]Claude Beaudoin, [2]Laurent Coutte, [1]Sonia Netzer, [2]Carmen Brenner, [2]Isabelle Huvent, [1]Jean-Luc Baert
*[1]UMR 8526 CNRS - Institut Pasteur, Institut de Biologie de Lille, France. [2]Laboratoire de Virologie Moléculaire – Faculté de Médecine, ULB, Brussels, Belgium. [3]Laboratoire de Biologie du Développement – UPRES EA 1033 - SN3, USTL, Villeneuve d'Ascq, France.*

Key words: Ets, Transcription Factor, Breast Cancer, PEA3

Abstract The *ets* genes encode eukaryotic transcription factors that are involved in tumorigenesis and developmental processes. The signature of the Ets family is the ETS-domain, which binds to sites containing a central 5'-GGAA/T-3' motif. They can be sub-classified primarily because of the high amino acid conservation in their ETS-domains and, in addition, in the conservation of other domains generally characterized as transactivating. This is the case for the PEA3 group, which is currently made up of three members, PEA3/E1AF, ER81/ETV1 and ERM, which are more than 95% identical in the ETS-domain and more than 85% in the transactivation acidic domain. The members of the PEA3 group are activated through both the Ras-dependent and other kinase pathways, a function which emphasizes their involvement in several oncogenic mechanisms. The expression pattern of the three PEA3 group genes during mouse embryogenesis suggests that they are differentially regulated, probably to serve important functions such as tissue interaction. Although the target genes of these transcription factors are multiple, their most frequently studied role concerns their involvement in the metastatic process. In fact, PEA3 group members are over-expressed in metastatic human breast cancer cells and mouse mammary tumors, a feature which suggests a function of these transcription factors in mammary oncogenesis. Moreover, when they are ectopically over-expressed in non-metastatic breast cancer cells, these latter become metastatic with the activation of transcription of matrix metalloproteinases or adhesion molecules, such as ICAM-1.

# 1. INTRODUCTION

The *ets* genes encode a family of eukaryotic transcription factors that includes more than 30 members from sponges to humans[1,2]. They have been involved both in tumorigenesis and in a number of developmental processes. Members of this family were originally identified on the basis of a region of primary sequence identity with the protein product of the *v-ets* oncogene encoded by the *E26* (*E twenty-six*) avian erythroblastosis virus. This signature is the ETS-domain[3], a domain of 85 amino acids structured as a winged helix-turn-helix structure and responsible for DNA-binding[1]. Many promoters have been characterized as containing active Ets-binding sites. For example, Ets proteins are involved in the regulation of the transcription of membrane receptors, growth factors or transcription factors[1,2]. Except for very limited examples, the specificity of an Ets protein to the regulation of gene transcription has not yet been established. However, this specificity could be at three levels: (1) the expression sites of the *ets* gene, (2) the DNA-binding specificity of the ETS-domain, and (3) the presence of specific domains required for protein-protein interactions. These factors can be sub-classified primarily because of the amino acid conservation in their ETS-domains and, in addition, in the conservation of other domains generally characterized as transactivating.

# 2. THE PEA3 GROUP MEMBERS

This is the case for the PEA3 group, which is currently made up of three members, PEA3 (also called E1AF in the human or ETV4)[4-6], ER81 (also called ETV1 in the human) [7-9] and ERM (also called ETV5) [10-12], which are more than 95% identical in the ETS-domain and more than 85% in the 32 residue acidic domain, and almost 50% identical in the final 61 residues corresponding to the carboxy-terminal tail of the proteins (Ct) [13]. ERM and ER81 are more closely related to each other than PEA3, suggesting that a common ancestor of the three genes has undergone two successive duplications. As illustrated in Fig. 1, human *erm* gene is composed of 14 exons split into at least 65 kbp of genomic DNA[14]. Human *etv1* and human *e1af* are each composed of 13 exons covering more than 85 kbp and 19 kbp, respectively[15,16]. The genomic organization of the ETS and the acidic domains of these genes is similar; i.e. they are both encoded by three different exons[14-16]. Concerning their chromosomal locations in the human, *erm* is situated at position 3q27-q29 [10,11], *e1af* at position 17q22 [17,18] and *etv1* at position 7q21 [8].

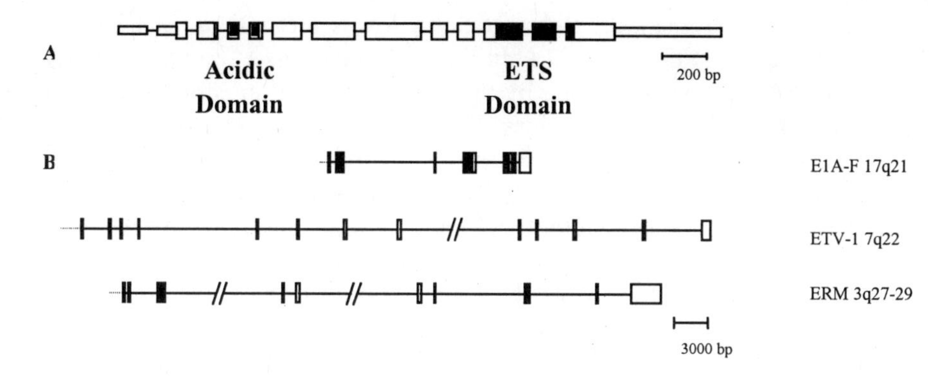


Fig 1. Organization of the three human genes of the PEA3 group. (A) Schematic representation of genomic DNA exons. The 13 exons are represented by the rectangles. The introns are indicated by horizontal lines. The protein-coding region is represented by the large box, which contains the acidic and the ETS-domains; the flanking 5'- and 3'-untranslated regions are shown as small boxes. (B) Comparison of the genomic organization of the three human PEA3 group members. Chomosomal localization is presented for each gene.

# 3. TRANSCRIPTION FACTOR CAPACITIES

All Ets transcription factors bind to sites containing a central "GGAA/T" motif. The residues flanking this motif dictate whether a particular ETS-domain will bind the site. It has been shown by means of gel shift analysis that the PEA3 group proteins bind to this DNA core consensus sequence[1,13,19]. In fact, *in vitro* target detection assay experiments have shown that the three PEA3 group proteins recognize similar sequences outside the core sequence. ERM contains two inhibitory domains for the DNA-binding activity which are adjacent to the ETS-domain. These are the Ct domain and a central region spanning residues 203 to 297 [20]. Only the Ct domain is conserved in the three PEA3 group members.

In transient cotransfection assays these three PEA3 group proteins increase transcription of a reporter plasmid which contains artificial multimerized Ets-responsive elements that may or may not be adjacent to an AP1 site [21], as well as reporter plasmids containing the functional promoter regions of the human metalloproteinases (MMP) 1, 3 and 9 [22], the human vimentin[23], and the human ICAM-I[24] genes. This transactivating activity is due to two conserved domains : the 32 residues of the acidic domain and the 61 residues in the Ct domain of human ERM [11,20,21], mouse ER81 [25] and mouse PEA3. The central region encompassing the DNA-binding inhibitory region decreases the transactivation potency of ERM and ER81 [11,20,22,25]. Structurally, the first 15 residues of the 32-residue amino terminal domain of these proteins form an alpha helix which contains the main transactivation potency[26]. Ets transcription factors exhibit low selectivity in binding site preference, suggesting that in addition to protein-DNA interactions, the specificity of promoter targeting by these factors relies on cooperation with other groups of transcription factors. This type of functional interactions has been demonstrated for almost all Ets proteins. The most studied cooperation is the ternary complex, in which the Ets proteins from the Elk group interact with the serum responsive factor (SRF) on the c-*fos* promoter[1]. As the other Ets proteins, the transcription factors from the PEA3 group do not functionally act solely to activate the transcription. ERM physically interacts with basal machinery elements from the $TF_{II}D$ complex, such as the TATA-binding protein (TBP) and the TAFII60 protein[26]. ERM also cooperates with other transcription factors, such as the androgen receptor, which negatively regulates MMP-1 expression probably following ERM fixation [27]. Recent data indicate that the PEA3 group members also functionally interact with the transcriptional co-activator CBP (personal communication).

## 4. ACTIVATION OF PEA3 GROUP MEMBERS BY TRANSDUCTION PATHWAYS

Differential phosphorylation of transcription factors by signal transduction pathways such as the mitogen activated protein kinase (MAPK) pathways plays a crucial role in the regulation of gene expression. The activation of MAPK cascades leads to changes in the activity of many Ets factors[1,28]. The transcription capacities of mouse and zebrafish PEA3, mouse and human ER81, and human ERM have been shown to be increased by components of these cascades, Ras, Raf-1, MEK, and MAPK ERK-1 and ERK-2; thus suggesting that these factors may contribute to the nuclear response to stimulation of cells and also to Ras-induced cell transformation [15,19,21,25,29]. The JNK/SAPK pathway is also involved in mouse PEA3 activation[29]. Moreover, protein kinase A (PKA) is also able to increase the transcriptional activity of the human ERM[21], the human ETV1 [15] and the zebrafish PEA3 [19] through a classical PKA consensus site, RRGS, present at the edge of the ETS-domain. In contrast, the mouse and human PEA3 proteins contain a RRGA sequence in place of the PKA site, thus avoiding these proteins to be activated by the PKA (personal communication).

# 5. EXPRESSION OF THE PEA3 GROUP MEMBERS

## 5.1 In the embryonic development

Recent data are yet available in regard to a role of these three genes during embryonic development. A prerequisite to investigations in this field is to obtain an accurate spatio-temporal expression map for the *erm*, *er81* and *pea3* genes. To this end, *in situ* hybridization used to compare their expression patterns during critical stages of murine embryogenesis shows that all three genes are expressed in numerous developing organs coming from different embryonic tissues. They appear co-expressed in different organs but present specific sites of expression, so that the resultant expression pattern could in fact reveals several distinct functions depending upon isolated and/or various combinations of the PEA3 member expression[12]. In developing dorsal root ganglia, *erm* is expressed both in satellite glia that express the NRG1 receptor ErbB3, and in neurons. However, this transcript is not detectable in presumptive Schwann cells along peripheral nerves. ERM represents thus the first mammalian marker that distinguishes satellite glia from presumptive Schwann cells at an early developmental stage [30]. In the Xenopus the homologue of ER81 (XER81) is expressed in the marginal zone at the onset of gastrulation. Over-expression of XER81 in Xenopus embryos results in the induction of ectopic, tail-like protrusions, or disturbed eye development. In later embryogenesis XER81 transcripts are found in neural crest cells, eyes, otic vesicles and pronephros and its expression requires active FGF signalling. The spatial overlap of eFGF and XER81 expression supports the idea that XER81 transcription could be a marker for regions with active FGF signalling in the embryo[31]. In the chick embryo, it has been demonstrated that motor neuron pools and subsets of muscle sensory afferents can be defined by the expression of *pea3* and *er81*. There is a matching in *pea3* and *er81* expression by functionally interconnected sensory and motor neurons. Expression of these genes by motor and sensory neurons fails to occur after limb ablation, suggesting that their expression is coordinated by signals from the periphery. These genes may therefore participate in the development of selective sensory-motor circuits in the spinal cord[32]. Altogether, these data suggest that these genes probably serve important functions as cell proliferation control, tissue interaction mediator or cell differentiation, all over successive steps of the mouse, chick, and Xenopus organogenesis

## 5.2 In the adult

At the mRNA level, ERM has been classified in adult human and mice as a ubiquitously expressed gene with its highest expression in the brain[10]. ER81/ETV1 displays an expression pattern with high expression levels in human and murine lung, heart and brain[7-10]. In contrast, PEA3/E1AF presents an expression pattern very restricted in normal adult tissues. It is almost exclusively expressed in the brain[4,12].

# 6. THE FUSION OF PEA3 GROUP GENES AND THE EWS GENE IN THE EWING'S SARCOMA

In almost all Ewing's sarcoma tumors, the RNA-binding protein gene *ews* is fused to an *ets* gene, either *fli* or *erg* by a t(11;22)(q24;q12) or t(21;22)(q22;q12) chromosome translocation[33,34]. *Etv1* and *e1af* have also been identified as translocation partners of *ews* in Ewing's sarcoma involving t(7;22)(p22;q12) and t(17;22)(q12;q12) translocations in these undifferentiated child sarcomas, leading to the synthesis of a chimeric protein which is formed by the transactivating domain of EWS and the ETS-domain of ETV1 [8] or E1AF [35-37]. The translocation breakpoint observed in the Ewing's sarcoma involving part of the *e1af* gene is situated between the exons coding the acidic domain and the ETS-domain. More precisely, the translocation breakpoint for the *e1af* gene is situated in intron 8, which is about 2.5 kbp long and contains repetitive Alu sequences, which were shown as being involved in the mechanisms of translocation with the *ews* gene[36]. At the present time, only the EWS-Fli chimera protein presents typical oncogenic properties, whereas EWS-Erg is less transforming and EWS-Etv1 is not at all[38]. Together with the putative cooperation with the activating domain of EWS, these capacities could be crucial in inducing the transcription of genes linked to transformation.

# 7. THE PEA3 GROUP MEMBERS AND THE BREAST CANCER METASTASIS

Although the target genes of these transcription factors are multiple, their most frequently studied role concerns their involvement in the metastatic process. In fact, it has been shown that PEA3 group members are over-expressed in metastatic human breast cancer cells[39] and mouse mammary tumors[40], a feature which suggests a function of these transcription factors in mammary oncogenesis. An initial experiment has shown that when E1AF is ectopically over-expressed in the non-metastatic MCF-7 human breast cancer cell line, the cells become metastatic in nude mice by activating the transcription of the matrix metalloprotease collagenase IV [41]. E1AF confers the invasive phenotype on cancer cells. E1AF is thus supposed to play an important role in cancer invasiveness/metastasis through transcription of metastasis-related genes. This role has been confirmed in another model, where non-metastatic mouse fibrosarcoma cells became metastatic when E1AF is ectopically expressed[42]. This E1AF over-expression contributes to invasiveness by activating MT1-MMP expression. By contrast, Hida *et al.* [43], showed in human metastatic oral squamous carcinoma cells that repression of E1AF by using a specific antisense RNA restrained the invasive phenotype. It however remains to be tested whether the effect obtained with E1AF is the same when the expression of the two other members of the PEA3 group is changed.

Several data now indicate that these factors are not *per se* oncogenic since we have generated transgenic lines specifically over-expressing these genes in the mammary gland and which do not develop tumors after two years (unpublished, Netzer and de Launoit). In contrast, crossing-over of these transgenic mice with mice developing tumors could add interesting data concerning the role of these factors in *in vivo* metastatic process.

Since these transcription factors are probably involved in the regulation of specific metastatic processes in the breast, as well as in other tissues, they can be used in the near future as metastatic markers. Moreover, gene therapies using antisense of these genes could be envisaged to treat metastatic breast cancer.

**Acknowledgements**

This work has been carried out on the basis of grants awarded by the "Centre National de la Recherche Scientifique" (France), the “Institut Pasteur de Lille”, the "Association pour la Recherche contre le Cancer" (France), the “Fonds National de la Recherche Scientifique” (Belgium) and the “Action de Recherche Concertée” from the “Communauté Française” (Belgium). LC is holder of a ARC fellowship. CB is holder of a MRC fellowship.

# 8. REFERENCES

1. Sharrocks A. D., Brown A. L., Ling Y. and Yates P. R., The ETS domain transcription factor family. Int. J. Biochem. Cell. Biol. 29, 12, 1371-1387, 1997
2. Dittmer J. and Nordheim A., Ets transcription factors and human disease. Biochem. Biophys. Acta. 1377, F1-F11, 1998

3. Karim F. D., Urness L. D., Thummel C. S., Klemsz M. J., McKercher S. R., Celada A., Van Beveren C., Maki R. A., Gunther C. V., Nye J. A. and Graves B. J., The ETS-domain: a new DNA-binding motif that recognizes a purine-rich core DNA sequence. Genes Dev. 4, 9, 1451-1453, 1990
4. Xin J. H., Cowie A., Lachance P. and Hassell J. A., Molecular cloning and characterization of PEA3, a new member of the *Ets* oncogene family that is differentially expressed in mouse embryonic cells. Genes Dev. 6, 481-496, 1992
5. Higashino F., Yoshida K., Kamio K. and Fujinaga K., Isolation of a cDNA encoding the adenovirus E1A enhancer binding protein: a new human member of the *ets* oncogene family. Nucl. Acid Res. 21, 547-553, 1993
6. Friedman L. S., Ostermeyer E. A., Lynch E. D., Szabo C. I., Anderson L. A., Dowd P., Lee M. K., Rowell S. E., Boyd J. and King M. C., The search for BRCA1. Cancer Res. 54, 24, 6374-82, 1994
7. Brown T. A. and McKnight S. L., Specificities of protein-protein and protein-DNA interaction of GABPα and two newly defined *ets*-related proteins. Genes Dev. 6, 2502-2512, 1992
8. Jeon I. S., Davis J. N., Braun B. S., Sublett J. E., Roussel M. F., Denny C. T. and Shapiro D. N., A variant Ewing's sarcoma translocation (7;22) fuses the EWS gene to the ETS gene ETV1. Oncogene 10, 6, 1229-1234, 1995
9. Monté D., Coutte L., Baert J.-L., Angeli I., Stéhelin D. and de Launoit Y., Molecular characterization of the Ets-related human transcription factor ER81. Oncogene 11, 771-780, 1995
10. Monté D., Baert J.-L., Defossez P.-A., de Launoit Y. and Stéhelin D., Molecular cloning and characterization of human ERM, a new member of the ETS family closely related to mouse PEA3 and ER81 transcription factors. Oncogene 9, 1397-1406, 1994
11. Nakae K., Nakajima K., Inazawa J., Kitaoka T. and Hirano T., ERM, a member of the PEA3 subfamily of Ets transcription factors, can cooperate with c-jun. J. Biol. Chem. 270, 40, 23795-23800, 1995
12. Chotteau-Lelièvre A., Desbiens X., Pelczar H., Defossez P.-A. and De Launoit Y., Differential expression patterns of the PEA3 group transcription factors through murine embryonic development. Oncogene 15, 937-952, 1997
13. de Launoit Y., Baert J.-L., Chotteau A., Monté D., Defossez P.-A., Coutte L., Pelczar H. and Leenders F., Structure-function relationships of the PEA3 group of Ets-related transcription factors. Biochem. Mol. Med. 61, 2, 127-35, 1997
14. Monté D., Coutte L., Dewitte F., Defossez P.-A., Le Coniat M., Stéhelin D., Berger R. and de Launoit Y., Genomic organization of the human ERM (ETV5) gene, a PEA3 group member of ETS transcription factors. Genomics 35, 1, 236-240, 1996
15. Coutte L., Monté D., Imai K., Pouilly L., Dewitte F., Vidaud M., Adamski J., Baert J.-L. and de Launoit Y., Characterization of the human and mouse ETV1/ER81 transcription factor genes: role of the two alternatively spliced isoforms in the human. Oncogene in press, 1999
16. Coutte L., Monté D., Baert J.-L. and de Launoit Y., Genomic organization of the human E1A-F gene, a member of ETS transcription factors. Gene in press, 1999
17. Isobe M., Yamagichi F., Yoshida K., Higashino F. and Fujinaga K., Assignment of the ets-related transcription factor E1AF gene (ETV4) to human chromosome region 17q21. Genomics 28, 357-359, 1995
18. Osborne-Lawrence S., Welcsh P. L., Spillman M., Chandrasekharappa S. C., Gallardo T. D., Lovett M. and Bowcock A. M., Direct selection of expressed sequences within a 1-

Mb region flanking BRCA1 on human chromosome 17q21. Genomics 25, 1, 248-55, 1995
19. Brown L. A., Amores A., Schilling T. F., Jowett T., Baert J.-L., Ingham P. W., de Launoit Y. and Sharrocks A. D., Molecular characterisation of the zebrafish PEA3 ETS-domain transcription factor. Oncogene 17, 1, 93-104, 1998
20. Laget M.-P., Defossez P.-A., Albagli O., Baert J.-L., Dewitte F., Stéhelin D. and de Launoit Y., Two functionally distinct domains cooperate for transactivation by the ETS family member ERM. Oncogene 12, 6, 1325-1336, 1996
21. Janknecht R., Monté D., Baert J.-L. and de Launoit Y., The ETS-related transcription factor ERM is a nuclear target of signaling cascades involving MAPK and PKA. Oncogene 13, 8, 1745-1754, 1996
22. Higashino F., Yoshida K., Noumi T., Seiki M. and Fujinaga K., Ets-related protein E1A-F can activate three different matrix metalloproteinase gene promoters. Oncogene 10, 1461-1463, 1995
23. Chen J. H., Vercamer C., Li Z., Paulin D., Vandenbunder B. and Stehelin D., PEA3 transactivates vimentin promoter in mammary epithelial and tumor cells. Oncogene 13, 1667-1675, 1996
24. de Launoit Y., Audette M., Pelczar H., Plaza S. and Baert J.-L., The transcription of the intercellular adhesion molecule-1 is regulated by Ets transcription factors. Oncogene 16, 2065-2073, 1998
25. Janknecht R., Analysis of the ERK-stimulated transcription factor ER81. Mol. Cell. Biol. 16, 1551-1556, 1996
26. Defossez P.-A., Baert J.-L., Monnot M. and de Launoit Y., The ETS family member ERM contains an alpha-helical acidic activation domain that contacts TAFII60. Nucleic Acids Res. 25, 22, 4455-4463, 1997
27. Schneikert J., Peterziel H., Defossez P.-A., Klocker H., de Launoit Y. and Cato A. C. B., Androgen receptor Ets protein interaction is a novel mechanism for steroid hormone mediated down regulation of matrix metalloproteinase expression. J. Biol. Chem. 271, 39, 23907-23913, 1996
28. Treisman R., Regulation of transcription by MAP kinase cascades. Curr. Opin. Cell Biol. 8, 2, 205-15, 1996
29. O'Hagan R. C., Tozer R. G., Symons M., McCormick F. and Hassell J. A., The activity of the Ets transcription factor PEA3 is regulated by two distinct MAPK cascades. Oncogene 13, 1323-1333, 1996
30. Hagedorn L., Paratore C., Mercader N., Brognoli G., Baert J.-L., Suter U. and Sommer L., The Ets domain transcription factor Erm distinguishes satellite glia from Schwann cells and is maintained in satellite cells by neuregulin signalling. submitted 1999
31. Munchberg S. R. and Steinbeisser H., The Xenopus Ets transcription factor XER81 is a target of the FGF signaling pathway. Mech. Dev. 80, 1, 53-65, 1999
32. Lin J. H., Saito T., Anderson D. J., Lance-Jones C., Jessell T. M. and Arber S., Functionally related motor neuron pool and muscle sensory afferent subtypes defined by coordinate ETS gene expression. Cell 95, 3, 393-407, 1998
33. Delattre O., Zucman J., Plougastel B., Desmaze C., Melot T., Peter M., Kovar H., Joubert I., De Jong P., Rouleau G., Aurias A. and Thomas G., Gene fusion with an ETS DNA-binding domain caused by chromosome translocation in human tumours. Nature 359, 6391, 162-165, 1992
34. Zucman J., Melot T., Desmaze C., Ghysdael J., Plougastel B., Peter M., Zucker J. M., Triche T. J., Sheer D., Turc-Carel C., Ambros P., Combaret V., Lenoir G., Aurias A.,

Thomas G. and Delattre O., Combinatorial generation of variable fusion proteins in the Ewing family of tumours. EMBO J. 12, 4481-4487, 1993

35. Kaneko Y., Yoshida K., Handa M., Toyoda Y., Nishihira H., Tanaka Y., Sasaki Y., Ishida S., Higashino F. and Fujinaga K., Fusion of an ETS-family gene, EIAF, to EWS by t(17;22)(q12;q12) chromosome translocation in an undifferentiated sarcoma of infancy. Genes Chromosomes Cancer 15, 2, 115-121, 1996
36. Ishida S., Yoshida K., Kaneko Y., Tanaka Y., Sasaki Y., Urano F., Umezawa A., Hata J. and Fujinaga K., The genomic breakpoint and chimeric transcripts in the EWSR1-ETV4/E1AF gene fusion in Ewing sarcoma. Cytogenet. Cell Genet. 82, 3-4, 278-83, 1998
37. Urano F., Umezawa A., Yabe H., Hong W., Yoshida K., Fujinaga K. and Hata J., Molecular analysis of Ewing's sarcoma: another fusion gene, EWS-E1AF, available for diagnosis. Jpn. J. Cancer Res. 89, 7, 703-11, 1998
38. Braun B. S., Frieden R., Lessnick S. L., May W. A. and Denny C. T., Identification of target genes for the ewing's sarcoma EWS/FLI fusion protein by representational difference analysis. Mol. Cell. Biol. 15, 8, 4623-4630, 1995
39. Baert J.-L., Monte D., Musgrove E. A., Albagli O., Sutherland R. L. and de Launoit Y., Expression of the PEA3 group of ETS-related transcription factors in human breast-cancer cells. Int. J. Cancer 70, 590-597, 1997
40. Trimble M. S., Xin J. H., Guy C. T., Muller W. J. and Hassell J. A., PEA3 is overexpressed in mouse metastatic mammary adenocarcinomas. Oncogene 8, 3037-3042, 1993
41. Kaya M., Yoshida K., Higashino F., Mitaka T., Ishii S. and Fujinaga K., A single ets-related transcription factor, E1AF, confers invasive phenotype on human cancer cells. Oncogene 12, 221-227, 1996
42. Habelhah H., Okada F., Kobayashi M., Nakai K., Choi S., Hamada J., Moriuchi T., Kaya M., Yoshida K., Fujinaga K. and Hosokawa M., Increased E1AF expression in mouse fibrosarcoma promotes metastasis through induction of MT1-MMP expression. Oncogene 18, 9, 1771-6, 1999
43. Hida K., Shindoh M., Yasuda M., Hanzawa M., Funaoka K., Kohgo T., Amemiya A., Totsuka Y., Yoshida K. and Fujinaga K., Antisense E1AF transfection restrains oral cancer invasion by reducing matrix metalloproteinase activities. Am. J. Pathol. 150, 6, 2125-2132, 1997

14.

# Transcription Factor NF 1 Expression in Involuting Mammary Gland

Rosemary Kane, Darren Finlay, Teresa Lamb and Finian Martin
*Dept. of Pharmacology and Biotechnology Centre, University College Dublin, Belfield, Dublin 4, Ireland*

Key words: Mammary Gland, Involution, Transcription Factor, NF1

Abstract: Transcripts of each of the four NF1 genes (NF1 A, B, C (CTF/NF1) and X) are expressed in both lactating and involuting mouse mammary gland but there is an indication that increased expression of an NF1 C (CTF/NF1) transcript accompanies early involution. The involution-associated 74 kD NF1 and the 114 kD lactation-associated NF1 are recognised by an anti-NF1 C–specific antibody that does not cross-react with other NF1 proteins. It is most likely that this lactation/involution switch in NF1 factors represents a change in expression of NF1 C (CTF/NF1) proteins.

## 1. INTRODUCTION:

Clusterin gene expression is rapidly induced in early involution of the mouse mammary gland after weaning. A search for involution enhanced DNaseI footprints in the proximal mouse TRPM-2/clusterin gene promoter led to the identification and characterization (by DNase I footprinting) of a twin NF1 binding element at -356/-309, relative to the proposed transcription start-site, with nuclear extracts from 2d-involuting mouse mammary gland showing an enhanced footprint over the proximal NF1 element, relative to extracts from lactating mammary gland. Subsequent EMSA and western analysis led to the detection of a 74 kD NF1 protein whose expression is triggered in early involution in the mouse mammary gland. This protein was not found in lactation where three other NF1 proteins of 114, 68 and 46 kD were detected. Reiteration of the epithelial cell apoptosis associated with early mammary gland involution, *in vitro*, in a primary cell culture system, triggered the appearance of the 74 kD NF1. Overlaying the cells with laminin-rich ECM suppressed the apoptosis and the expression of the 74 kD NF1 and, in the presence of lactogenic hormones, initiated milk protein gene expression and the expression of two of the lactation associated NF1 proteins (68 and 46 kD)[1]. Below we describe the transcripts of NF1 genes in mouse mammary gland which potentially encode the involution-associated 74 kD NF1.

## 2. MATERIALS AND METHODS

The basic procedures employed are described by Furlong *et al.*[1]. Specific PCR analyses employed hot start and suitable cycling conditions established for each primer pair.

## 3. RESULTS

NF1 (nuclear factor 1) and CTF/NF1 (CCAAT-binding transcription factors/NF1) proteins are a heterogeneous family of transcription factors which have a highly conserved N-terminal DNA binding and dimerisation domain[2,3,4] and proline-rich carboxy-terminal transactivating domains that are highly divergent in size and sequence. Thus, related products encoded by four different genes have been characterized (NF1-A, B, C and X) and multiple splice variants which differ in the carboxy terminus of these have been described[5,6,7]. Based on the sequence of known transcripts of the four mouse NF-1 genes we devised an RT-PCR strategy that would allow us clone the NF1 transcripts expressed in lactating and involuting mouse mammary gland. Table 1 shows that multiple transcripts of each of the four NF1 genes (NF1 A, B, C (CTF/NF1) and X) are expressed in both lactating and involuting mouse mammary gland. Three transcripts of NF1 A, two of NF1 B and two of NF1 X were cloned and characterised (by DNA sequencing) from both lactating and early involuting mammary gland. Only in the case of NF1 C (CTF/NF1) were transcripts found exclusively in either lactating or involuting mammary tissue. However, these were seemingly rare and previously undescribed transcripts: NF1 C-Δ-5^8,9,10* was detected in lactating tissue; and, NF1 C-Δ-8,9 and NF1 C-Δ-5^8,9 were detected in involuting tissue (numbers denote the spliced-out exons; 5^8, spliced from within exon 5 to within exon 8).

Table I.: NF1 Transcripts Expressed in the Mouse Mammary Gland

| Transcript | Lactation | Involution |
|---|---|---|
| **NF1A** | | |
| mNF1-A1 (Y07690*) | + | + |
| mNF1-A2 (Y07691) | + | + |
| **NF1-B** | | |
| mNF1-B1 (Y07685) | + | + |
| mNF1-B2 (Y07687) | + | + |
| mNF1-B3 (Y07686) | + | + |
| **NF1-X** | | |
| mNF1-X1 (Y07688) | + | + |
| mNF1-X2 (Y07689) | + | + |
| mNF1-X3 (S81451) | - | - |
| **NF1-C** | | |
| (h)CTF-1(X12492) | - | - |
| mNF1-C2, (U57635, Y07692, Y07693) | + | + |
| mCTF-5 (X92857)** | + | + |
| **Novel mouse NF1-C** | | |
| mNF1-C3 (Exons 7&9 spliced) | + | + |
| mNF1-C4 (Exons 8&9 spliced) | - | + |
| mNF1-C5 (Exons 9&10 spliced)** | + | + |
| mNF1-C6 (Exons 5^8&9 spliced) | - | + |
| mNF1-C7 (Exons 5^8,9&10 spliced) | + | - |

*Genbank accession numbers are provided for all previously cloned NF1s **Described previously for human not for mouse

We next designed specific primer sets for PCR analysis of total NF1 A, NF1 B and NF1 X transcripts and the predominant individual NF1 A, B and X transcripts in lactating (L), involuting (2 di) and re-suckled 2-day involuting mammary gland (36h s). There was no striking difference in total NF1 A, B or C transcript levels between lactating, involuting or re-differentiated mammary tissue; however, analysis of the individual NF1 A transcripts suggested that there was greater expression of NF1 A1 in lactation and the resuckled gland than in involution (Fig 1). In the case of the NF1 B family members a bias towards increased expression in involution was seen; no discernible differences could be established for the individual NF1 X transcripts (Fig. 1).

While these experiments suggested that the involution-associated 74 kD NF1 was possibly encoded by an NF1 B gene we obtained more compelling evidence that it was encoded by an NF1 C/CTF-NF1 gene: The 74kD involution-associated-NF1 was detectable in western analysis by a range of antibodies raised against the conserved N-terminal NF1 DNA binding domain[1] but also by an anti-CTF/NF1 C-specific antibody[8]. We established that this latter antibody is seemingly NF1 C-specific by its ability to detect HA-epitope-tagged mouse NF1 C2 transiently expressed in JEG 3 cells (which have low endogenous NF1 expression[9]; the antibody did not detect expressed epitope-tagged NF1 A, B or X. An anti-HA antibody detected all four expressed proteins in transiently transfected JEG 3 cells (results not shown).

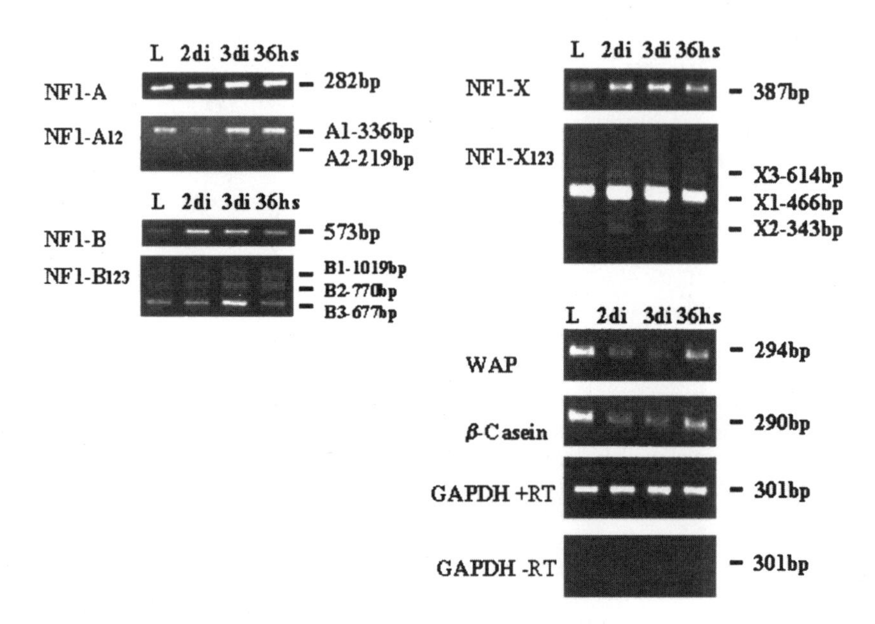


Figure 1. *NF1 transcript levels in lactating and involuting mouse mammary gland*

The dominant NF1 C transcripts detectable in mammary epithelial cells are NF1 C2 and NF1 C5. We established an RT PCR assay for these two transcripts (Fig. 2). We found a definite tendency towards higher expression of these transcripts in involuting mammary gland than in lactating and re-differentiated mammary gland (Fig. 2). However, this increased NF1 C transcript expression tends to be higher on day 3 of involution while expression of the 74 kD involution-associated protein tends to be higher on day 2 of involution[1].

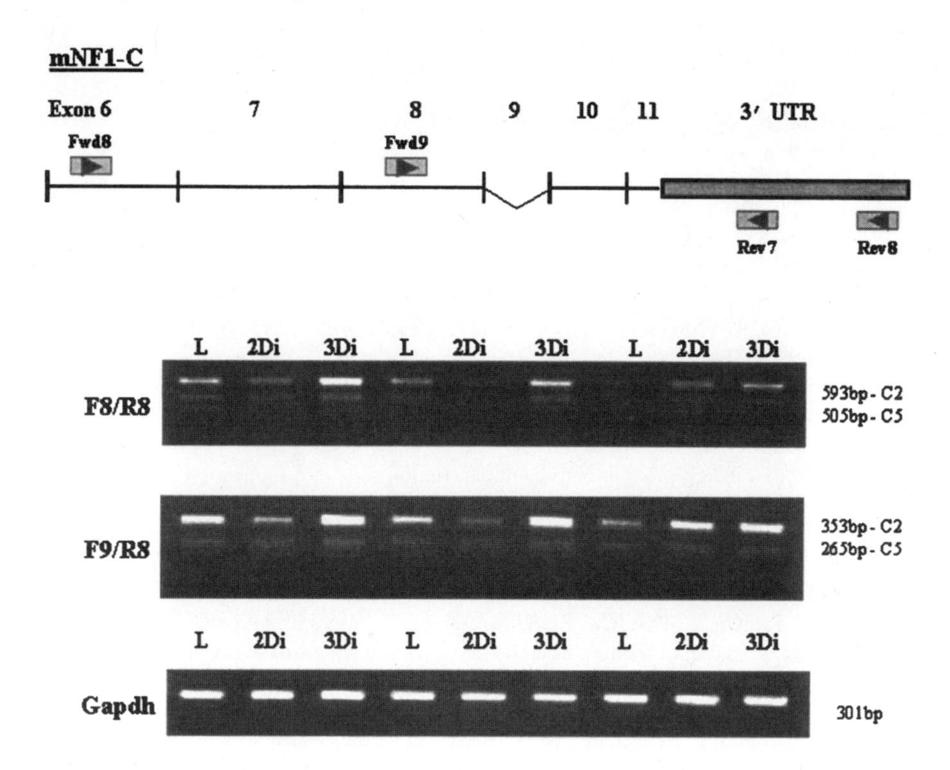


Figure 2. *NF1 C transcripts in lactating and involuting mouse mammary gland.*

## 4. CONCLUSION

The involution-associated 74 kD NF1 (and the 114 kD lactation-associated NF1 are recognised by an anti-NF1 C–specific antibody (see above). It is most likely, therefore, that this lactation/involution switch in NF1 factors represents a change in expression of NF1 C (CTF/NF1) proteins. We present preliminary evidence here that there is also an involution-associated increase in NF1 C transcript levels but highest levels are detectable slightly later in involution (day 3), after highest levels of the 74 kD protein are seen (day 2). Of significant consequence, also, is the fact that the known mouse NF1 C transcripts encode proteins of approximately < 60 kD. Thus, the 74 kD NF 1 is a post-translationally modified NF1 C or encoded by an NF1 transcript of novel exonic structure. We are presently endeavouring to answer these questions.

**Acknowledgements.**
This work was supported by grants from Enterprise Ireland and The Health Research Board, Ireland.

# 5. REFERENCES


1. E. Furlong, N. Keon, F. Thornton, T. Rein and F. Martin , 1996, Expression of a 74 kD NF-1 protein is induced in mouse mammary gland involution. Involution enhanced occupation of a twin NF-1 binding element in the TRPM-2/clusterin promoter. *J. Biol. Chem.*, **271**, 29688 – 29697.
2. Roy, R.J. and Guerin, S.L., 1994, The 30-Kda rat liver transcription factor Nuclear Factor 1 binds the rat growth-hormone proximal silencer. *Eur J Biochem* **219**:799-806.
3. Reifel-Miller, A.E., Calnek, D.S. and Grinnell, B.W., 1994, Tyrosine phosphorylation regulates the DNA binding activity of a Nuclear Factor 1-like repressor protein. *J Biol Chem* **269**:23861-4.
4. Hay, R.T., 1985, The origin of adenovirus DNA replication: Minimal DNA sequence requirement in vivo. *EMBO J.* **4**:421-426.
5. Santoro, C., Mermod, N., Andrews, P.C. and Tjian, R., 1988, A family of human CCAAT-box-binding proteins active in transcription and DNA replication: cloning and expression of multiple cDNAs. *Nature* **334**:218-24.
6. Apt, D., Liu, Y. and Bernard, H.U., 1994, Cloning and functional analysis of spliced isoforms of human Nuclear Factor I-X: interference with transcriptional activation by NF1/CTF in a cell-type specific manner. *Nucleic Acids Res* **22**:3825-3833.
7. Inoue, T., Tamura, T., Furuichi, T. and Mikoshiba, K. , 1990, Isolation of complementary DNAs encoding a cerebellum-enriched Nuclear Factor I family that activates transcription from the mouse myelin basic protein promoter. *J Biol Chem* **265**:19065-70.
8. Szabo, P., Moitra, J., Rencendorj, A., Rakhely, G., Rauch, T. and Kiss, I., 1995, Identification of a Nuclear Factor-I family protein-binding site in the silencer region of the cartilage matrix protein gene. *J Biol Chem* **270**:10212-219.
9. Chaudry A., Lyons G. E. and Gronostajski R.M., 1997, Expression patterns of the four nuclear factor 1 genes during mouse embryogenesis indicate a potential role in development. *Developmental Dynamics*, **208**, 313-325.

15.

# The Elf Group of Ets-Related Transcription Factors

*ELF3 and ELF5*

Ross S. Thomas[2], Annie N.Y. Ng[1], Jiong Zhou[1], Martin J. Tymms[3], Wolfgang Doppler[4] and Ismail Kola[1]

[1] Centre for Functional Genomics and Human Disease, Monash University, Melbourne, Australia: [2] Transcription Laboratory, Imperial Cancer Research Fund, London, UK: [3] National Vision Research Institute of Australia, Melbourne, Australia: [4] Institut für Medizinische Chemie und Biochemie, Innsbruck, Austria

Key words  ELF3, ELF5, transcription factors

## 1. INTRODUCTION

We provide an initial characterisation of two new mammalian ETS transcription factors, ELF3 and ELF5. Expression of ELF3 and ELF5 appears to be restricted to the epithelial cells of multiple organs, and we are examining their role in normal mammary differentiation and function, and in mammary neoplasia. We show evidence that both ELF3 and ELF5 are able to positively regulate transcription of the whey acidic protein (WAP) promoter in mammary epithelial cells, independently of hormone treatment.

## 1.1 Background

One of the major milk proteins produced by mammary epithelial cells during pregnancy and lactation is whey acidic protein (WAP). WAP expression is very low until mid-gestation, after which time the combined actions of the glucocorticoid and prolactin pathways produce a several thousand-fold increase in transcription. Transcription is also partly dependent upon the binding of factors to the mammary-cell activating factor (MAF) site[1,2], a conserved ETS-like element present in the WAP proximal promoter and in the regulatory regions of other mammary specific genes.

The ETS family of transcription factors regulate gene expression during normal biological processes such as haemopoiesis, angiogenesis and cartilage/skeletal development, and aberrant forms have also been implicated with various types of neoplasia. Despite the large variety of ETS factors, functions in epithelial cells, such as those in the mammary gland, have not been examined extensively. Overexpression of PEA3 is, however, associated with aggressive subsets of breast carcinoma that also overexpress erbB2[3]. PEA3 does not appear to be involved in normal mammary function, but may be required for early mammary development.

We have recently cloned two new ETS transcription factors, ELF3[4] and ELF5[5]. The HUGO Nomenclature Committee has upheld the ELF naming convention despite ELF3 being cloned and differently named by other groups (ESX[6], ESE-1[7], jen[8], ERT[9]).

## 2. EXPRESSION OF ELF3 AND ELF5

We examined the expression of ELF3 and ELF5 in mouse organs by Northern blot analysis of poly $A^+$ RNA. The expression patterns of these two genes were similar, but both differ significantly to the expression of other ETS family members. Most other ETS factors are expressed strongly in haemopoietic compartments, but ELF3 and ELF5 levels were undetectable in adult thymus or spleen (Figs 1a and 2a), nor in peripheral lymphocytes of leukaemic cell lines (data not shown). Strong expression, however, was observed in a subset of organs that contain secretory epithelial cells, such as the lung, stomach, prostate and mammary gland.

The mammary gland undergoes distinct phases of growth, differentiation and regression; from virginal mice to pregnancy, parturition and weaning. Examination of expression levels through these stages revealed that ELF3 is present in virginal mammary gland and during early pregnancy, when cells are proliferating, but drops in late pregnancy when epithelial cells undergo terminal differentiation and begin to produce milk components (Fig 1b). Expression remains low during lactation, but reappears upon weaning, when most of the epithelial cells in the mammary gland are undergoing apoptosis. Another group's study made similar findings[10]. *In situ* expression analysis in human mammary gland confirms that ELF3 is expressed specifically in the epithelial cells of the ductules and lobular structures (Fig 1c).

One possible explanation for the pattern of ELF3 expression is that it is associated with a small proportion of epithelial stem cells, rather than the differentiated functional cells, and that these cells survive apoptosis to re-colonise the mammary gland during the next pregnancy.

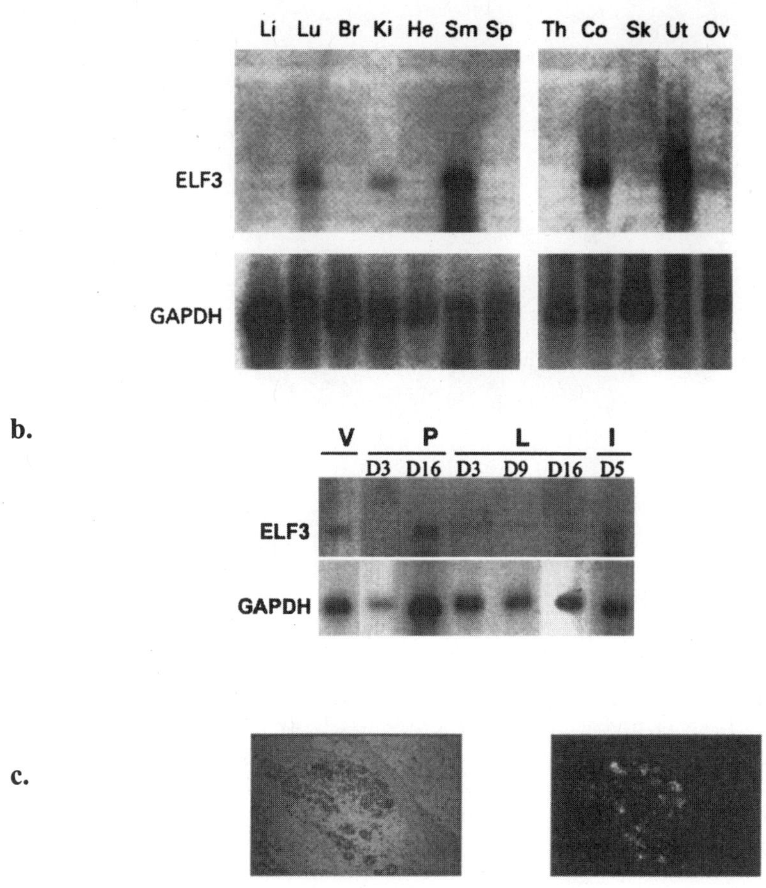


*Fig. 1.* Expression of ELF3. **a.** Northern blot of mouse organs probed with labelled ELF3 and GAPDH cDNAs. Li; liver, Lu; lung, Br; brain, Ki kidney, He; heart, Sm; small intestine, Sp; spleen, Th; thymus, Co; colon, Sk; skin, Ut; uterus, Ov; ovary. **b.** Northern blot of mouse mammary gland. V; virgin, P; pregnant, L; lactating, I; involuting. **c.** Human mammary gland section probed with labelled ELF3 cDNA (light field and exposure).

ELF5 is also expressed specifically in human mammary gland epithelial cells (Fig 2b), and, in two time points, mammary ELF5 expression is increased during pregnancy in the mouse (Fig 2a), suggesting that the temporal pattern of ELF5 expression may be very different to that of ELF3.

**a.**

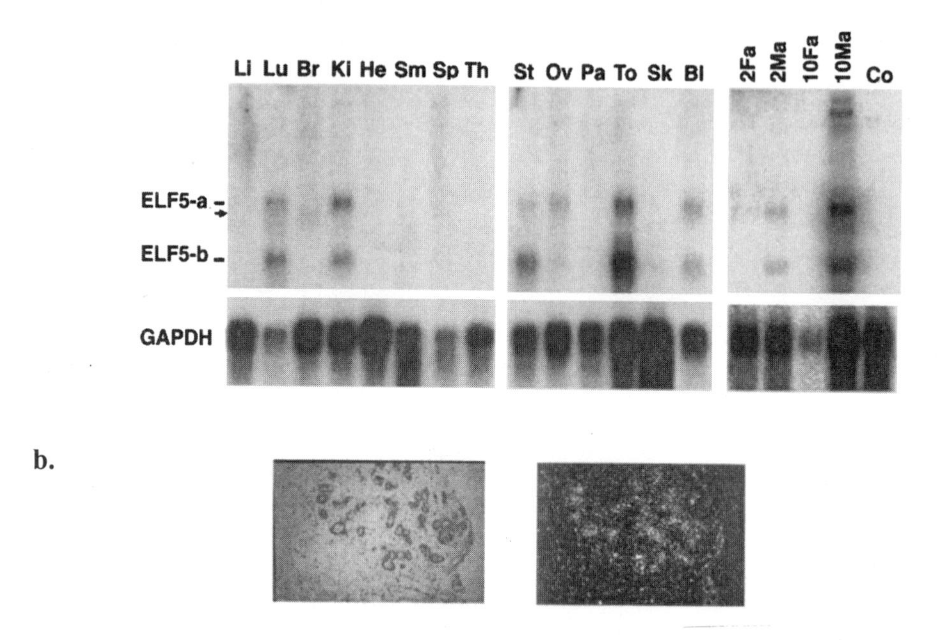


*Fig. 2.* Expression of ELF5. **a.** Northern blot of mouse organs probed with labelled ELF5 and GAPDH cDNAs. As above, and additional organs St; stomach, Pa; pancreas, To; tongue, Bl; bladder, 2Fa; fat, 2Ma; mammary gland day2 pregnancy, 10Fa; fat, 10Ma; mammary gland day 10 pregnancy. **b.** Human mammary gland section probed with labelled ELF5 cDNA (light field and exposure).

## 3. FUNCTION OF ELF3 AND ELF5 IN MAMMARY GLAND

Expression of the WAP gene has been studied extensively as it is highly specific to mammary epithelial cells and is induced several thousand fold during late pregnancy and lactation. The WAP promoter responds to lactogenic hormones, but also contains other important transcription factor elements that are essential for expression. *In vitro* studies indicate that the MAF site is critical for maximal WAP expression, independently of hormone treatment[1]. Mouse transgenic studies indicate that the site is important for expression of WAP during pregnancy, but not critical during lactation, when the promoter is most active and under the control of hormones[2].

The MAF site is similar to others described as binding sites for the ETS family of transcription factors. We have investigated the ability of ELF3 and

ELF5 to control WAP transcription through the MAF site (herein called ETS/1) and another adjacent ETS site (ETS/2).

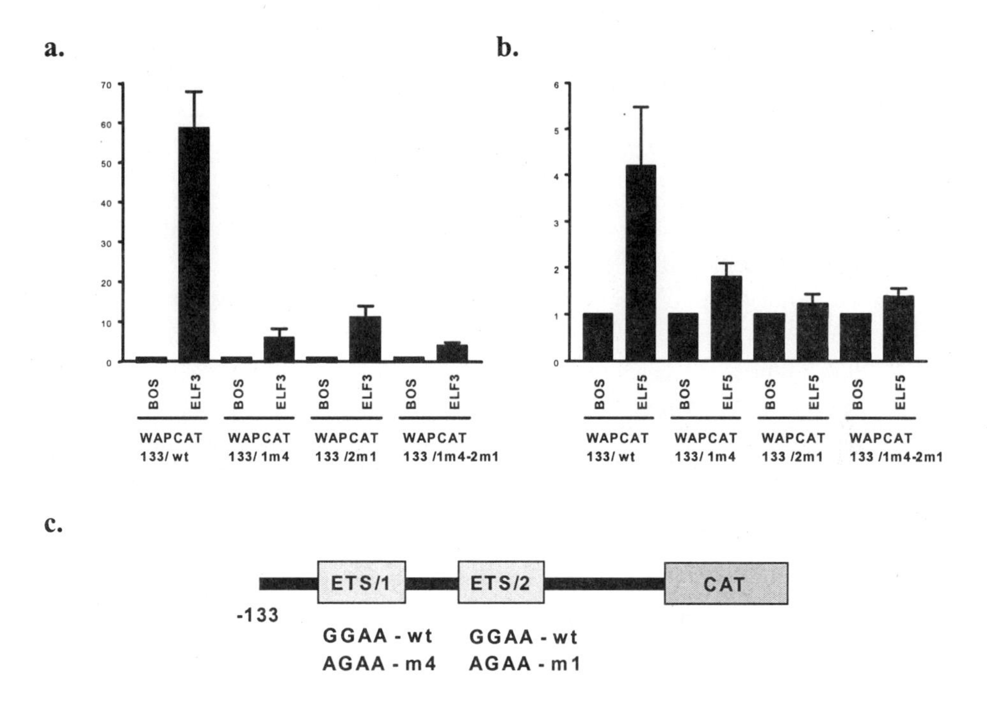


*Figure 3.* Transactivation of the WAP promoter by ELF3 and ELF5. **a.** Co-transfection of an ELF3 expression vector with WAP promoter-reporter constructs in HC-11 cells. Transcriptional activity is measured by CAT assay. BOS is an empty expression vector control. **b.** As above, but with ELF5. **c.** Schematic of the 133 bp WAP promoter attached to the CAT reporter gene. ETS/1 (MAF) and ETS/2 are binding sites for ELFs. Mutations m4 and m1 of the binding sites are indicated.

Recombinant ELF3 and ELF5 proteins are able to strongly interact with the ETS/1 site, and also more weakly with the ETS/2 site, in electrophoretic mobility shift assay (data not shown). Consistent with the binding, we found that both ELF3 (fig 3a) and ELF5 (fig 3b) were able to transactivate a WAP promoter construct in a mammary epithelial cell line, although ELF3 was much more potent than ELF5. Both factors were able to function independently of lactogenic hormones. A longer promoter construct, containing more distal hormone response elements, behaved similarly (data not shown).

Point mutations (fig 3c) of either ETS/1 or ETS/2 alone were enough to disable most transactivation, indicating the ELF3 and ELF5 operate through these sites. Both mutations together had only a slightly greater effect, suggesting that the ETS/1 and ETS/2 sites work cooperatively, both being required for maximal promoter activity.

## 4. CONCLUSION

Expression of mammary-specific genes is under the control of both hormone dependent and independent mechanisms, neither of which may be totally mammary-specific. It seems that a complex interplay of transcription factors may be responsible for mammary expression. Our challenge is to determine whether ELF3 and/or ELF5 are important for the expression of mammary-specific genes *in vivo*, and whether they cooperate with other stimuli, such as prolactin. Because the mammary gland is not a static tissue, but can continually undergo cycles of growth, differentiation and function, it will be interesting to examine whether ELF3 and ELF5 operate cooperatively or differentially during these different phases.

## REFERENCES

1. Welte T, Garimorth K, Philipp S, Jennewein P, Huck C, Cato ACB and Doppler W, 1994, *Eur. J. Biochem.* **223**: 997-1006.
2. McKnight RA, Spencer M, Dittmer J, Brady JN, Wall RJ and Hennighausen L, 1995, *Mol. Endo.* **9**: 717-24.
3. Benz CC, O'Hagan RC, Richter B, Scott GK, Chang CH, Xiong X, Chew K, Ljung BM, Edgerton S, Thor A and Hassell JA, 1997, *Oncogene* **15**: 1513-25.
4. Tymms MJ, Ng AY, Thomas RS, Schutte BC, Zhou J, Eyre HJ, Sutherland GR, Seth A, Rosenberg M, Papas T, Debouck C, Kola I, 1997, *Oncogene* **15**: 2449-62.
5. Zhou J, NG AY, Tymms MJ, Jermiin LS, Seth AK, Thomas RS, Kola I, 1998 *Oncogene* **17**: 2719-32.
6. Chang CH, Scott GK, Kuo WL, Xiong X, Suzdaltseva Y, Park JW, Sayre P, Erny K, Collins C, Gray JW, Benz CC, 1997, *Oncogene* **14**: 1617-22.
7. Oettgen P, Alani RM, Barcinski MA, Brown L, Akbarali Y, Boltax J, Kunsch C, Munger K, Libermann TA, 1997, *Mol. Cell. Biol.* **17**: 4419-33.
8. Andreoli JM, Jang SI, Chung E, Coticchia CM, Steinert PM, Markova NG, 1997, *Nuc. Acids Res.* **25**: 4287-95.
9. Choi SG, Yi Y, Kim YS, Kato M, Chang J, Chung HW, Hahm KB, Yang HK, Rhee HH, Bang YJ, Kim SJ, 1998, *J. Biol. Chem.* **273**: 110-7.
10. Neve R, Chang, CH, Scott, GK, Wong, A, Friis RR, Hynes, NE and Benz CC, 1998, Faseb J. 12;1541-50.

16.

# The Role of Stat3 in Apoptosis and Mammary Gland Involution

*Conditional deletion of Stat3*

[1]Rachel S Chapman, [1]Paula Lourenco, [2]Elizabeth Tonner, [2]David Flint, [1]Stefan Selbert, [3]Kyoshi Takeda, [3]Shizuo Akira, [1]Alan R Clarke, and [4]Christine J Watson

*[1]Dept. of Pathology, University of Edinburgh, UK, [2]Hannah Research Institute, Ayr, UK, [3]Dept. of Host Defence, Research Institute for Microbial Diseases, Osaka University, Japan, and [4]Dept. of Pathology, University of Cambridge, UK*

Key words: Stat3, apoptosis, mammary gland, lox/Cre, involution

Abstract: STATs (signal transducer and activator of transcription) are a family of latent transcription factors which are activated in response to a variety of cytokines and growth factors. This family of signalling molecules have been implicated in growth, differentiation, survival and apoptosis. In this article, we will review work which highlights the role of individual STAT factors in mammary gland and demonstrate the value of genetically modified mice in defining the function of STAT3. Involution of the mouse mammary gland is characterised by extensive apoptosis of the epithelial cells and the activation of STAT3. STATs 3 and 5 have reciprocal patterns of activation throughout a mammary developmental cycle suggesting that STAT5 may be a survival factor and STAT3 a death factor for differentiated mammary epithelium. To clarify the role of STAT3 in mammary epithelial apoptosis, we have generated a conditional knockout using the lox/Cre recombination system. Mammary glands from crosses of transgenic mice expressing Cre recombinase under the control of the β-lactoglobulin milk protein gene promoter with mice harbouring one floxed STAT3 allele and one null STAT3 allele, showed a decrease in epithelial apoptosis and a dramatic delay of the involution process upon forced weaning. This was accompanied by precocious activation of STAT1 and increases in p53 and p21 levels - these may act as a compensatory mechanism for initiating the eventual involution which occurs in STAT3 null mammary glands. This demonstrates for the first time the importance of STAT factors in signalling the initiation of physiological apoptosis *in vivo* and highlights the utility of the lox/Cre system for addressing the function of genes, which have an embryonic lethal phenotype, specifically in mammary gland.

# 1. INTRODUCTION

Involution of the mammary gland is the final phase in the developmental cycle initiated by pregnancy. The removal of the lobuloalveolar compartment is a dramatic process which is characterised by apoptosis of these epithelial cells. The signals which trigger this physiological cell death have interested mammary gland biologists for many years but it is only recently that specific signalling molecules have been shown to have a role in the remodelling process.

One of the reasons for the slow progress has been the difficulty of modelling involution *in vitro* since lactation is a prerequisite for normal involution and full lactation requires a complex hormonal and extracellular environment. Elegant cell culture models have been developed, particularly in the laboratory of Mina Bissell[1]. However, understanding the regulatory mechanism of mammary gland involution and the role of specific genes in this process requires an *in vivo* approach using transgenesis. Overexpression of genes, or expression at the wrong time in development, by introducing fusion gene constructs has been very informative. For example, precocious expression of the matrix metalloproteinase stromelysin-1 caused unscheduled apoptosis during pregnancy which was abolished by crossing with transgenic mice overexpressing TIMP-1 [2] whilst involution was delayed by overexpressing IGF-1 [3]. A more subtle approach is to delete the gene of interest by gene targeting. This circumvents problems which arise from perturbation of other signalling pathways by overexpression of a particular component of one pathway which can complicate interpretation of the phenotype.

Gene targeting is now a relatively straightforward procedure and thousands of genes have been knocked out. Gene deletion in the mammary gland is complicated by the fact that it is one of the few tissues which undergoes most of its development in the adult. Genes which have functions in other tissues or at earlier stages of development are difficult therefore to investigate by conventional targeting. An example is the prolactin receptor which has been knocked out and shown to have a multitude of effects[4]. Many genes are required during embryonic development thus precluding analysis of the function of these genes in the adult. This problem can be overcome by conditional gene targeting, that is, knocking out a gene in a tissue-specific manner.

We have developed a conditional gene targeting strategy to investigate the function of specific transcription factors and signalling molecules in mammary development. This article will focus on the use of this approach to determine the role of a member of the Jak/STAT signalling pathway, STAT3, in mammary development particularly during the involution phase.

## 2. THE JAK/STAT SIGNALLING PATHWAY.

About 8 years ago, a mammary gland transcription factor was described independently in both our laboratory and that of Berndt Groner and named MPBF (for milk protein binding factor) or MGF (for mammary gland factor) respectively[5,6]. Subsequently, cloning of MPBF/MGF[7] revealed it to be a member of the family of STAT (signal transducer and activator of transcription) factors originally identified in Darnell's laboratory[8]. This family of transcription factors comprises seven members which are localised in pairs on three different chromosomes (counting STAT5a and STAT5b, which are highly similar, as one gene). In evolutionary terms, STAT5 appears to be the ancestral gene having homologues in other species such as Drosophila and Dictyostelium[8]. STATs are latent transcription factors which are sequestered in the cytoplasm in an inactive form. Upon ligand engagement of a cytokine (or in some cases) a growth factor receptor, the receptor subunits aggregate and initiate a cascade of tyrosine phosphorylation events in which the receptor-associated Jaks (Janus protein tyrosine kinases) become activated. This creates a docking site for the STAT which then becomes phosphorylated on a single tyrosine residue, dimerises and dissociates from the receptor and translocates to the nucleus where is interacts with its recognition site in the promoters of target genes. The specificity of STATs for receptors appears to depend on the cell type and a single ligand can activate a number of different STATs. The target genes are just beginning to be characterised but the milk protein gene promoters were among the earliest targets to be identified. Indeed, the original identification of STAT5 in the mammary gland was through its interaction with the β-lactoglobulin (BLG) and β-casein promoters.

## 3. STATS IN THE MAMMARY GLAND

The first clue that STAT5 was important for mammary function came from mutational analyses of the BLG and whey acidic protein (WAP) promoters. Transgenic mice were generated which had mutations in combinations of the three STAT binding sites in the sheep BLG promoter (9) or the two sites in the rat WAP promoter[10]. Abolishing STAT binding reduced expression of BLG by 80% and WAP by 90% but did not affect the tissue-specificity of expression. Mutation of a single site in the BLG promoter had little effect whereas mutation of any 2 sites dramatically reduced expression[9]. It has recently been shown that binding of 2 STAT dimers to adjacent sites, where they form tetramers, induces maximal expression[11]. The next step was to determine if other members of the STAT family were expressed in mammary tissue and whether their expression pattern mimicked that of STAT5. Interestingly, STAT5a/b are unique in their induction during pregnancy since both STATs 1 and 3 are expressed at fairly constant levels and STAT4 is barely detectable[12]. STAT2 DNA binding activity could not be detected in extracts from lactating tissue although there is some evidence that STAT6 may have a role in mammary gland.

Analysis of protein levels and DNA binding activities showed that STATs 3 and 5 have a reciprocal relationship suggesting that they play very different roles in mammary gland[12,13]. Of particular interest is the activation of STAT3 at the beginning of involution indicating that it may be a death signal for the epithelial cells. The subsequent knockout of both STAT5 genes showed unequivocally that STAT5a, at least, is required for mammopoeisis and not suprisingly, given its requirement for efficient milk protein gene expression, for lactogenesis[14,15]. The role of STAT5b is less clear although it is worth noting that a high proportion of STAT5 DNA binding activity in mammary tissue is composed of heterodimers and 5a and 5b despite the lower levels of STAT5b protein[12]. However, the knockout of the STAT3 gene proved to have an embryonic lethal phenotype, with death occurring around day 5 of gestation[16]. The demonstration, in ES cells, that STAT3 is required for stem cell renewal explains this very early lethality[17]. In order to address the role of STAT3 in mammary involution, it was therefore necessary to generate a conditional knockout.

## 4. THE LOX/CRE RECOMBINATION SYSTEM AND GENERATION OF BLG/CRE TRANSGENIC MICE

A number of approaches have been devised to facilitate conditional gene targeting. These include the lox/Cre recombination system from P1 bacteriophage, and the yeast Flp system. The most extensively used of these is lox/Cre. This is summarised in Fig. 1 and reviewed[18]. The essential components of this binary system are a transgenic line harbouring a Cre recombinase/tissue-specific promoter fusion construct and a floxed knockout line containing ideally one null (non-functional) allele of the chosen gene (in this case STAT3) and one allele with 2 loxP sites inserted within introns a suitable distance apart. We first generated a transgenic line with mammary-restricted Cre expression.

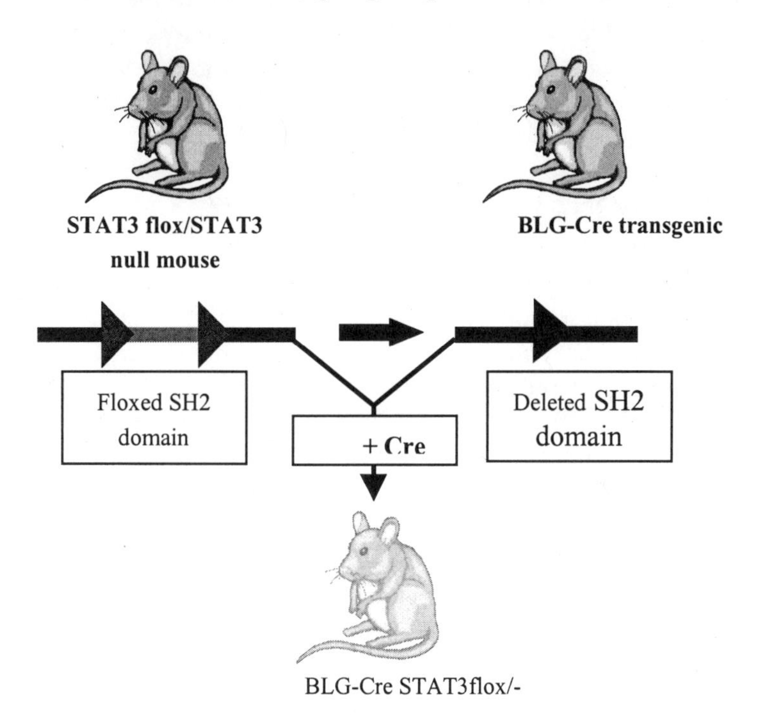


**Fig. 1 STAT3 deleted only in mammary epithelial cells**

The BLG promoter has been used extensively, primarily in John Clark's laboratory, to drive expression of genes specifically to the secretory epithelial cells of mammary gland during late pregnancy and lactation. This promoter was chosen to drive expression of Cre recombinase to mammary gland. Variable expression in the 15 Cre-positive transgenic lines generated was observed and a subset of these were analysed further.

The specificity of expression of the Cre transgene was examined by RT-PCR and mammary-restricted expression observed in a number of these. The two lines with the highest, mammary restricted expression were crossed to a reporter strain harbouring a floxed DNA ligase I (LigI) allele to determine functionality and specificity of the Cre recombinase. Using Southern blotting and a quantitative PCR based approach to detect the excision event, we observed 70-80% recombination in the lactating mammary gland in crosses with one of the lines (line BLG-Cre 74) with less than 1.1% in all other tissues. Lower levels of recombination occurred in mammary glands of virgin animals (7%) and during pregnancy (20-30%).

Immunohistochemistry and *in situ* PCR show that Cre mediated recombination is very likely restricted to the epithelial cells of the mammary gland which are embedded therefore in a wild-type stroma[19]. Conditional gene targeting using this BLG-Cre line will allow thus the role of genes in lactogenesis or apoptosis of the epithelial compartment to be investigated.

## 5. CONDITIONAL STAT3 KNOCKOUT

Due to the early embryonic lethality of STAT3, a floxed strain was generated, in the laboratory of Shizuo Akira, in which an exon of the essential SH2 domain was flanked by loxP sites[20]. Both floxed and null STAT3 animals were crossed with the BLG-Cre transgenic line 74 to generate animals of the desired genotype. These were then analysed for phenotype during lactation and involution.

Given the activation profile of STAT3, it was anticipated that the most dramatic phenotype would be observed during involution and indeed this was the case. Fig. 2 shows H & E staining of mammary sections from day 3 involution time points. It is striking that involution is not initiated in the conditional knockouts for at least 3 days following the removal of pups and that the extent of apoptosis in the epithelial cells is reduced. Mammary glands in conditional knockout animals do eventually undergo remodelling demonstrating that a compensatory mechanism(s) exists.

STAT3 floxed/BLG-Cre 3 day involution

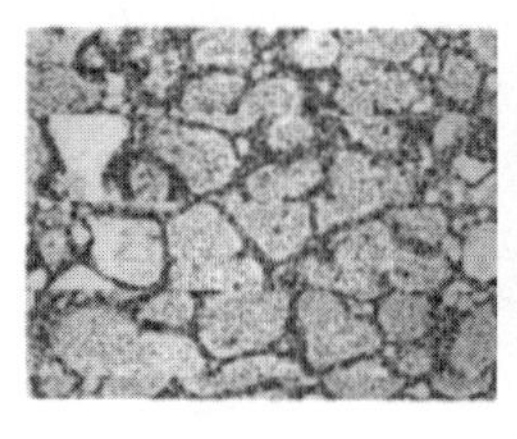

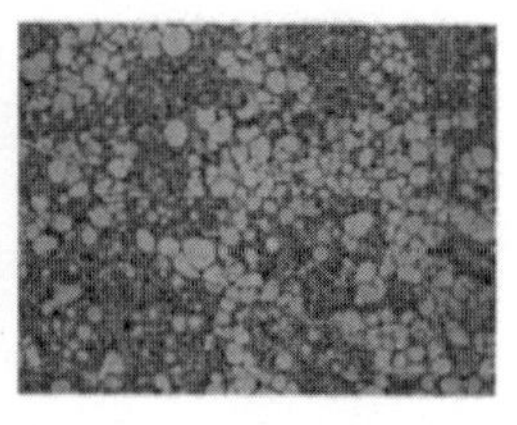

Floxed / -        Floxed / +

Fig. 2. Mammary phenotype of Stat3 knockout mice. H & E stained sections of mammary glands from control (right) and conditional STAT3 knockout (left) mice at day 3 of involution.

Analysis of changes at the molecular level shed some light on possible mechanisms for initiating this delayed involution process and apoptosis[21]. Of particular interest is the precocious activation of STAT1, a dimerisation partner for STAT3, which is often activated in concert with it (Fig. 3). Perhaps STAT1 can activate a similar set of target genes. This speculation can be addressed by generating double knockouts of both STATs 1 and 3 and by using microarray analysis to identify downstream target genes of these transcription factors.

## 6. STATS AND BREAST CANCER

What is the relevance of this work for breast cancer? We demonstrated a number of years ago that constitutive activation of STATs 1 and 3, but not STAT5, is a feature of invasive breast carcinoma but not *in situ* or benign disease[22]. Constitutive STAT activation has subsequently been shown to be associated with transformation by Src, Abl, and other oncoproteins and tumour viruses[23] in a variety of cell types. More recently, STAT factors have been shown to be involved in apoptosis. STAT3 is required for cytokine-induced apoptosis in myeloid leukaemia cells[24] but can suppress apoptosis in a pro-B cell line[25]. Similarly, STATs 1 and 5 have been implicated in cell survival and death[26]. Truncated, dominant-negative STAT mutants can block these effects[27] and in this context it is worth noting that dominant-negative mutants can occur naturally.

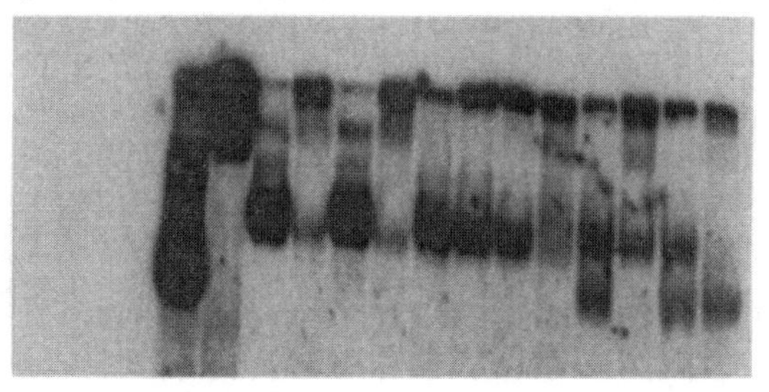


Fig. 3 Precocious activation of STAT1 in STAT3 knockout mammary glands. Electrophoretic mobility shift assays of nuclear extracts prepared from mammary tissue from either control (C) or STAT3 null (KO) mice at day 10 of lactation (L) or day 3 of involution (I). The 2 tracks on the left are from KIM-2 mammary epithelial cells stimulated with IFN-γ which activates only STAT1. Supershift of the complexes with antibodies to either STAT 1, 3, or 5 is indicated above each lane. The highest mobility band is the STAT1 complex which is present in the STAT3 null glands, but not in the control glands, during involution.

Cancer may be viewed as a failure of apoptosis. Our observations on the conditional STAT3 knockout demonstrate that STAT3 is required for apoptosis. It will be interesting to determine if these mice are more susceptible to tumour development following pregnancy or if deletion of STAT3 has a protective effect. In light of the opposing roles of different STATs in the mammary epithelial cell, it may be necessary to perturb the function of more than one member of this family to cause cancer. Crosses of knockouts of STATs 1 and 3 will be particularly interesting. The undoubted role of the Jak/STAT pathway in apoptosis and transformation places STAT factors in the range of therapeutic targets for breast cancer and high throughput analysis is being used by drug companies to discover inhibitors of STAT function.

## 7. THE WAY AHEAD

We have presented an example of conditional gene targeting in the mammary gland using the lox/Cre recombination system and a BLG-Cre transgenic line. This approach is applicable to any gene of interest and we are currently generating floxed Jak2 animals to overcome the embryonic lethal phenotype and investigate its role in mammary development. Since Jak2 lies upstream of both STATs 3 and 5 and is implicated in other signalling pathways, a mammary conditional Jak2 knockout can be expected to have a different phenotype from both the STAT3 and STAT5 knockouts.

The BLG-Cre transgenics from our laboratory and the WAP-Cre transgenics from the Henninghausen laboratory[28] will be a valuable resource to those interested in the conditional knockout of any one of the number of genes which have already been implicated in breast cancer. However, a further level of sophistication in conditional gene targeting is desirable to provide both tissue-specificity and temporal regulation. Strategies to achieve this have been devised and the new millennium should herald the arrival of finely tuned gene deletion.

**Acknowledgements**

We would like to thank John Clark for providing the BLG promoter. This work was supported by grants to Christine Watson and Alan Clarke from the Biotechnology and Biological Sciences Research Council, the Association for International Cancer Research and The Cancer Research Campaign.

# 8. REFERENCES

1. Weaver, VM and MJ Bissell 1999. Functional culture models to study mechanisms governing apoptosis in normal and malignant mammary epithelial cells *J. Mam Gland Biol. & Neoplasia* **4:** 193-201.
2. Alexander CM, EW Howard, MJ Bissell, and Z Werb. 1996. Rescue of mammary epithelial cell apoptosis and entactin degradation by a tissue inhibitor of metalloproteinases-1 transgene. *J. Cell Biol.* **135:** 1669-1677.
3. Neuenschwander, S., A. Schwart, T.L. Wood, C.T. Roberts Jr, L. Hennighausen and D. Le Roith. 1996. Involution of the lactating mammary gland is inhibited by the IGF system in a transgenic mouse model. *J. Clin. Invest.* **97:** 2225-2232.
4. Ormandy, CJ, A Camus, J Barra, JD Damotte, B Lucas, H Buteau, M Edery, N Brousse, C Babinet, N Binart, and PA Kelly 1997. Null mutation of the prolactin receptor gene produces multiple reproductive defects in the mouse *Genes & Dev.* **11**:167-178.
5. Watson, C.J., K.E. Gordon, M. Robertson, and A.J. Clarke 1991. Interaction of DNA-binding proteins with a milk protein gene promoter in vitro : identification of a mammary-specific factor. *Nuc. Acids Res.* **19:** 6603-6610.
6. Schmitt-Ney, M., W. Doppler, R.K. Ball, and B. Groner 1991. Beta-casein gene promoter activity is regulated by the hormone-mediated relief of transcriptional repression and a mammary gland-specific nuclear factor. *Mol. Cell. Biol.* **11:** 3745-3755.
7. Wakao, H., F. Gouilleux, and B. Groner 1994. Mammary gland factor (MGF) is a novel member of the cytokine regulated transcription factor gene family and confers the prolactin response. *EMBO J.* **13:** 2182-2191.
8. Ihle, J.N. 1996. STATs: Signal transducers and activators of transcription. *Cell* **84:** 331-334.
9. Burdon, T.G., K. A. Maitland, A.J. Clark, R. Wallace and C. J. Watson 1994. Regulation of the sheep β-lactoglobulin gene by lactogenic hormones is mediated by a transcription factor that binds to an IFN-γ activation site-related element. *Mol. Endo.* **8** 1528-1536.
10. Li, S. and J. M. Rosen 1995. Nuclear factor-1 and mammary gland factor (STAT5) play a critical role in regulating rat whey acidic protein gene expression in transgenic mice. *Mol.Cell. Biol* **15**: 2063-2070.
11. John, S, U. Vinkemeier, E Soldaini, JE Darnell, and WJ Leonard 1999. The significance of tetramerization in promoter recruitment by Stat5. *Mol. Cell. Biol.* **19:** 1910-1918.
12. Philp, J.C., T.G. Burdon, and C.J. Watson. 1996. Differential activation of STATs 3 and 5 during mammary gland development. *FEBS Lett.* **396:** 77-80.
13. Liu, X.W., G.W. Robinson, and L. Hennighausen. 1996. Activation of Stat5a and Stat5b by tyrosine phosphorylation is tightly linked to mammary gland differentiation. *Mol.Endocrinol.* **10:** 1496-1506.
14. Liu, X.W., G.W. Robinson, K.U. Wagner, L. Garrett, A. WynshawBoris, and L. Hennighausen. 1997. Stat5a is mandatory for adult mammary gland development and lactogenesis. *Genes & Dev.* **11:** 179-186.
15. Teglund, S., C. Mckay, E. Schuetz, J.M. vanDeursen, D. Stravopodis, D.M. Wang, M. Brown, S. Bodner, G. Grosveld, and J.N. Ihle. 1998. Stat5a and Stat5b proteins have essential and nonessential, or redundant, roles in cytokine responses. *Cell* **93:** 841-850.
16. Takeda, K., K. Noguchi, W. Shi, T. Tanaka, M. Matsumoto, N. Yoshida, T. Kishimoto, and S. Akira. 1997. Targeted disruption of the mouse Stat3 gene leads to early embryonic lethality. *Proc.Natl.Acad.Sci.U.S.A.* **94:** 3801-3804.

17. Niwa, H., T.G. Burdon, I. Chambers and A. G. Smith 1998. Self-renewal of pluripotent embryonic stem cells is mediated via activation of STAT3. *Genes & Dev.* **12:** 2048-2060.
18. Rossant,J and A McMahon 1999. "Cre"-ating mouse mutants - a meeting review on conditional mouse genetics *Genes & Dev.* **13:** 142-145.
19. Selbert, S., D.J. Bentley, D.W. Melton, D. Rannie, P. Lourenco, C.J. Watson, and A.R. Clarke. 1998. Efficient BLG-Cre mediated gene deletion in the mammary gland. *Transgenic Res.* **7:** 387-396.
20. Takeda, K., T. Kaisho, N. Yoshida, J. Takeda, T. Kishimoto, and S. Akira. 1998. Stat3 activation is responsible for IL-6-dependent T cell proliferation through preventing apoptosis: Generation and characterization of T cell-specific Stat3-deficient mice. *J.Immunol.* **161:** 4652-4660.
21. Chapman, R.S., P.Lourenco, E. Tonner, D. Flint, S. Selbert, K. Takeda, S. Akira, A.R. Clarke, and C.J. Watson 1999. Suppression of epithelial apoptosis and delayed mammary gland involution in mice with a conditional knockout of STAT3. *Genes & Dev.* **13:** 2604-2616.
22. Watson, C.J. and W.R. Miller 1995. Elevated levels of members of the STAT family of transcription factors in breast carcinoma nuclear extracts. *British J. Cancer* **71:** 840-844.
23. Garcia, R. and R. Jove 1998. Activation of STAT transcription factors in oncogenic tyrosine kinase signaling. *J. Biomed. Sci.* **5**: 79-85.
24. Minami, M., M. Inoue, S. Wei, K. Takeda, M. Matsumoto, T. Kishimoto, and S. Akira. 1996. STAT3 activation is a critical step in gp130-mediated terminal differentiation and growth arrest of a myeloid cell line. *Proc.Natl.Acad.Sci. U.S.A.* **93:** 3963-3966
25. Fukada, T., M. Hibi, Y. Yamanaka, M. Takahashi, Y. Tezuka, T. Fujitani, T. Yamaguchi, K. Nakajima, and T. Hirano 1996. Two signals are necessary for cell. by a cytokine receptor gp130: Involvement of STAT3 in anti-apoptosis. *Immunity* -460.
26. Zamorano, J., H.Y. Wang, R. Wang, Y. Shi, G.D. Longmore, and A.D. Keegan. 1998. Regulation of cell growth by IL-2: Role of STAT5 in protection from apoptosis but not in cell cycle progression. *J.Immunol.* **160:** 3502-3512.
27. Bovolenta, C., L. Testolin, L. Benussi, P.J. Lievens, and E. Liboi. 1998. Positive selection of apoptosis resistant cells correlates with activation of dominant negative STAT5. *J.Biol.Chem.* **273:** 20779-20784.
28. Xu, X., K. Wagner, D. Larson, Z. Weaver, C. Li, T. Reid, L. Hennighausen, A. Wynshaw-Boris and C. Deng. 1999. Conditional mutation of *Brca1* in mammary epithelial cells results in blunted ductal morphogenesis and tumour formation. *Nature Gen.* **22:** 37-43.

# 17.

# Synergistic and Antagonistic Interactions of Transcription Factors in the Regulation of Milk Protein Gene Expression

*Mechanisms of cross-talk between signalling pathways*

Wolfgang Doppler, Sibylle Geymayer, and Harald G. Weirich
*Insitut for Medical Chemistry and Biochemistry, University of Innsbruck, Austria*

Key words: STAT5, glucocorticoid receptor, NF-κB, prolactin, beta casein

Abstract: The stage and tissue specific expression of milk protein genes in the mammary gland is controlled by modular response regions with multiple binding sites for distinct classes of transcription factors, which either co-operate or are antagonistic. In addition, the activity of some of these factors is individually control-led by diverse extracellular signals. A well studied paradigm for a synergistic co-operation is the activation of beta-casein gene transcription by prolactin and glucocorticoids mediated by the signal transducer and activator of transcription STAT5 and the glucocorticoid receptor (GR). As an example for an antagonistic interaction we can demonstrate inhibition of prolactin signalling by TNF-α, which is mediated by NF-κB. In both cases, the interactions occur at several levels: For GR and STAT5, the synergy is discussed to be promoted by protein-protein interactions. Furthermore, we can demonstrate a co-operation between GR and STAT5 in DNA binding by a mechanism, which is dependent on the integrity of the DNA binding domain of the GR and on the existence of half-palindromic GR binding sites in the hormone response region. Indirect effects of glucocorticoids by modulation of the expression of secondary genes are also important. They might account for the observed enhancement of prolactin induced tyrosine phosphorylation of STAT5 by glucocorticoids. For NF-κB and STAT5, one component of the antagonism is the inhibition of STAT5 tyrosine phosphorylation by activation of NF-κB. Another potential mechanism is the inhibition of DNA binding of STAT5 due to overlapping binding sites for STAT5 and NF-κB in the beta-casein gene promoter. Thus, synergistic and antagonistic interactions between GR, NF-κB, and STAT5 involve (a) cross-talk mechanisms influencing the activation of STAT5 and (b) promoter-dependent interactions modulating the DNA binding activity of the transcription factors

# 1. INTRODUCTION

The activation of milk protein gene expression is an extensively studied example for the interaction of several signalling pathways. Long before the discovery of molecular details concerning the operation of the individual pathways, endocrinologists could demonstrate a synergism involving three hormones, namely the lactogenic hormones prolactin, glucocorticoids and insulin (reviewed[1]). Later it became evident, that signals elicited by extracellular matrix components are also necessary for the activation of milk protein gene expression in mammary epithelial cells[2]. Numerous other extracellular stimuli, most of them triggering the inhibition of milk protein gene expression, were discovered subsequently. Current knowledge indicates that many if not all of these extracellular signals influence milk protein gene expression at the level of transcription initiation. A summary of extracellular signals and transcription factors (TFs) implicated in the regulation of milk protein gene expression is given in Table 1. For the signals triggered by steroid hormones, prolactin, insulin, interleukin-4 (IL-4), and TNF-α, the transcription factors mediating the response to these stimuli have been identified. For other signals, such as insulin and TGF-β, the targeted transcription factor (TF) is either not known or the response to the signal is not dependent on the activation or repression of a TF. It should be noted that for several TFs with an established function in milk protein gene transcription such as NF-1, Oct-1, CTF/NF-1 and Ets domain proteins, it has still to be established whether and how they are regulated by extracellular stimuli.

# 2. MODULAR ARRANGEMENT OF TRANSCRIPTION FACTOR BINDING SITES IN THE ß-CASEIN GENE PROMOTER

An important principle in directing the interaction between TFs is the modular arrangement of their binding sites in the DNA response regions of the regulated genes. Depending on the relative position of binding sites and the affinity of the TFs to these sites, cooperative binding or competition for binding sites can be observed, resulting in either synergy or antagonism. The structure of TF binding sites in the two lactogenic hormone response region (LHRR) of the β-casein gene has been studied in detail. Fig. 1 shows a schematic overview of the principle binding sites mapped in the rat[18,21-24], the bovine[25] and the human[26] gene and the conservation of these binding sites between species. A common feature of all LHRRs is the presence of multiple binding sites for CCAAT enhancer binding proteins and at least one binding site for STAT5. The human and ruminant proximal LHRR exhibits impaired functional activity in comparison to the LHRR of rodents[12,26].

*Table 1.* Extracellular signals and targeted TFs regulating milk protein gene transcription. So far, the activation of STAT5 by insulin has only been described in non-mammary tissues[3]. Epidermal growth factor has a dual function on milk protein gene expression: it promotes the competence of mammary epithelial cells to express the β-casein gene, but represses the action of lactogenic hormones in terminal differentiated cells[4]. A truncated version of the CCAAT enhancer protein C/EBP β is expressed in mammary epithelial cells (LIP) and is discussed to block the action of full length C/EBP β (LAP)[5].

| Extracellular Signal | Targeted TF | Effect on Transcription | Ref. |
|---|---|---|---|
| Prolactin | STAT5 | activation | 6 |
| interleukin-4 | STAT6 | activation | 7 |
| Glucocorticoids | Glucocorticoid receptor | activation | 1 |
| Progestins | Progesterone receptor | repression | 1 |
| Insulin | STAT5? | activation | 3 |
| Epidermal growth factor | STAT5 | repression, (activation) | 4,8,9 |
| Tumor necrosis factor α | NF-κB | repression | 10 |
| Transforming growth factor β | ? | repression | 11 |
| extracellular matrix | STAT5, NF-1 ? | activation | 12,13 |
| ? | NF-1 | activation | 14-17 |
| ? | CCAAT enhancer binding proteins | activation, repression (?) | 5,18 |
| ? | Ets domain proteins | activation | 19 |
| ? | octameric factor | activation | 20 |
| ? | YY-1 | repression | 21,22 |

This might relate to the lack of a second STAT5 binding site, which is required for the function of the LHRR in rodents (H. Weirich, unpublished results). In these species, activation of milk protein gene expression depends on a distal enhancer region with multiple C/EBP and STAT5 sites (Fig. 1).

YY1 has been described as a factor participating in the repression of β-casein gene transcription by its binding to a site in the proximal LHRR. Due to the close proximity of binding sites, STAT5 inhibits binding of YY1. Such a STAT5 mediated relief of repression by YY1 has been postulated to be important for stimulation of β-casein gene expression[21,22]. In the next sections two other examples for interactions of TFs with STAT5 are discussed. In both cases, a full understanding of their action on STAT5 signalling has to consider in addition mechanisms other than interactions mediated by the LHRR.

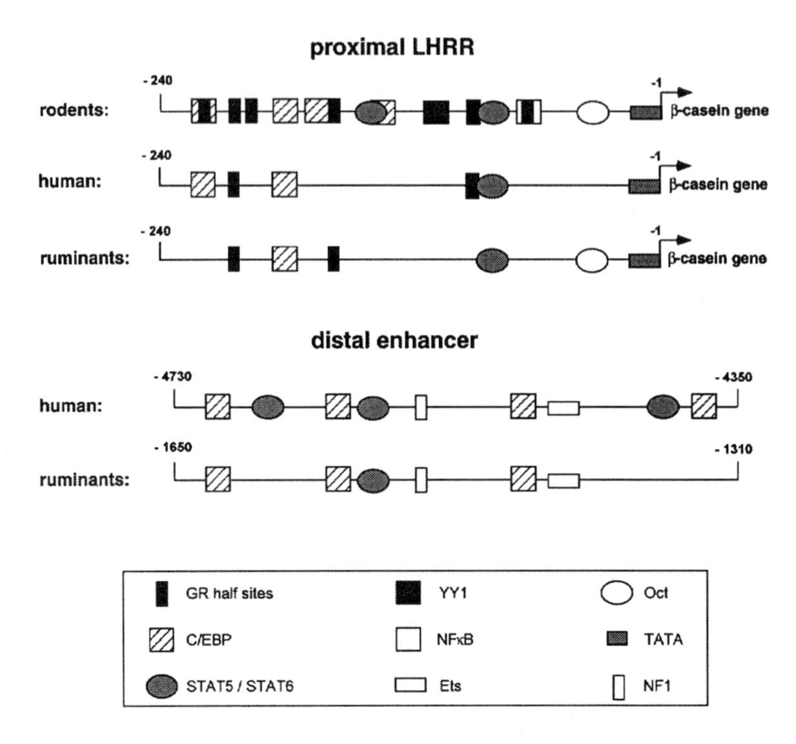


Fig. 1 Structural organisation of TF binding sites in the two defined regulatory regions of the β-casein gene. For rodents, the conserved transcription factor binding sites mapped in the sequence of the mouse and rat gene are shown. For ruminants, the depicted binding sites are identical for sheep, goat and cattle.

# 3. SYNERGY BETWEEN GLUCOCORTICOID RECEPTOR (GR) AND STAT5

The proximal rat LHRR of the β-casein gene has been shown to direct the synergy between the GR and STAT5 [27]. Evidence obtained from immunoprecipitation experiments indicate protein-protein interactions between the GR and STAT5, which do not depend on the binding of the GR to DNA[28,29]. These interactions appear to contribute to the functional synergy between GR and STAT5 [30]. However, mutational analysis of the GR binding sites in the β-casein gene promoter revealed that the structural integrity of GR binding sites is a prerequisite to achieve a synergy between these two TFs[31], indicating that binding of the GR to DNA is also important. A particularity of the binding of GR to the LHRR of milk protein genes is the recognition of half-palindromic sites, which do not function in the absence of activated STAT5 to mediate the action of GR on transcription. This mode of interaction provides an efficient means to restrict gene expression to conditions were both STAT5 and the GR are active[32].

In addition to the direct interactions between GR and STAT5 at the β-casein gene promoter, activated GR promotes STAT5 tyrosine phosphorylation[33]. Indirect actions of the GR, such as inhibition of the expression of cytokine inducible SH-2 domain proteins by glucocorticoids might be important for this effect and might explain the delayed effects and slow kinetics of glucocorticoid action on β-casein gene transcription[34]. The left part of Fig. 2 illustrates the two levels of interaction between GR and STAT5.

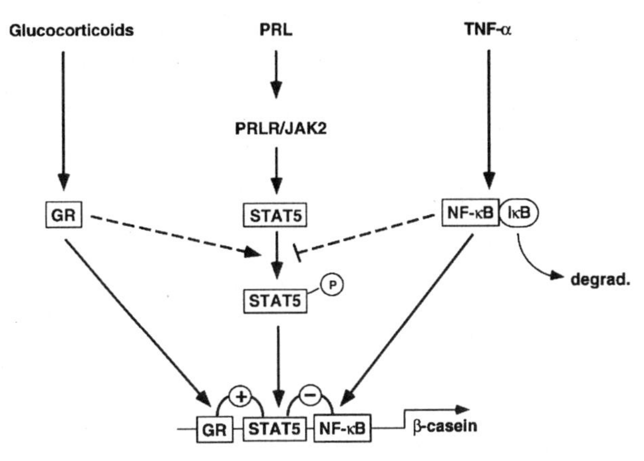


Fig. 2 Cross-talk mechanisms between glucocorticoid, prolactin and tumor necrosis-factor α induced signalling pathways in mammary epithelial cells.

## 4. ANTAGONISM BETWEEN NF-κB AND STAT5

Tumor necrosis factor α (TNF-α) has been shown to be a multifunctional regulator of mammary gland development with an inhibitory action on milk protein gene expression[35,36]. It is an activator of NF-κB. We and others (C.J.Watson, personal communication) could demonstrate developmental regulated nuclear NF- κB p50/p65 in the mammary gland of mice, with peak levels between mid- and late-pregnancy. During the lactation period, DNA binding activity of NF- κB p50/p65 was low due to inhibited nuclear translocation (S.Geymayer et al., submitted). Co-transfection experiments performed with 293 cells revealed an inhibition of STAT5 dependent activation of β-casein gene expression by NF- κB p50/p65. Similar as in the case of the above described synergism between STAT5 and GR, more than one mechanism is potentially involved in the antagonism between the

STAT5 and NF-κB signalling pathways: (a) competition of STAT5 and NF-κB for binding in the β-casein gene promoter, mediated by binding sites for NF-κB in the proximal LHRR (Fig. 1); and (b) inhibition of STAT5 tyrosine phosphorylation, which was found to be inversely correlated to the NF-κB activation status in the mammary epithelial cell line HC11 (S.Geymayer et al., submitted).

In conclusion, for the interaction of TFs involved in the regulation of milk-protein gene transcription, two different types of cross-talk mechanism are operative. The structural organization of the TF binding sites in the LHRR determines the gene specific response to simultaneous activation of signalling pathways. Thereby, species specific arrangement of sites are likely to contribute to the differences in strength and stage dependence of expression in different species. A second level of interaction, which does not necessarily require binding of interacting TF to DNA, is promoted by cross-talk between the signalling pathways leading to the activation of the TFs. A concise analysis of the co-operation or antagonism between TF has to consider both modes of interaction.

## 5. REFERENCES

1. Topper YJ, Freeman CS. Multiple hormone interactions in the developmental biology of the mammary gland. Physiol Rev 60, 1049-1106, 1980.
2. Lee YHP, Lee WH, Kaetzel CS, Parry G, Bissell MJ. Interaction of mouse mammary epithelial cells with collagen substrata: regulation of casein gene expression and secretion. Proc Natl Acad Sci U S A 82, 1419-1423, 1985.
3. Chen J, Sadowski HB, Kohanski RA, Wang LH. Stat5 is a physiological substrate of the insulin receptor. Proc Natl Acad Sci U S A 94, 2295-2300, 1997.
4. Hynes NE, Taverna D, Harwerth IM, Ciardiello F, Salomon DS, Yamamota T, Groner B. Epidermal growth factor receptor, but not c-*erb*B-2 activation, prevents lactogenic hormone induction of the β-casein gene in mouse mammary epithelial cells in culture. Mol Cell Biol 10, 4027-4034, 1990.
5. Raught B, Liao WSL, Rosen JM. Developmentally and hormonally regulated CCAAT/enhancer- binding protein isoforms influence beta-casein gene expression. Mol Endocrinol 9, 1223-1232, 1995.
6. Wakao H, Gouilleux F, Groner B. Mammary gland factor (MGF) is a novel member of the cytokine regulated transcription factor gene family and confers the prolactin response. EMBO J 13, 2182-2191, 1994.
7. Moriggl R, Berchtold S, Friedrich K, Standke GJ, Kammer W, Heim M, Wissler M, Stocklin E, Gouilleux F, Groner B. Comparison of the transactivation domains of Stat5 and Stat6 in lymphoid cells and mammary epithelial cells. Mol Cell Biol 17, 3663-3678, 1997.
8. Welte T, Garimorth K, Philipp S, Doppler W. Prolactin-dependent activation of a tyrosine phosphorylated DNA binding factor in mouse mammary epithelial cells. Mol Endocrinol 8, 1091-1102, 1994.

9. Doppler W, Villunger A, Jennewein P, Brduscha K, Groner B, Ball RK. Lactogenic hormone and cell type-specific control of the whey acidic protein gene promoter in transfected mouse cells. Mol Endocrinol 5, 1624-1632, 1991.
10. Varela LM, Ip MM. Tumor necrosis factor-α: a multifunctional regulator of mammary gland development. Endocrinology 137, 4915-4924, 1996.
11. Mieth M, Boehmer FK, Ball R, Groner B, Grosse R. Transforming growth factor-β inhibits lactogenic hormone induction of β-casein expression in HC11 mouse mammary epithelial cells. Growth Factors 4, 9-15, 1990.
12. Schmidhauser C, Bissell MJ, Myers CA, Casperson GF. Extracellular matrix and hormones transcriptionally regulate bovine β-casein 5' sequences in stably transfected mouse mammary cells. Proc Natl Acad Sci U S A 87, 9118-9122, 1990.
13. Streuli CH, Edwards GM, Delcommenne M, Whitelaw CBA, Burdon TG, Schindler C, Watson CJ. Stat5 as a target for regulation by extracellular matrix. J Biol Chem 270, 21639-21644, 1995.
14. Lubon H, Hennighausen L. Conserved region in the rat α-lactalbumin promoter is a target site for protein binding *in vitro*. Biochem J 256, 391-396, 1988.
15. Watson CJ, Gordon KE, Robertson M, Clark AJ. Interaction of DNA-binding proteins with a milk protein gene promoter in vitro: identification of a mammary gland-specific factor. Nucleic Acids Res 19, 6603-6610, 1991.
16. Mink S, Härtig E, Jennewein P, Doppler W, Cato ACB. A mammary cell-specific enhancer in mouse mammary tumor virus DNA is composed of multiple regulatory elements including binding sites for CTF/NFI and a novel transcription factor, mammary cell-activating factor. Mol Cell Biol 12, 4906-4918, 1992.
17. Li S, Rosen JM. Nuclear factor I and mammary gland factor (STAT5) play a critical role in regulating rat whey acidic protein gene expression in transgenic mice. Mol Cell Biol 15, 2063-2070, 1995.
18. Doppler W, Welte T, Philipp S. CCAAT/enhancer-binding protein isoforms β and δ are expressed in mammary epithelial cells and bind to multiple sites in the β-casein gene promoter. J Biol Chem 270, 17962-17969, 1995.
19. Alam T, An MR, Mifflin RC, Hsieh C-C, Ge X, Papaconstantinou J. *trans*-Activation of the $\alpha_1$-acid glycoprotein gene acute phase responsive element by multiple isoforms of C/EBP and glucocorticoid receptor. J Biol Chem 268, 15681-15688, 1993.
20. Groner B, Altiok S, Meier V. Hormonal Regulation of Transcription Factor Activity in Mammary Epithelial Cells. Mol Cell Endocrinol 100, 109-114, 1994.
21. Raught B, Khursheed B, Kazanasky A, Rosen J. YY1 represses β-casein gene expression by preventing the formation of a lactation-associated complex. Mol Cell Biol 14, 1752-1763, 1994.
22. Meier VS, Groner B. The nuclear factor YY1 participates in repression of the β-casein gene promoter in mammary epithelial cells and is counteracted by mammary gland factor during lactogenic hormone induction. Mol Cell Biol 14, 128-137, 1994.
23. Schmitt-Ney M, Doppler W, Ball RK, Groner B. β-Casein gene promoter activity is regulated by the hormone-mediated relief of transcriptional repression and a mammary-gland-specific nuclear factor. Mol Cell Biol 11, 3745-3755, 1991.
24. Welte T, Philipp S, Cairns C, Gustafsson J-Å, Doppler W. Glucocorticoid receptor binding sites in the promoter region of milk protein genes. J Steroid Biochem Molec Biol 47, 75-81, 1993.
25. Myers CA, Schmidhauser C, Mellentinmichelotti J, Fragoso G, Roskelley CD, Casperson G, Mossi R, Pujuguet P, Hager G, Bissell MJ. Characterization of BCE-1, a

transcriptional enhancer regulated by prolactin and extracellular matrix and modulated by the state of histone acetylation. Mol Cell Biol 18, 2184-2195, 1998.

26. Winklehner-Jennewein P, Geymayer S, Lechner J, Welte T, Hansson L, Geley S, Doppler W. A distal enhancer region in the human beta casein gene mediates the response to prolactin and glucocorticoid hormones. Gene 217, 127-139, 1998.
27. Doppler W, Groner B, Ball RK. Prolactin and glucocorticoid hormones synergistically induce expression of transfected rat β-casein gene promoter constructs in a mammary epithelial cell line. Proc Natl Acad Sci U S A 86, 104-108, 1989.
28. Stöcklin E, Wissler M, Gouilleux F, Groner B. Functional interactions between Stat5 and the glucocorticoid receptor. Nature 383, 726-728, 1996.
29. Cella N, Groner B, Hynes NE. Characterization of Stat5a and Stat5b homodimers and heterodimers and their association with the glucocorticoid receptor in mammary cells. Mol Cell Biol 18, 1783-1792, 1998.
30. Stöcklin E, Wissler M, Moriggl R, Groner B. Specific DNA binding of Stat5, but not of glucocorticoid receptor, is required for their functional cooperation in the regulation of gene transcription. Mol Cell Biol 17, 6708-6716, 1997.
31. Lechner J, Welte T, Tomasi JK, Bruno P, Cairns C, Gustafsson J-Å, Doppler W. Promoter-dependent synergy between glucocorticoid receptor and Stat5 in the activation of β-casein gene transcription. J Biol Chem 272, 20954-20960, 1997.
32. Lechner J, Welte T, Doppler W. Mechanism of interaction between the glucocorticoid receptor and Stat5: role of DNA-binding. Immunobiol 197, 175-186, 1997.
33. Wyszomierski SL, Yeh J, Rosen JM. Glucocorticoid receptor/signal transducer and activator of transcription 5 (STAT5) interactions enhance STAT5 activation by prolonging STAT5 DNA binding and tyrosine phosphorylation. Mol Endocrinol 13, 330-343, 1999.
34. Doppler W, Höck W, Hofer P, Groner B, Ball RK. Prolactin and glucocorticoid hormones control transcription of the β-casein gene by kinetically distinct mechanisms. Mol Endocrinol 4, 912-919, 1990.
35. Ip MM, Shoemaker SF, Darcy K. Regulation of rat mammary epithelial cell proliferation and differentiation by tumor necrosis factor-α. Endocrinology 130, 2833-2844, 1992.

18.

# Nucleosome Organisation of the β-Lactoglobulin Gene

*Transcription complex formation*

C. Bruce A. Whitelaw

*Roslin Institute (Edinburgh), Division of Molecular Biology, Roslin, Midlothian, UK*

Key words: Chromatin, mammary, ovine, STAT5

Abstract: In most mammals, the major whey protein beta-lactoglobulin (βlg) represents a marker for tissue-specific, temporally regulated gene expression in the mammary gland. Prolactin, the major lactogenic stimulus which activates βlg expression acts through a cytoplasmic signalling cascade ending in the activation of the transcription factor STAT5. Although much is known about the complexities of this signalling cascade, little is known about how this transcription factor functions within the context of chromatin. Using DNaseI as a probe of chromatin structure we have identified temporally regulated elements within the ovine βlg gene domain. The appearance of these hypersensitive sites accompanies changes in expression state of the βlg gene. Changes in DNaseI hypersensitivity at the proximal promoter region, while reflecting STAT5 activation, is not dependent upon STAT5 interaction at this site. We have mapped the nucleosome positions over the entire βlg gene, both *in vitro* using the monomer extension assay and *in vivo* using cuprous phenanthroline to probe for nucleosome-linker positions. The specific positioning pattern detected, which reflects strong sequence-directed positioning over the proximal promoter, complement the STAT5 consensus sites within this region. The comparison of both the functional and chromatin data enables a model for βlg gene transcription to be developed.

## 1. INTRODUCTION

The accurate switching 'on' and 'off' of gene expression is central for life with each gene having its own specific spatial and temporal expression pattern. To ensure this pattern is faithfully adhered to several layers of regulation are imposed within the cell. Over the past two decades there have been great advances in our understanding of how soluble transcription factors regulate gene expression. With this knowledge has come an increased awareness of the importance of the whole nuclear environment within which genes are regulated. Gene expression takes place in the context of chromatin, a condition that many think contributes to the developmental and spatial control of transcription. Clearly a full understanding of how a gene is regulated must incorporate knowledge of the chromosomal environment in which it is contained.

We have focussed our effort on investigating mammary gene expression. The mammary gland represents a differentiated tissue which expresses a specific set of genes in response to a variety of hormonal, cellular and extracellular matrix derived signals. It is also the only organ in which the majority of development takes place in the adult and which can undergo successive cycles of development and regression. In addition, the mammary gland is an important organ from an agricultural perspective (milk production) and can be harnessed for new genetic biotechnologies[1].

## 2. β-LACTOGLOBULIN GENE

In most mammals, the major whey protein is β-lactoglobulin (βlg). The function of βlg is unconfirmed although it shares considerable similarity to retinol-binding protein and other lipocalins[2], and therefore may transport small hydrophobic molecules[3]. In ruminants, this protein represents a marker for tissue-specific, temporally regulated gene expression in the mammary gland[4]. Numerous studies in a variety of cell-based systems have shown that βlg expression is regulated by a complex interaction of hormones and growth factors in association with cell-cell and cell-extracellular matrix interactions[5]. The major lactogenic stimulus prolactin[6], activates βlg expression through a cytoplasmic signalling cascade ending in the activation of the transcription factor STAT5 [7,8]. There are three binding sites for STAT5[9] within the ovine βlg gene promoter (Fig. 1).

Activation of βlg gene expression is accompanied by changes in the chromatin structure surrounding the promoter region. No DNaseI hypersensitive sites are present in the liver where the gene is not expressed, while distinct sites are present in the lactating sheep mammary gland[10]. Additional mammary specific hypersensitive sites are detected within the first two introns prior to over βlg transcription[11]. Analysis, both in cell culture[12,13] and transgenic mice[10,14], indicates that only the hypersensitive site located at the promoter, termed HSIII, is essential for expression in mammary cells.

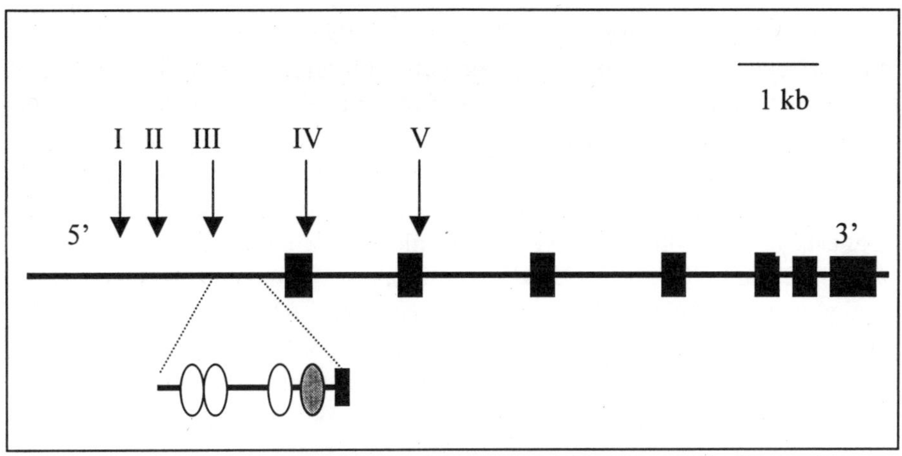


Fig. 1. The ovine βlg gene contains 7 exons (black boxes) within about 5 kb of genomic sequence. The entire sequence is known (X12817). Mammary specific DNaseI hypersensitive sites are depicted as arrows with roman numerals[8]. Specific transcription binding sites are shown in the enlarged promoter region; open circles, STAT5; filled circle, TATA-box.

This region (Fig. 1) contains several binding sites for the transcription factors NF-1 and STAT5[9], the latter of which is activated by the intracellular signaling pathway induced by prolactin[7,8], the main lactogenic hormone.

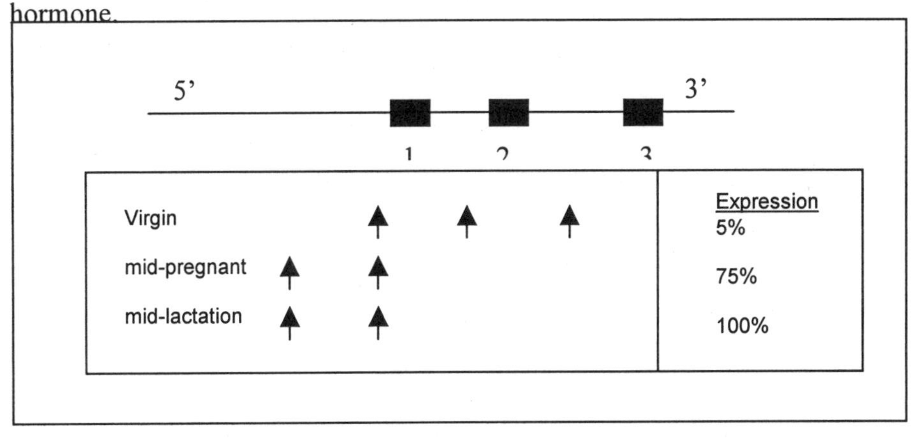


Fig. 2. DNaseI hypersensitive site formation on the ovine βlg gene during pregnancy and lactation in sheep[11]. The ovine βlg gene is shown with exons as black boxes and hypersensitive sites are shown as arrows. Relative mammary expression levels as a percentage of maximal levels obtained during lactation are shown.

In the virgin animal βlg expression is barely detectable. Then, as pregnancy proceeds, βlg expression gradually increases, with the first major increase occurring at the 110th day of pregnancy. Expression continues to increase through parturition to reach maximal levels during lactation[4].

DNaseI hypersensitivity at HSIII parallels this expression profile, with minimal cutting before mid-pregnancy and substantial cutting thereafter and during lactation[15]. This temporal pattern of DNaseI hypersensitivity parallels the differential activation of complexes containing the prolactin induced transcription factors STAT5a and STAT5b [7]. Specifically, there is a marked increase in STAT5a relative to STAT5b during late pregnancy[16] which suggests a major role for STAT5a in regulating βlg expression. In support of this, STAT5a null-mice indicate that this transcription factor is the principal mediator of the lactogenic signal[17]. Nevertheless, STAT5a interaction with the βlg promoter is not essential for tissue-specific expression of βlg[12]. Therefore, other mechanisms must be involved in establishing a mammary-restricted transcription factor complex upon the βlg promoter.

## 3. NUCLEOSOME MAPPING OF β-LACTOGLOBULIN PROMOTER

The positioning of nucleosomes contributes to both the structure and function of chromatin and has a decisive role in controlling gene expression[18]. A major determinant of nucleosome positioning, and therefore distribution, is the DNA sequence itself. Other components may modulate DNA-directed nucleosome positioning, but must act in conjunction with, rather than independently of the DNA sequences. In most studies of sequence-directed nucleosome positioning, only short regions of DNA have been analysed in any detail. Understandably, the tendency has been to focus in the immediate vicinity of regulatory DNA sequences. Only one study has addressed an entire gene[19].

We have mapped the *in vivo* nucleosomal profile for about 10 kb of contiguous genomic sequence encompassing the ovine βlg gene. Two probes of nucleosomal organisation have been used. The first relies on the ability of the enzyme MNase to preferentially digest DNA in the linker region between two nucleosomes. The second approach involves cuprous phenanthroline probed mercaptopropionic acid cleaved chromatin. Again cleavage is predominately in the linker region between nucleosomes. Both methods clearly indicate that a periodic nucleosome pattern based on 180 bp is present on both the inactive and active βlg gene. Furthermore, reflecting gene expression state, there are distinct differences between the active and inactive chromatin. In addition, a specific relationship between nucleosome positioning over the promoter region and the location of known transcription factor binding sites is evident (Fig. 3).

To further define the nucleosomal profile of βlg we have mapped *in vitro* reconstituted, sequence directed nucleosomal positioning. This has been performed using the monomer extension assay[19] developed by Dr. Jim Allan (University of Edinburgh) and optimised for βlg by Simon Boa. The *in vitro* map essentially supports the *in vivo* map and offers the opportunity to identify sequence determinates of nucleosomal positioning.

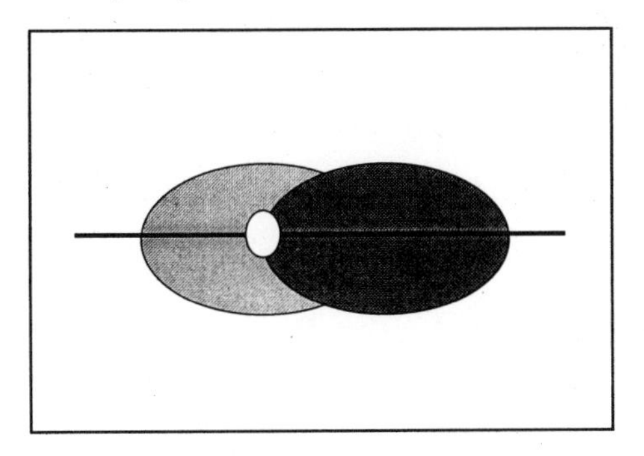

Fig. 3. Alternative nucleosome positions relative to STAT5 binding site (open circle). Thus, the STAT5 site resides either within the linker region between two adjacent nucleosomes or near the nucleosome dyad.

# 4. PROPOSED MODEL FOR TRANSCRIPTION COMPLEX FORMATION AT β-LACTOGLOBULIN PROMOTER

From the DNaseI and nucleosomal mapping data we propose a co-operative model for transcription complex formation at the ovine βlg promoter. This model revolves around the precise positioning of a nucleosome within the promoter (Boa et al., manuscript in preparation). Furthermore, STAT5 probably fulfils a late function to stabilise the transcription factor complex (Whitelaw et al., manuscript in preparation), thus ensuring efficient transcription.

**Acknowledgements**

I gratefully acknowledge that much of this work was performed by Simon Boa and John Webster; and through my on-going collaboration with Dr. Jim Allan (University of Edinburgh). The initial STAT5 studies were performed by Tom Burdon, Jerome Demmer and Christine Watson. This work was supported by the BBSRC (UK).

# 5. REFERENCES


1. Wilmut I, Whitelaw, CBA. Strategies for production of pharmaceutical proteins in milk. Reprod Fert Dev 6, 625-630, 1994.
2. Ali S, Clark AJ. Characterization of the gene encoding ovine beta-lactoglobulin. Similarity to the genes for retinol binding protein and other secretory proteins. J Mol Biol 199, 415-426, 1988.
3. Flower DR. The lipocalin protein family: structure and function. Biochem J 318, 1-14, 1996.
4. Harris S, McClenaghan M, Simons JP, Ali S, Clark AJ. Developmental regulation of the sheep beta-lactoglobulin gene in the mammary gland of transgenic mice. Dev Genet 12, 299-307, 1991.
5. Streuli CH, Edwards GM, Delcommenne M, Whitelaw CBA, Burdon TG, Schindler, C, Watson CJ. Stat5 as a target for regulation by extracellular matrix. J Biol Chem 270, 21639-21644, 1995.
6. Osborne R, Howell M, Clark AJ, Nicholas KR. Hormone-dependent expression of the ovine beta-lactoglobulin gene. J Dairy Res 62,321-319, 1995.
7. Hennighausen L, Robinson WG, Wagner KU, Liu X. Prolactin signaling in mammary gland development. J Biol Chem 272, 757-7569, 1997.
8. Watson CJ, Burdon TG. Prolactin signal transduction mechanisms in the mammary gland: the role of the Jak/Stat pathway. Rev Reprod 1-5, 1996.
9. Watson CJ, Gordon KE, Robertson M, Clark AJ. Interaction of DNA-binding proteins with a milk protein gene promoter in vitro: identification of a mammary gland-specific factor. Nucl Acids Res 19, 6603-6610, 1991.
10. Whitelaw CBA, Harris S, Archibald AL, McClenaghan M, Simons JP, Clark AJ. Position-independent expression of the ovine beta-lactoglobulin gene in transgenic mice. Biochem J 286, 31-39, 1992.
11. Whitelaw CBA, Webster J. Temporal profiles of appearance of DNase I hypersensitive sites associated with the ovine beta-lactoglobulin gene differ in sheep and transgenic mice. Mol Gen Genet 257, 649-654, 1998.
12. Burdon TG, Maitland KA, Clark AJ, Watson CJ. Regulation of the sheep beta-lactoglobulin gene by lactogenic hormones is mediated by a transcription factor that binds an interferon-gamma activation site-related element. Mol Endocrinol 8, 1528-1536, 1994.
13. Demmer J, Burdon TG, Djian J, Watson CJ, Clark AJ. The proximal milk protein binding factor binding site is required for the prolactin responsiveness of the sheep beta-lactoglobulin promoter in Chinese hamster ovary cells. Mol Cell Endocrinol 107, 113-121, 1995.
14. Webster J, Wallace R, Clark A., Whitelaw CBA. Tissue-specific, temporally regulated expression mediated by the proximal ovine beta-lactoglobulin promoter in transgenic mice. Cell Mol Biol Res 41, 11-15, 1995.
15. Whitelaw CBA. Regulation of ovine beta-lactoglobulin gene expression during the first stage of lactogenesis. Biophys Biochem Res Comm 209, 1089-1093, 1995
16. Philp JAC, Burdon TG, Watson CJ. Differential activation of STATs 3 and 5 during mammary gland development. FEBS Lett 396, 77-80, 1996.
17. Liu X, Robinson GW, Wagner KU, Garret L, Wynshaw-Boris A, Hennighausen L. Stat5a is mandatory for adult mammary gland development and lactogenesis. Genes Dev 11, 179-186, 1997.

18. Kornberg RD, Lorch Y. Twenty-five years of the nucleosome, fundamental particle in the eukaryotic chromosome. Cell 98, 285-294, 1999.
19. Davey C, Pennings S, Meersseman G, Wess TJ, Allan J. Periodicity of strong nucleosome positioning sites around the chicken adult beta-globin gene may encode regularly spaced chromatin. Proc Natl Acad Sci U S A 92, 11210-11214, 1995.

19.

# Chromatin Remodeling in Hormone-Dependent and - Independent Breast Cancer Cell Lines

[2]Claire Giamarchi, [1]Catherine Chailleux, [1]Hélène Richard-Foy.
*[1]Laboratoire de Biologie Moléculaire Eucaryote, CNRS, Toulouse, France and [2] Institut Claudius Regaud, Laboratoire de Biochimie, Toulouse, FRANCE*

Key words: breast cancer, oestrogens, chromatin structure, gene expression

Abstract: Chromatin restricts the accessibility of DNA to regulatory factors; its remodelling over the regulatory regions contributes to the control of gene expression. An increasing number of evidence links defects in chromatin remodelling machinery and cancer. Our aim is to elucidate the role of chromatin structure in the control of the expression of hormone-induced genes in breast cell lines oestrogen-dependent or -independent for growth. Mammary tumour growth is controlled by steroid hormones via their nuclear receptor and by growth factors via tyrosine kinase receptors. 50% of these tumours elude to hormonal control. This limits the anti-oestrogen therapy. As a model, we have analysed in several cell lines the chromatin organisation of the regulatory regions of two genes, pS2 that is associated with a good prognostic, and cathepsin D (catD) that is a bad prognostic marker. The expression of the two genes is oestrogen-regulated in oestrogen-dependent cell line MCF7. In contrast in the hormone-independent cell line MDA MB 231, pS2 is not expressed and catD is constitutively expressed. Within the regulatory regions of pS2 gene, we have localised two regions that undergo a hormone-dependent change in chromatin structure in MCF7 cells but not in MDA MB 231 and that can be correlated with gene expression. In contrast catD regulatory regions did not display hormone-dependent changes in chromatin structure, suggesting that hormone regulation takes place within regions with a constitutively open chromatin structure.

# 1. INTRODUCTION

Chromatin structure plays a major role in processes such as DNA transcription, replication and repair. This structure is dynamic and changes can be correlated with transcriptional repression or activation. Cell differentiation is accompanied by changes in chromatin structure of large domains of the genome that become transcriptionally competent (Fig. 1). Transcription of the genes located within these domains involves additional changes in chromatin structure localised within small regulatory regions.

In breast cancer cells, cell proliferation is controlled by estrogens and by growth factors such as EGF and IGFI. This is achieved through changes in the level of expression of groups of genes. Transcriptional activation of a gene by a steroid, via its cognate receptor, or by growth factors, via their tyrosine-kinase receptors, both result in changes in composition and/or activity of large multi-protein complexes containing CBP/p300 [1].This process is accompanied by changes in chromatin structure over the proximal promoter and regulatory regions.

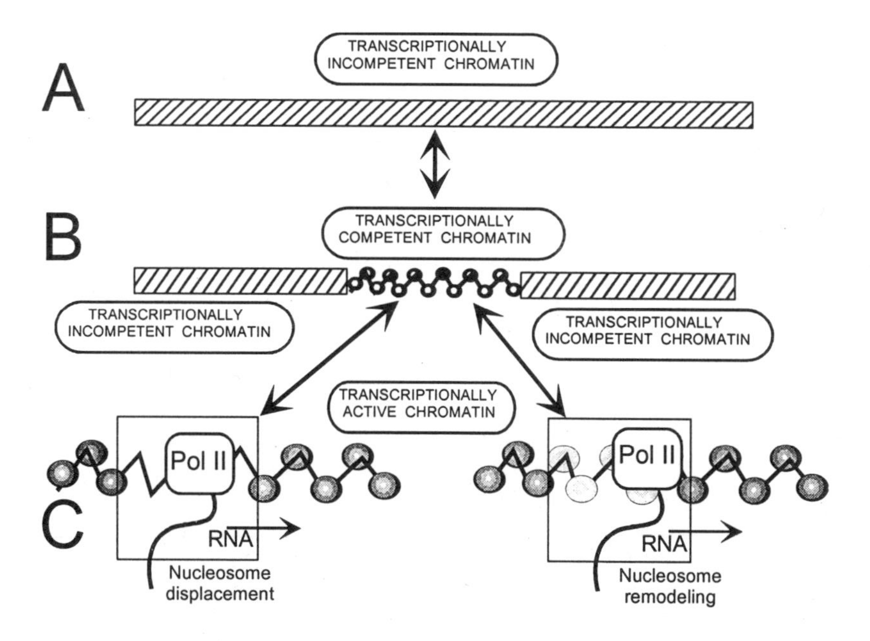


**Fig. 1: Chromatin structure and transcription.** A - Transcriptionally incompetent chromatin. B - The opening of chromatin domains generates transcriptionally competent chromatin. C - Within these domains, transcription may be triggered. A localized change in chromatin structure (nucleosome displacement or nucleosome remodelling?) accompanies transcription.

One major problem in breast cancer biology concerns the cellular changes leading to breast cell population whose growth is hormone-independent. The mechanisms involved in such a cell progression remain unclear. Establishment of the hormone-independent phenotype is accompanied by a loss of the expression of the estrogen receptor (ERα). However the lack of ERα is not the cause of the hormone independence[2].

To get new insights on the mechanisms involved in the establishment of the hormone-independent phenotype of breast cancer cells, we have analysed the chromatin structure of the regulatory regions of two model genes: pS2 and cathepsin D (CatD) in three cell lines. The MCF7 cells are hormone-dependent, MDA MB 231 are hormone independent and HE5 are MDA MB231 cells in which the expression of ERα has been restored[3]. We have mapped the DNase I hypersensitive sites (these sites mark regions with an "open" chromatin structure) et compared the results with transcription of the genes, assayed by Northern blot analysis.

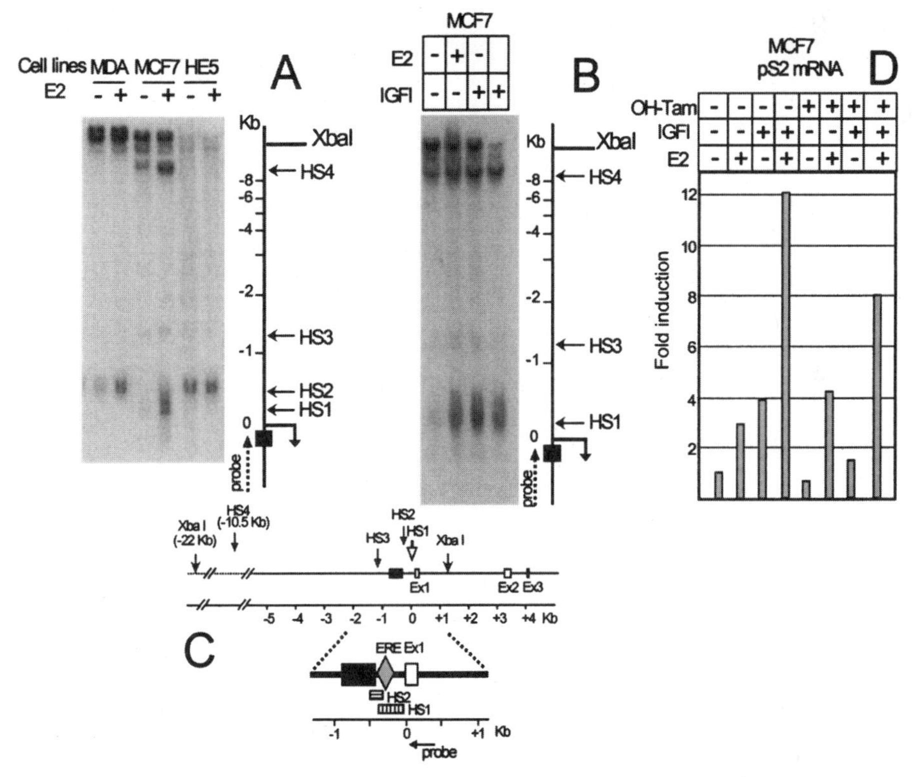


**Fig. 2. Chromatin structure of pS2 regulatory regions and gene expression.** A - Comparison of DNase I hypersensitivity of pS2 gene in the 3 cell lines, untreated or treated with hormone (E2). B - Effect of estrogen and IGFI on DNase I hypersensitivity of pS2 regulatory regions in MCF7 cells. C - Mapping of the DNase I hypersensitive sites in the 5'-flanking sequences of pS2 gene. D - Influence of estradiol (E2), IGFI, and an estrogen antagonist (OH-Tam) on pS2 expression in MCF7 cells. For the experimental details see Giamarchi et al., 1999.

## 2. THE pS2 GENE

We have investigated the chromatin structure of the regulatory region of pS2 gene in MCF-7, MDA MB 231, and HE5 cells treated or not with estradiol. Fig 2A shows the DNase I hypersensitivity profiles of pS2 gene in the 3 cell lines, treated or not with estradiol. In MCF7, in which pS2 transcription is hormone-induced, there were two DNase I hypersensitive sites (pS2-HS1 and pS2-HS4) induced by estrogen treatment. In contrast, in MDA MB 231 and HE5 cells, in which pS2 is not expressed whether or not

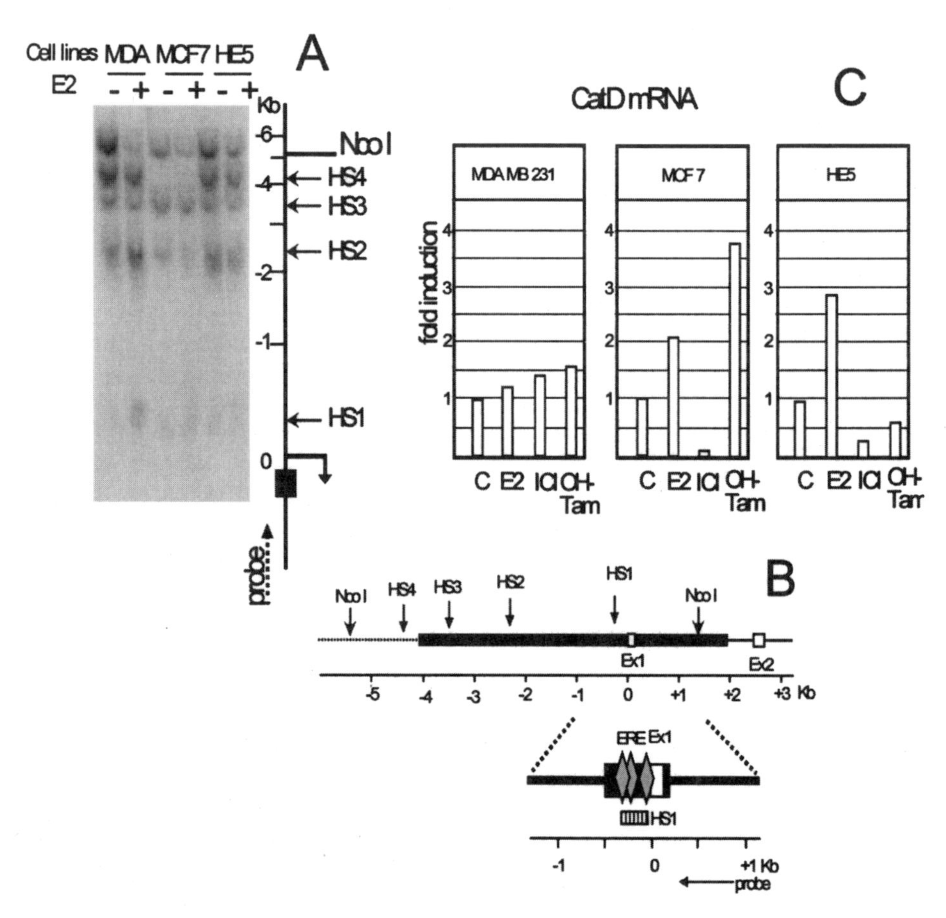


**Fig. 3. Chromatin structure of CatD regulatory regions and gene expression.** A – Comparison of DNase I hypersensitivity of CatD gene in the 3 cell lines, untreated or treated with hormone (E2). B - Mapping of the DNase I hypersensitive sites in the 5'-flanking sequences of CatD gene. C – Influence of estradiol (E2) and two anti-estrogens (ICI and OH-Tam) on CatD mRNA expression in the 3 cell lines.

hormone is present, there was only one hormone-independent DNase I hypersensitive site (pS2-HS2). This site maps immediately upstream of site pS2-HS1, and encompasses a region containing an AP1 binding site (Fig 2C). A synergistic effect of estrogens and IGFI on pS2 gene transcription has been described[4]. We have compared the chromatin structure of pS2 gene in MCF7 cells untreated or treated with estradiol, IGFI or both. Estrogen or IGF treatment induced DNase I hypersensitive sites that are indistinguishable (pS2-HS1 and pS2-HS4, Fig 2B). Treatment with both IGFI and estradiol did not increase the intensity of the hypersensitive sites, contrasting with the effect on RNA synthesis (Fig 2D). This is the first description of a DNase I hypersensitive site induced by a growth factor[5]. The fact that hypersensitive sites induced by estradiol and IGFI are indistinguishable may be due to the recruitment of the partners from different signal transduction pathways in a large coactivator complex[1,6].

## 3. THE CatD GENE

We have performed the same chromatin structure analysis as for the pS2 gene in the three cell lines (Fig. 3). Contrasting with the results obtained for pS2 gene, there is no hormone-dependent DNase I hypersensitive site in the regulatory regions of CatD gene (Fig 3A), although RNA synthesis is induced by estradiol in the hormone dependent cell line MCF7 (Fig 3C). In HE5 cells the expression of estrogen receptor allows recovering the hormonal regulation of CatD while in the parental cell line MDA MB 231 CatD was constitutively expressed at a high level (Fig 3C). In these two cell lines the DNase I hypersensitivity profiles are indistinguishable. Both cell lines display an additional site compared to MCF7 cells (CatD-HS4, Fig. 3A). Similar studies in other hormone-independent cell lines are in progress to determine if the presence of this DNase I hypersensitive site is characteristic of the hormone-independent status, or of MDA MB 231 cells.

## 4. CONCLUSION

The promoters of pS2 [7,8] and CatD[9,10] genes have been extensively studied with classical approaches. The comparison of chromatin structure changes in the regulatory regions of the two genes suggests that the mechanisms controlling their expression are different. The lack of expression of pS2 in cell lines MDA MB 231 and HE5 correlates with a closure of the hormone-dependent hypersensitive sites pS2-HS1 and pS2-HS4. As it should be expected from mRNA measurements, the presence of ERα in HE5 cells has no effect on the chromatin structure. This indicates that in hormone-independent cells there is a closure of the chromatin over large domains (Fig. 1 - A), one of them containing the pS2 gene. Extinction of ERα gene in hormone-independent cells may obey to similar mechanisms. It has been established that factors other than ERα are involved in the progression of the cells towards hormone-independent growth[2,11]. Such factors could also play a critical role in modulating chromatin structure. The situation for CatD gene is different. In hormone-dependent cells, CatD transcription is controlled by estrogens. However changes in transcription level are not accompanied by chromatin structure changes. Contrasting with pS2 gene, in hormone-independent cells, cathepsin D is constitutively expressed at a high level; in these cells, re-expression of ERα allows recovering the hormonal control of CatD gene expression. This indicates that in hormone-independent cells, it is the lack of ERα that results in an abnormally high expression of CatD. The chromatin structure of the regulatory regions of CatD does not undergo a closure in hormone-independent cells. DNase I hypersensitive sites are constitutive, allowing gene expression (actively transcribing chromatin). In the presence of unliganded ERα there is a decrease in CatD expression. This decrease is not accompanied by detectable changes in chromatin structure probably because the chromatin being already open, proteins can fully interact with each other and regulate transcription. In this context, the exact role of histone acetyltransferases, histone deacetylases[12] and/or chromatin remodelling complexes remain to elucidate.

## 5. REFERENCES


1. Montminy M. Something new to hang your hat on. Nature, 387, 654-655,1997.
2. Levenson A and Jordan C. Transfection of the human estrogen receptor (ER) cDNA into ER-negative mammalian cell lines. J. Steroid Biochem. Molec. Biol., 51, 229-239,1994.
3. Touitou I, Vignon F, Cavailles V and Rochefort H. Hormonal regulation of cathepsin D following transfection of the estrogen or progesterone receptor into three sex steroid hormone resistant cancer cell lines. J Steroid Biochem Mol Biol, 40, 231-237,1991.
4. Chalbos D, Philips A, Galtier F and Rochefort H. Synthetic antiestrogens modulate induction of pS2 and cathepsin-D messenger ribonucleic acid by growth factors and adenosine 3',5'- monophosphate in MCF7 cells. Endocrinology, 133, 571-576,1993.
5. Giamarchi C, Solanas M, Chailleux C, Augereau P, Vignon F, Rochefort H and Richard-Foy H. Chromatin structure of the regulatory regions of pS2 and cathepsin D genes in hormone-dependent and -independent breast cancer cell lines. Oncogene, 18, 533-541,1999.
6. Torchia J, Rose DW, Inostroza J, Kamel Y, Westin S, Glass CK and Rosenfeld MG. The transcriptional co-activator p/CIP binds CBP and mediates nuclear-receptor function. Nature, 387, 677-684,1997.

7. Berry M, Nunez AM and Chambon P. Estrogen-responsive element of the human pS2 gene is an imperfectly palindromic sequence. Proc Natl Acad Sci U S A, 86, 1218-1222,1989.
8. Nunez AM, Berry M, Imler JL and Chambon P. The 5' flanking region of the pS2 gene contains a complex enhancer region responsive to estrogens, epidermal growth factor, a tumor promoter (TPA), the c-Ha-ras oncoprotein and the c-jun protein. EMBO J, 8, 823-829,1989.
9. Augereau P, Miralles F, Cavailles V, Gaudelet C, Parker M and Rochefort H. Characterisation of the proximal estrogen-responsive element of human cathepsin D gene. Molecular endocrinology, 8, 693-703,1994.
10. Cavailles V, Augereau P and Rochefort H. Cathepsin D gene is controlled by a mixed promoter, and estrogens stimulate only TATA-dependent transcription in breast cancer cells. Proc. Natl. Acad. Sci. USA, 90, 203-207,1993.
11. Garcia M, Derocq D, Freiss G and Rochefort H. activation of estrogen receptor transfected into a receptor-negative breast cancer cell line decreases the metastatic and invasive potential of the cells. Proc. Natl. Acad. Sci USA, 89, 11538-11542,1992.
12. Utley RT, Ikeda K, Grant PA, Cote J, Steger DJ, Eberharter A, John S and Workman JL. Transcriptional activators direct histone acetyltransferase complexes to nucleosomes. Nature, 394, 498-502,1998.

20.

# The Use of Transplanted Mammary Gland to Study Cancer Signalling Pathways

Paul A.W.Edwards
*Dept. of Pathology, University of Cambridge, U.K*

Key words: Mammary, breast cancer

Abstract: Mammary epithelium can be genetically manipulated by reconstituting a mammary gland, in an animal, from epithelium and a mammary fat pad from which the endogenous epithelium has been removed at 3 weeks of age. Genes can be introduced into the epithelium before transplantation using retrovirus vectors. To remove genes from the epithelium at present requires epithelium to be transplanted from knockout donor mice, but this is a valuable extension of knockout technology, as (a) it creates knockout epithelium in a normal stromal and systemic environment, or vice versa, and (b) where the knockout mouse does not survive into adulthood, epithelium can be rescued from embryos after about 12 days of gestation, and grown to form mature mammary epithelium in a normal recipient mammary fat pad.

## 1. MAMMARY TRANSPLANTS

A powerful way to investigate the mammary gland is to genetically manipulate the epithelium by reconstructing the mammary gland, in vivo, from transplanted epithelial cells. Mammary epithelial cells are isolated from one mouse mammary gland and put into a mammary fat pad in another mouse, from which the natural epithelium has been removed. Remarkably, transplanted epithelial cells grow into the fat pad and form a mammary epithelium almost exactly like a naturally-occurring one, except that it is not connected to a nipple[1]. To introduce a gene into the mammary epithelium, the mammary epithelial cells to be transplanted can be infected with nonreplicating retrovirus vector between isolation from one mouse and transplantation into the host[2,3] (Fig. 1).

This approach has a number of advantages over the more familiar approach of creating a transgenic mouse with an exogenous gene driven by a mammary-specific promoter. The first advantage is that the gene is entirely confined to mammary epithelial cells, so any promoter can be used - typically actin, SV40 or retrovirus LTR promoters have been used. In contrast, transgenics have to be made with mammary-specific promoters, the MMTV promoter, the WAP (Whey acidic protein) promoter or the beta-lactoglobulin promoter. All three are much more active in lactation than in virgin animals and the MMTV promoter at least is active in tissues other than mammary gland. A major difference between the approaches is that in the transgenics all mammary epithelial cells contain the introduced gene, and all are capable of expressing it, though in practice expression can be patchy. In the transplants infected with retrovirus, only a few percent of the transplanted cells express the gene, unless those cells have a selective advantage after transplantation[3]. However, while this is a weakness of transplantation where the whole gland is to be altered, it is a unique strength of the approach for modelling cancer development, as individual cells can be genetically altered among an excess of normal cells, a true model of the cancer situation.

Examples of the use of this approach include the expression of mutant forms of the EGF-receptor family of receptors, such as *neu*, a point-mutated form of rat *erbB2,* and *v-erbB*, the original retroviral truncated EGF-receptor mutant. These gave focal abnormalities of growth pattern, as expected if clones or small groups of cells were expressing the introduced gene. There were striking differences in the patterns of abnormal growth. *neu* gave a spectrum of lesions, including DCIS and occasional tumours, that resembled histologically a range of lesions found in human breasts[4]. *v-erbB* on the other hand gave enlarged and distorted ducts with loose cells in the lumens[5]. Thus two closely-related receptors gave different alterations to growth pattern. Whether the difference was between *erbB2* and *erbB*, or was due to the type of activating mutation, or even to the species of origin of the receptor (rat versus chicken) remains to be determined. A very different result was obtained when the genes *myc*, *Wnt-1* and *Wnt-4* were expressed this way[2,6,7]. In all three cases, large parts or even the whole of a transplanted mammary epithelium were hyperplastic, suggesting that cells expressing the exogenous gene had a selective advantage over their normal neighbours during outgrowth of the transplant. This may be important as a model of cancer development[3,8]. Also, the pattern of growth in the virgin animal induced by expression of *Wnt-1* and *Wnt-4* was remarkably similar to the pattern induced early in pregnancy, suggesting that the normal change of pattern in pregnancy may be signalled locally by a member of the *Wnt* family, perhaps *Wnt-4* itself, which is expressed naturally in early pregnancy[7]. Other genes expressed this way include *ras*[9,10] and the marker gene beta-galactosidase[11,12].

Transplantation can be used not only to introduce genes into epithelium, but also to manipulate mammary epithelium lacking a gene, and here the method solves different problems of the transgenic approach. In transgenic knockout mice where mammary epithelium is altered, it is not in general possible to tell whether the effect on the epithelium is due to alteration in the behaviour of the epithelial cells themselves, or is an indirect effect of a systemic change, or changes to the environment in which the epithelium grows. And, of course, in some cases no knockout mammary epithelium can be obtained because the mutation is lethal before the mice reach maturity. Both problems can often be solved by transplanting fragments of mammary epithelium between mice.

For example, mice with null mutations in the cyclinD1 gene develop more or less normally, but the mammary epithelium fails to respond properly to pregnancy. This could be because the epithelial cells were unable to respond to signals, but it was at least as likely that they were not receiving signals. In pregnancy the mammary epithelium proliferates in response to systemic signals coming from, for example, the pituitary and the placenta, and these signals may well act on the stroma of the mammary fat pad to provoke local signals to the epithelium, as for many epithelial-mesenchymal signals. To distinguish these possibilities we transplanted mammary epithelium from knockout mice into normal, histocompatible (an important issue as discussed below) recipients. The resulting mice had knockout epithelium in a normal fat pad and mouse, but they continued to show the same attenuated response to pregnancy, showing that most or all of the reduction in response to pregnancy was a failure within the epithelial cells themselves[13]. In principle, we could also have transplanted normal epithelium into knockout recipients to see if their mammary epithelium had a normal response, but in the mixed-background transgenic line we used, this combination would not have been histocompatible. Transplantation of, or into, transgenic mice is often complicated by the fact that transgenic lines are frequently made by crossing two strains of mice. The resulting transgenics individually have an unknown and variable combination of histocompatibility alleles from the two backgrounds. Such tissue can be transplanted into an F1 mouse from a cross between the two strains, as it has all the alleles, but nothing can be transplanted into these transgenics as they will almost invariably lack at least one allele from any donor chosen. The solution to this problem is for the transgenic knockout line to be made on a pure 129 background.

To solve the problem of lethality of mutations, it may often be possible to rescue mammary epithelium from embryonic or neonatal mice and transplant it into adult normal mice, where it grows to form epithelium apparently equivalent to mature adult epithelium. Transplantation can be done from about 12 or 13 days of gestation, and indeed the developing mammary glands, or rather the nipples, are easiest to find at or soon after this point, before the presence of hair follicles and folds in the skin make them more difficult to recognise.

In the future it will probably be possible to create clones of knockout epithelial cells in a normal fat pad, to model clonal loss of a tumour suppressor gene. This could be done by transplanting epithelium from conditional knockout mice – in which the gene of interest is flanked by *lox* sites[14] – and inducing knockout of the genes in clones by infecting them with a *cre*-expressing retrovirus. (Cre protein catalyses recombination between *lox* sites).

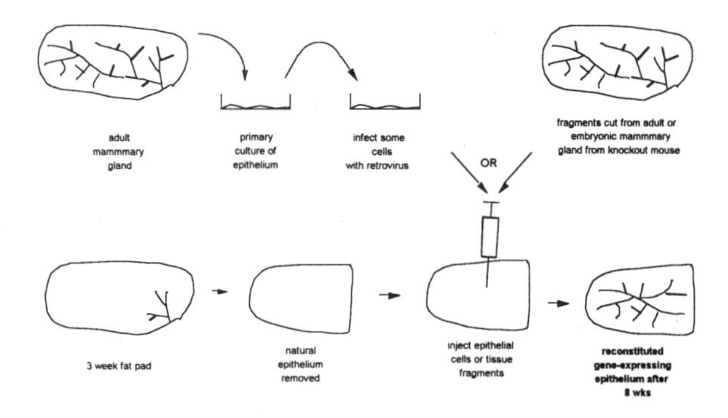


Fig. 1. Creating genetically-manipulated mammary epithelium by transplantation

## 2. REFERENCES

1. DeOme KB, Faulkin, LJ, Bern HA, Blair P B (1959). Development of mammary tumors from hyperplastic alveolar nodules transplanted into gland-free mammary fat pads of female C3H mice. Cancer Research 19, 515-520.
2. Edwards PAW, Ward JL, Bradbury JM (1988). Alteration of morphogenesis by the v-myc oncogene in transplants of mouse mammary gland. Oncogene 2, 407-412.
3. Edwards PAW (1996). Tissue reconstitution or transgenic mammary gland technique for modelling breast cancer development. In Mammary tumour cell cycle, differentiation and metastasis: advances in cellular and molecular biology of breast cancer, R. B. Dickson and M. E. Lippman, eds. (Boston: Kluwer Academic), pp. 23-36.
4. Bradbury JM, Arno J, Edwards PAW (1993). Induction of epithelial abnormalities that resemble human breast lesions by the expression of the neu/erbB2 oncogene in reconstituted mouse mammary gland. Oncogene 8, 1551-1558.

5. Abram CL, Bradbury JM, Page MJ, Edwards PAW (1995). v-erbB induces abnormal patterns of growth in mammary epithelium. In Intercellular Signalling in the Mammary Gland, C. J. Wilde, M. Peaker and C. H. Knight, eds. (New York: Plenum), pp. 67-68.
6. Edwards PAW, Hiby SE, Papkoff J, Bradbury JM (1992). Hyperplasia of mouse mammary epithelium induced by expression of the Wnt-1 (int-1) oncogene in reconstituted mammary gland. Oncogene 7, 2041-2051.
7. Bradbury JM, Edwards PAW, Niemeyer CC, Dale TC (1995). Wnt-4 expression induces a pregnancy-like growth pattern in reconstituted mammary glands in virgin mice. Dev Biol 170, 553-563.
8. Edwards PAW (1993). Tissue reconstitution models of breast cancer. Cancer Surveys 16, 79-96.
9. Aguilar-Cordova E, Strange R, Young LJT, Billy HT, Gumerlock PH, Cardiff RD (1991). Viral Ha-ras mediated mammary tumour progression. Oncogene 6, 1601-1607.
10. Bradbury JM, Sykes H, Edwards PAW (1991). Induction of mouse mammary tumours in a transplantation system by sequential introduction of the myc and ras oncogenes. Int J Cancer 48, 908-915.
11. Smith GH, Gallaghan D, Zweibel JA, Freeman SM, Bassin RH, Callaghan R (1991). Long-term in vivo expression of genes introduced by retrovirus-mediated transfer into mammary epithelial cells. Journal of Virology 65, 6365-6370.
12. Edwards PAW, Abram CL Bradbury JM (1996). Genetic manipulation of mammary epithelium by transplantation. Journal of Mammary Gland Biology and Neoplasia 1, 75-89.
13. Fantl F, Edwards PAW, Steel JH, Vonderhaar BK, Dickson C (1999) The Impaired Mammary Gland Development in *Cyl-1*$^{-/-}$ mice during pregnancy and Lactation is Epithelial Cell Autonomous. Developmental Biology, 212: 1-11
14. Xu X, Wagner KU, Larson D, Weaver Z, Li C, Ried T, Hennighausen L, Wynshaw-Boris A, Deng CX (1999) Conditional mutation of Brca1 in mammary epithelial cells results in blunted ductal morphogenesis and tumour formation. Nat Genet 22, 37-43

21.

# Development of Mammary Gland Requires Normal β1-Integrin Function

Marisa M. Faraldo, Marie-Ange Deugnier, Jean Paul Thiery and Marina A. Glukhova
*UMR 144, CNRS-Institut Curie, Section de Recherche, Paris, France.*

Key words: Mammary epithelium, Integrins, Transgenic mice

Abstract: To study the role of β1-integrins in mammary gland development we have generated transgenic mice expressing a dominant negative mutant of the β1-integrin chain in the mammary epithelium. The transgenic glands presented a delayed development in pregnancy and lactation due to decreased epithelial cell proliferation and increased apoptosis, whereas at the beginning of lactation, expression of milk proteins, WAP and β-casein was diminished. In correlation with transgene expression, the basement membrane component, laminin, and the β4 integrin were accumulated at the lateral surface of luminal epithelial cells, revealing defects in polarization. Our data show that β1-integrins are involved *in vivo* in the control of proliferation, apoptosis, differentiation, and maintenance of baso-apical polarity of mammary epithelium.

## 1. INTRODUCTION

The mammary gland consists of secretory alveoli interconnected by a system of branching ducts. Its development is tightly regulated and needs the concerted action of soluble and extracellular matrix bound factors as well as the correct cell-cell and cell-matrix interactions.

A number of studies in different cell systems, including the mammary gland, have shown that cell-matrix interactions are important regulators of cell growth and programmed cell death[1]. In addition, extracellular matrix (ECM) can regulate the phenotype of mammary epithelium. Particularly, adhesion of mammary epithelial cells to laminin is believed to mediate β-casein gene expression[2], whereas activity of STAT5 transcription factor, an essential regulator of milk gene transcription, is also controlled by cell-ECM interactions[3].

Integrins are the major cellular ECM receptors. They are transmembrane heterodimers constituted by non-covalently associated α and β subunits. The large extracellular domain of integrins binds to various ligands, i.e. to the ECM proteins or to other cell surface receptors. The cytoplasmic domain of the receptor interacts with the cytoskeletal complexes triggering a signal transduction cascade in response to ligand binding[4,5,6].

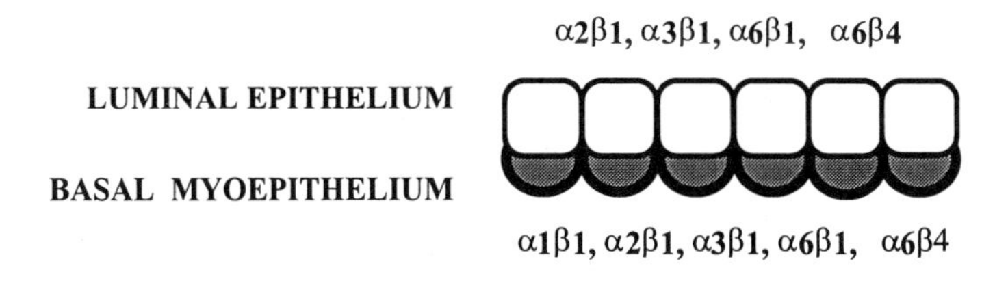


*Figure 1.* Integrins in the mammary gland bilayer.

In the mammary gland bilayer, luminal epithelial cells express α2β1, α3β1, α6β1 and α6β4 integrin dimers and basal myoepithelial cells present, in addition, α1β1[7, 8] (Fig.1). To study the involvement of β1-integrins in the cellular functions of mammary epithelium *in vivo*, we have targeted the expression of a transgene coding for a chimeric molecule containing the cytoplasmic and the transmembrane domains of the β1-integrin subunit and the extracellular domain of the T-cell differentiation antigen, CD4, to the luminal mammary epithelium using MMTV promoter. Such chimera neither binds to α-integrin subunits, nor interacts with the ECM integrin ligands. However, *in vitro*, it was shown to be delivered to focal contacts of the adherent cells and, if expressed at high level, to interfere with integrin functions such as adhesion to ECM proteins and FAK phosphorylation following integrin clustering[9]. Thus the chimeric molecule can uncouple adhesion from the intracellular integrin-associated events and acts as a dominant inhibitor of integrin function (Fig. 2A).

## 2. GENERATION OF TRANSGENIC ANIMALS

Four founders (F0) expressing the β1-chimera under the control of MMTV promoter were generated. The females from 3 transgenic lines were able to feed their litters of normal size, whereas 25% of the females of line 17 lost some or all the pups during the 1st and 2nd days of lactation.

Northern blot analysis of glands from different developmental stages has shown that the transgene was already expressed in 8-week-old virgin mice, although its expression was significantly up-regulated in pregnancy, and reach the maximal levels in lactation. The expression of the transgene, was high in line 17, moderate, in line 42, and weak, in lines 44 and 46. Consistent with the transgene expression levels, the morphological differences described below were more drastic in females from line 17, much less pronounced in line 42, and hardly detectable in lines 44 and 46.

## 3. EFFECTS OF TRANSGENE EXPRESSION ON MAMMARY EPITHELIUM

In the transgenic animals, we have analysed the mammary gland morphology, proliferation and apoptosis rates and differentiation of luminal epithelial cells at different stages of the mammary gland development.

### 3.1 Delay in the development of mammary glands expressing β1-chimera

Morphological analysis of the wild-type and transgenic glands at different developmental stages has shown the first alterations at mid-pregnancy stage. By day 10-12 of pregnancy the transgenic glands appeared smaller with a branching pattern less complex than the one observed in normal glands (Fig. 2B). Histological analysis of 2-day-lactating wild-type glands has shown a fat pad completely occupied by the secretory epithelium, with small fat stroma islets between the lobules. On the contrary, in the transgenic glands, the alveoli appeared sparse and were surrounded by vast areas of fat stroma.

Later in lactation no significant morphological differences between wild-type and transgenic glands were detected.

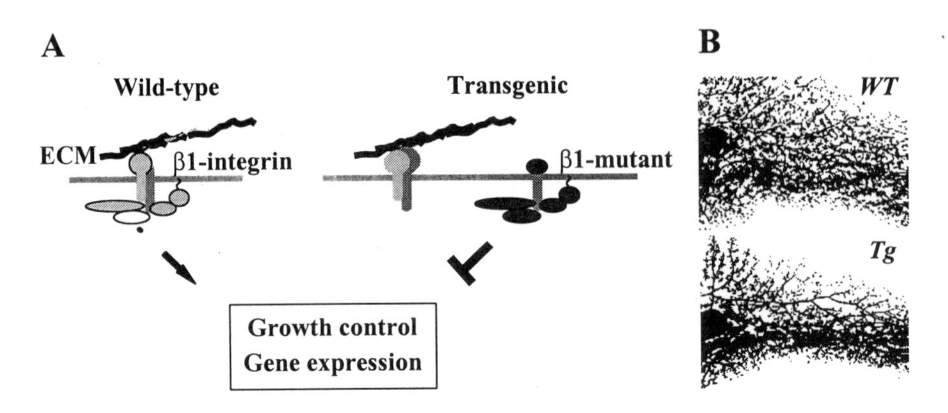


*Figure 2.* A) Perturbation of β1-integrin functions in the presence of transgene. B) Whole mount staining of mammary glands from wild-type (WT) and β1-transgenic (Tg) animals at 12 days of pregnancy.

The delayed development of the mammary glands of the transgenic mice might be due to decreased proliferation and/or increased apoptosis rates. We have found that proliferation was significantly diminished in transgenic glands at mid-pregnancy as well as at the second day of lactation (Table 1).

On the other hand, TUNNEL analysis has revealed an increase in the apoptosis rates in 12 and 18-day pregnant and in 2-day lactating transgenic glands whereas later in lactation, similar to wild-type glands, apoptosis rates were low and did not exceed 0.1% (Table 2).

*Table 1.* BrdU incorporation in normal and transgenic mammary glands

| Stages | Wild-type | Transgenic |
|---|---|---|
| 12-day-pregnant | 19.91 ± 1.47 | 12.10 ± 2.35 |
| 2-day-lactating | 7.96 ± 0.64 | 3.21 ± 2.08 |

Values presented as mean ± SD.

*Table 2.* Apoptosis in normal and transgenic mammary glands

| Stages | Wild-type | Transgenic |
|---|---|---|
| 12-day-pregnant | 1.11 ± 0.11 | 2.46 ± 1.19 |
| 18-day-pregnant | 0.42 ± 0.10 | 0.90 ± 0.18 |
| 2-day-lactating | 0.28 ± 0.10 | 1.15 ± 0.41 |
| 4-day-lactating | < 0.1 | < 0.1 |
| 10-day-lactating | < 0.1 | < 0.1 |

Values presented as mean ± SD.

## 3.2 β1-chimera expression affects differentiation of secretory epithelium during lactation

To determine whether the transgene expression affected the differentiation of mammary secretory epithelium, we analysed by Northern blot the expression of WAP and β-casein genes in 2-day-lactating normal and transgenic mammary glands. The amount of mRNA for both proteins are reduced in transgenic glands, however the expression levels varied in different animals. We have observed 40-80% and 20-60% reduction in the amount of mRNA for WAP and β-casein, respectively. These data are in keeping with the results of the *in vitro* studies suggesting that expression of β-casein gene in mammary epithelial cells requires interaction with laminin mediated by β1-integrin[2] and that DNA-binding activity of a transcription factor STAT5, regulator of milk protein gene expression, depends on adhesion to basement membrane[3].

## 3.3 β1-chimera expression in luminal epithelial cells affects polarity

Polarity is an important property of all organized epithelia and its establishment is mediated by the interaction of the cells with their basement membrane. We have detected evidences of altered cell polarity in the transgenic mammary epithelium as well as in cultured mammary epithelial cells expressing the transgene.

Overall, the mammary epithelium bilayer organization in the transgenic animals appeared normal, and the basement membrane around alveoli appeared continuous as revealed by staining with an anti-laminin antibody. However, surprisingly, in addition to basal localization, at the sites of transgene expression, laminin was accumulated at the lateral surface of luminal epithelial cells. Similar to laminin, distribution of the β4-integrin chain was altered and it was colocalized with laminin at lateral cell surfaces. These observations reveal defects in cell polarization and suggest malformation of cell-cell junctions.

We have isolated epithelial cells from mammary glands of wild-type and transgenic mice and cultured them in Matrigel, in order to compare their ability to establish polarity when interacting with the basement membrane components *in vitro*. The cells isolated from wild-type animals formed cysts if cultured in Matrigel, while the cells obtained from the transgenic glands remained in aggregates without lumen, failing to establish polarity and to get organized into cysts. In agreement with our results, polarization and organization of normal mammary epithelial cells into cysts in collagen gels as well as in Matrigel were reported to require function of β1-integrins[10].

In conclusion, our data prove that β1-integrins are involved in growth control and differentiation of mammary epithelium and are essential for normal mammary gland development and function. We are currently searching for the intracellular signalling pathways responsible for the impaired growth and differentiation control in the transgenic glands during pregnancy and lactation.

**Acknowledgements**

This work was supported by the Association pour la Recherche contre le Cancer (ARC 1661).

# 4. REFERENCES

1. Frisch, S.M. and Ruoslahti, E., 1997, Integrins and anoikis. *Curr. Op. Cell Biol.*, **9**: 701-706.
2. Streuli, C.H., Schmidhauser, C., Bailey, N., Yurchenko, P., Skubitz, A.P.N., Rockelley, C., Bissell, M.J. ,1995, Laminin mediates tissue-specific gene expression in mammary epithelia. *J. Cell Biol.*, **129**: 591-603.
3. Streuli, C.H., Edwards, G.M., Delcommenne, M., Whitelaw, B.A., Burdon, T.G., Schindler, C. and Watson, C.,1995, Stat5 as a target for regulation by ECM. *J. Biol. Chem.*, **270**: 21639-21644.
4. Dedhar, S. and Hannigan, G.E., 1996, Integrin cytoplasmic interactions and bidirectional transmembrane signalling. *Curr. Op. Cell Biol.*, **8**: 657-669.
5. Giancotti, F.G., 1997, Integrin signaling: specificity and control of cell survival and cell cycle progression. *Curr. Op. Cell Biol.*, **9**: 691-700.
6. Meredith, J.E., Winitz, S., McArthur Lewis, J., Hess, S., Ren, X.-D., Renshaw, M.W. and Schwartz, M.A., 1996, The regulation of growth and intracellular signaling by integrins. *Endocrine Rev.* **17**: 207-220.
7. Sonnenberg, A., Daams, H., van der Valk, M.A., Hilken, J. and Hilgers, G., 1986, Development of mouse mammary gland: identification of stages of differentiation of luminal and myoepithelial cells using monoclonal antibodies and polyvalent antiserum against keratin. *J. Histochem. Cytochem.* **34**: 1037-1046.
8. Glukhova, M., Koteliansky, V., Sastre, X. and Thiery, J.P., 1995, Adhesion systems in normal breast and in invasive breast carcinoma. *Am. J. Pathol.* **146**: 706-716.
9. Lukashev, M.E., Sheppard, D. and Pytela, R., 1994, Disruption of integrin function and induction of tyrosine phosphorylation by the autonomously expressed β1 integrin cytoplasmic domain. *J. Biol. Chem.*, **269**: 18311-18314.
10. Howlett A.F., Bailey N., Damsky C., Petersen O.W. and Bissell M.J., 1995, Cellular growth and survival are mediated by β1 integrins in normal human breast epithelium but not in breast carcinoma. *J. Cell Sci.*, **108**: 1945-1957.

22.

# Repression of the Putative Tumor Suppressor Gene Bard1 or Expression of Notch4(Int-3) Oncogene Subvert the Morphogenetic Properties of Mammary Epithelial Cells

J.V. Soriano[1], I. Irminger-Finger[2], H. Uyttendaele[3], G. Vaudan[2], J. Kitajewski[3], A.-P. Sappino[2] and R. Montesano[1].
[1]Dept. of Morphology, and [2]Division of Oncology, University Medical Center, Geneva, Switzerland. [3]Dept. of Pathology, Columbia University, New York, USA

Key words: *Bard1*, Brca1, breast cancer, int-3, morphogenesis, Notch, oncogene, tumor invasion

Abstract: We have investigated whether repression of the putative tumor suppressor gene BARD1 or expression of the Notch4(int-3) oncogene in non-tumorigenic mammary epithelial cells affects their in vitro morphogenetic properties. Bard1 (Brca1-associated ring domain) is a protein interacting with Brca1 and thought to be involved in Brca1-mediated tumor suppression. To investigate the potential role of Bard1 in mammary gland development, we repressed its expression in TAC-2 cells, a murine mammary epithelial cell line which, when grown in three-dimensional collagen gels, forms branching ducts in response to hepatocyte growth factor (HGF) and alveolar-like cysts in response to hydrocortisone. Whereas Bard1 repression did not markedly modify the tubulogenic response of TAC-2 cells to HGF, it dramatically altered cyst development, resulting in the formation of compact cell aggregates devoid of central lumen. In addition, when grown to post-confluence in two-dimensional cultures, Bard1-suppressed TAC-2 cells overcame contact-inhibition of cell proliferation and formed multiple cell layers. The Notch4(int-3) oncogene, which codes for a constitutively activated form of the Notch4 receptor, has been reported to induce undifferentiated carcinomas when expressed in the mammary gland. The potential effect of activated Notch4 on mammary gland morphogenesis was investigated by retroviral expression of the oncogene in TAC-2 cells. Notch4(int-3) expression was found to significantly reduce HGF-induced tubulogenesis and to markedly inhibit hydrocortisone-induced cyst formation. In addition, Notch4(int-3) expressing TAC-2 cells formed multilayers in post-confluent cultures and exhibited an invasive behavior when grown on the surface of collagen gels. Taken together, these results indicate that both repression of Bard1 and expression of Notch4(int-3) disrupt cyst morphogenesis and induce an invasive phenotype in TAC-2 mammary epithelial cells.

# 1. INTRODUCTION

During the multistage process of mammary carcinogenesis, stepwise accumulation of genetic changes causes uncontrolled growth, disruption of normal glandular architecture, and invasion of epithelial cells into the adjacent stroma. While unscheduled proliferation also occurs in benign adenomas, subversion of orderly epithelial tissue organization is a hallmark of malignancy and plays a crucial role in tumor progression. Interestingly, the inability of carcinoma cells to correctly polarize and to generate histotypic three-dimensional structures is reflected in *in vitro* assays of morphogenesis. Thus, unlike cells derived from normal mammary epithelium, breast carcinoma cells are unable to form duct-like or alveolar-like structures when grown in collagen or basement membrane gels[6,9]. Overexpression of normal or mutated proto-oncogenes or inactivation of tumor suppressor genes has been shown to result in mammary carcinogenesis and/or transformation of mammary epithelial cells *in vitro*. However, the potential role of oncogenes and tumor suppressor genes in the control of epithelial architecture has only recently begun to be elucidated[2,8]. To address this question, we have investigated whether repression of the putative tumor suppressor gene *Bard1* or expression of the Notch4 (int-3) oncogene in non-tumorigenic mammary epithelial cells affects their *in vitro* morphogenetic properties. For this purpose, we utilized TAC-2 clonal mammary epithelial cells, which when grown in three-dimensional collagen gels, have the ability to form branching tubules in response to HGF and alveolar-like cysts in response to hydrocortisone[11]. The studies summarized below indicate that both repression of *Bard1* and expression of Notch4(int-3) disrupt cyst morphogenesis and induce an invasive phenotype in TAC-2 mammary epithelial cells.

# 2. REPRESSION OF BARD1 (BRCA1-ASSOCIATED RING DOMAIN) GENE IN TAC-2 CELLS

BARD1 is a protein that interacts with the product of the breast cancer susceptibility gene *BRCA1*. The identification of *BARD1* mutations in human breast carcinomas has suggested that this protein participate in BRCA1-mediated tumor suppression[12]. However, whether *BARD1* plays a role in mammary gland morphogenesis is not known. To address this question, we reduced levels of *Bard1* expression in TAC-2 mammary epithelial cells. By transfecting TAC-2 cells with an antisense sequence of murine *Bard1*, we obtained a partial reduction of *Bard1* mRNA and protein expression levels. This resulted in marked phenotypic changes consisting of increased cell size and high frequency of multinucleated cells, when compared to mock-transfected TAC-2 cells.

Since disruption of normal glandular architecture is an early step in the process of mammary tumorigenesis, it was relevant to assess whether *Bard1* repression would affect the morphogenetic properties of TAC-2 cells. When grown in collagen gels under control conditions, mock-transfected TAC-2 cells gave rise to small colonies with a morphology ranging from irregular cell aggregates to poorly branched structures.

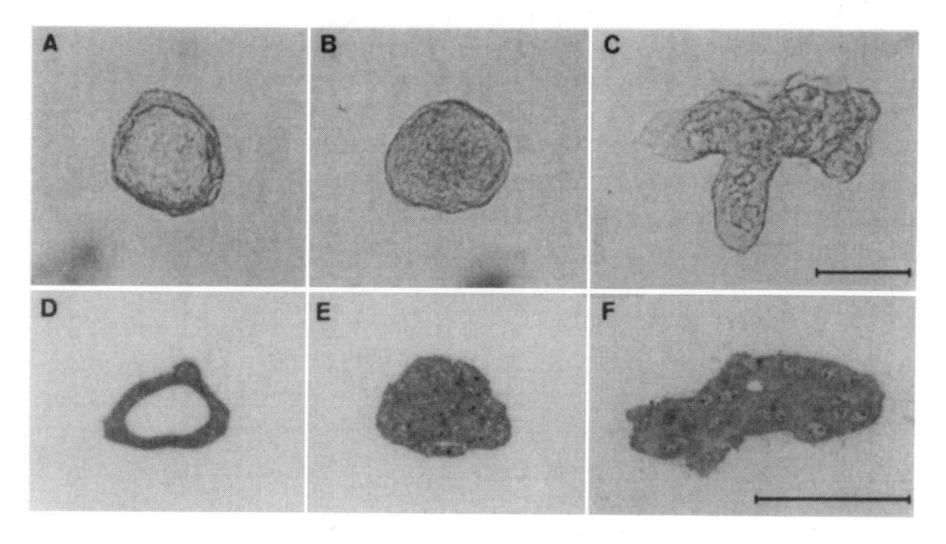


Fig. 1. Repression of *Bard1* abrogates hydrocortisone-induced lumen formation by TAC-2 cells. Cells suspended in collagen gels were incubated with 1 µg/ml hydrocortisone and 10 µg/ml insulin for 8 d. Under these conditions, mock-transfected cells form alveolar-like cystic structures containing a wide lumen (A) delimited by a palisade of cubic epithelial cells (D). In marked contrast, *Bard1* repressed cells form solid ball-like structures (B) devoid of lumen (E), or irregular aggregates (C) containing small focal lumina (F). A-C, Bright-field microscopy; D-F, semi thin sections. Bars, 100 µm. [Reproduced from the J. of Cell Biol. 143, 1329-1339, by copyright permission of the Rockefeller University Press.]

Addition of HGF to the cultures induced the formation of highly arborized branching cords, as previously observed with non-transfected TAC-2 cells[11]. Addition of HGF to collagen gel cultures of *Bard1* repressed cells induced branching tubulogenesis to a similar extent as in mock-transfected cells. These findings suggest that *Bard1* does not play a major role in the elongation and branching of duct-like structures by TAC-2 cells[4]. We next analyzed whether *Bard1* repression could influence lumen formation by TAC-2 cells. When grown in collagen gels in the presence of hydrocortisone, mock-transfected cells formed spheroidal cysts enclosing a widely patent lumen (Fig. 1A) delimited by a palisade of cubic epithelial cells (Fig. 1D), as previously observed with non-transfected TAC-2 cells[11]. In contrast, under the same experimental conditions, two different clones of *Bard1* repressed cells formed either solid aggregates devoid of lumen (Fig. 1B,E) or irregularly shaped cell aggregates occasionally containing small focal lumina (Fig. 1C,F). A quantitative analysis demonstrated a marked inhibition of lumen formation in *Bard1* repressed cells, as evidenced by a significant 6-fold and 8-fold ($p= 0.001$) decrease in the mean percentage of cysts, when compared to mock-transfected cells[4]. Taken together, these results demonstrate that repression of *Bard1* expression inhibits the formation of alveolar-like cystic structures by TAC-2 cells.

Since loss of contact-mediated inhibition of proliferation is a hallmark of malignant transformation, we analyzed the growth properties of mock-transfected and *Bard1* repressed TAC-2 cells. Although both cell lines showed a similar proliferation rate in subconfluent cultures, they had a markedly different behavior when grown to post-confluence. Whereas postconfluent mock-transfected cells formed regular, contact-inhibited monolayer, *Bard1* repressed cells overlapped each other and continued to grow in a disorganized criss-cross pattern. This differential behavior was further enhanced in the presence of exogenous growth factors. Thus, addition of either epidermal growth factor (EGF) or insulin-like growth factor-I (IGF-I) did not induce multilayering in mock-transfected cells, but resulted in a marked degree of stratification in *Bard1* repressed cells (Fig. 2).

This indicates that *Bard1* repressed cells have lost sensitivity to contact-inhibition of growth. However, *Bard1* repressed cells lacked the ability to grow under anchorage-independent conditions, and were not tumorigenic in vivo[4]. Altogether, our findings suggest that repression of *Bard1* induces a partially transformed phenotype in TAC-2 cells, consisting of disruption of normal alveolar-like morphogenesis and loss of contact-inhibition of cell growth. It remains to be established whether complete inhibition of *Bard1* expression could induce a fully malignant phenotype.

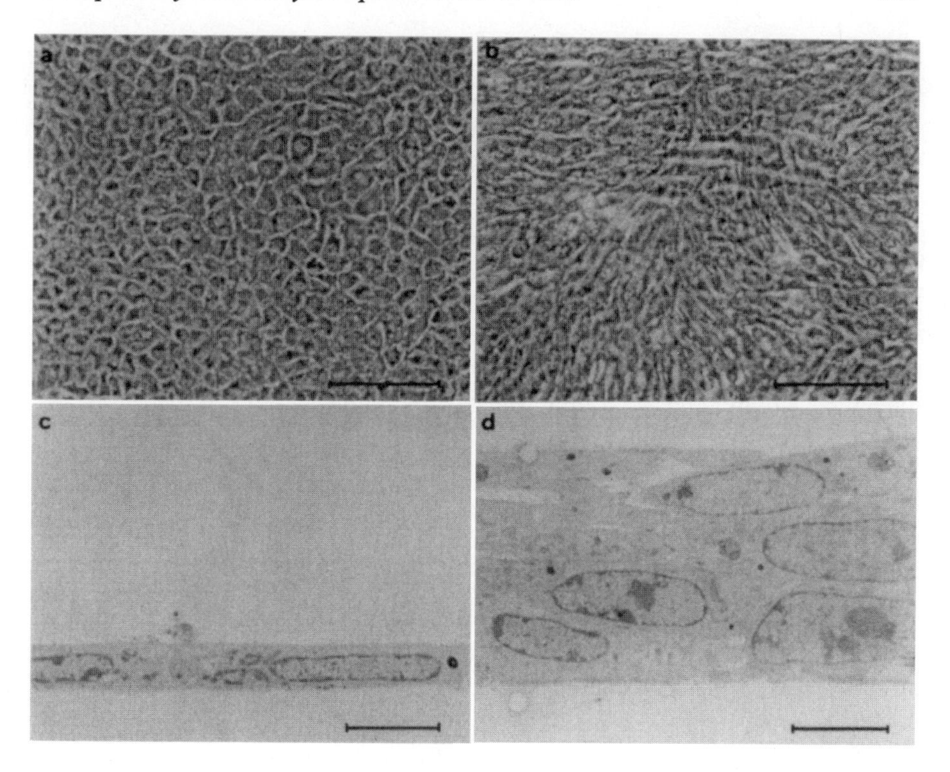


Fig. 2. Loss of contact-mediated inhibition of cell proliferation in *Bard1*-repressed cells. Cells were seeded in collagen-coated wells at saturating cell density and grown for 2 d, at which time the cultures were incubated in the absence or the presence of 20 ng/ml EGF for a further 5 d. In EGF-supplemented cultures, mock-transfected cells form a contact-inhibited cobblestone-like monolayer (a), whereas *Bard1* repressed cells (b) grow in a disordered crisscross pattern. (c and d) Thin sections perpendicular to the plane of cultures shown in a and b demonstrate the lack of stratification in mock-transfected cells and the obvious multilayering in *Bard1* repressed cells. Bars: (a and b) 100 μm; (c and d) 5 μm. [Reproduced from the Journal of Cell Biology, 143, 1329-1339, by copyright permission of the Rockefeller University Press.]

# 3. EXPRESSION OF AN ACTIVATED NOTCH4 (INT-3) ONCOPROTEIN IN TAC-2 CELLS

Genes of the Notch family encode transmembrane receptor proteins that regulate cell differentiation in a variety of cell types in both vertebrates and invertebrates[1]. Notch4 (int-3) is an oncogenic, truncated form of Notch4 which has most of its extracellular domain deleted and behaves as a constitutively activated receptor[7,13]. Expression of Notch4(int-3) as a transgene in the mouse impairs mammary gland ductal morphogenesis and promotes the formation of undifferentiated carcinomas[3,5,10]. To gain insight into the mechanisms that might be responsible for the tumorigenicity of this oncogene, we studied the biological consequences of Notch4(int-3) expression on the *in vitro* morphogenetic properties of TAC-2 cells. We found that retroviral expression of Notch4(int-3) in TAC-2 mammary epithelial cells impairs HGF-induced formation of branching tubules in collagen gels[13]. Interestingly, the structures formed by Notch4(int-3) expressing cells (TAC-2 int-3 cells) in collagen gels were not only less branched than those generated by mock-transfected TAC-2 cells, but were also frequently devoid of a central lumen. These observations led us to investigate whether expression of Notch4(int-3), in addition to inhibiting branching tubulogenesis, might specifically interfere with the process of lumen formation. When grown in collagen gels in the presence of hydrocortisone, mock-transfected TAC-2 cells formed alveolar-like cystic structures containing a wide lumen (cf. Fig. 1A,D), as previously observed with wild type TAC-2 cells[4,11]. In contrast, TAC-2 int-3 cells were virtually unable to form alveolar-like cystic colonies, and generated instead solid cell aggregates devoid of a central lumen. Thin section electron microscopy revealed that the wall of the cysts formed by mock-transfected cells consisted of closely apposed, well polarized epithelial cells. Intercellular junctions were located between the apical portions of adjacent cells, and microvilli were mostly restricted to the surface of the cell facing the lumen of the cyst (Fig. 3A). In marked contrast, the colonies formed by TAC-2 int-3 cells consisted of loosely associated epithelial cells lacking obvious cell polarity. The cells were separated by a wide intercellular space bridged by cytoplasmic processes establishing focal cell-cell contacts. Abundant polymorphic microvilli projected from the entire cell surface (except for regions of cell-collagen interactions), and no clearly defined apical pole could be identified (Fig. 3B)[14]. Taken together with the finding that TAC-2 int-3 cells fail to undergo branching tubulogenesis in response to HGF, these results demonstrate that activated Notch4(int-3) subverts the normal morphogenetic properties of TAC-2 cells. They also suggest that formation of disorganized cell aggregates by TAC-2 int-3 cells recapitulates the architectural destructuring observed in anaplastic carcinomas induced by transgenic expression of Notch4(int-3)[3,5,10].

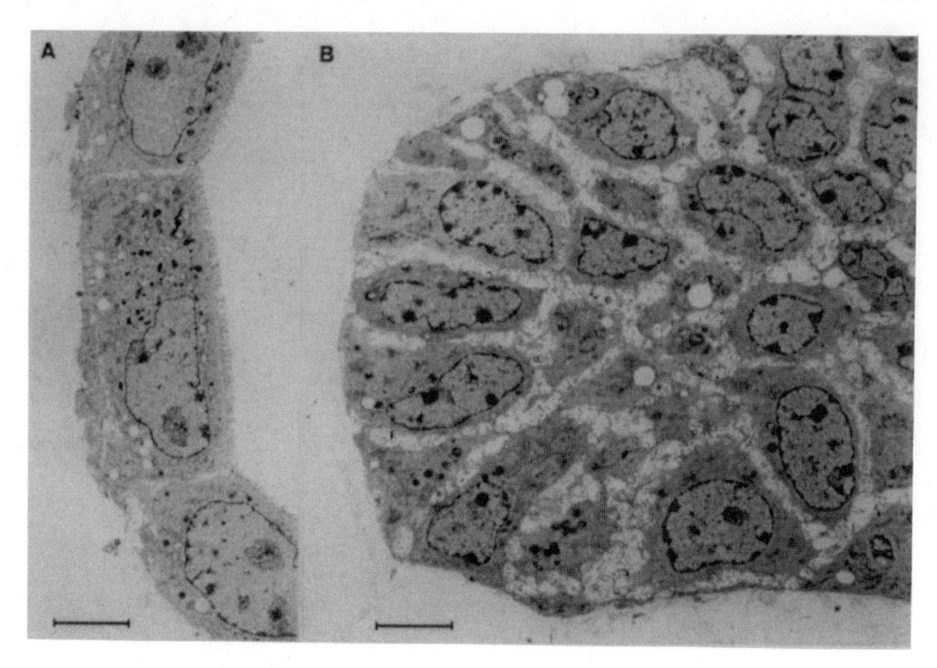


Fig. 3. Disruption of cyst morphogenesis and of apico-basal polarity by Notch4 (int-3). Thin sections of colonies formed in collagen gels by mock- and Notch4(int-3)-transfected TAC-2 cells incubated with hydrocortisone. (A) Wall of a cyst formed by mock-transfected cells. (B) Portion of a colony formed by TAC-2 int-3 cells. Bars, 5 μm. See text for details.

To determine whether expression of Notch4 (int-3) in TAC-2 cells induces additional changes that may be relevant to the process of mammary carcinogenesis, we evaluated the ability of TAC-2 cells to invade three-dimensional collagen gels. When grown on a collagen gel under control conditions, mock-transfected cells formed a cobblestone-like monolayer and remained exclusively confined to the surface of the gel. Addition of EGF to the cultures did not induce appreciable invasion (Fig. 4A,C). In contrast, TAC-2 int-3 cells invaded the underlying collagen matrix as thick branched cords. Invasion was enhanced by the addition of EGF to the cultures (Fig. 4B,D) and was suppressed in a dose-dependent manner by a synthetic inhibitor of matrix metalloproteinases. These results demonstrate that expression of Notch4(int-3) in TAC-2 cells is associated with the acquisition of invasive properties.

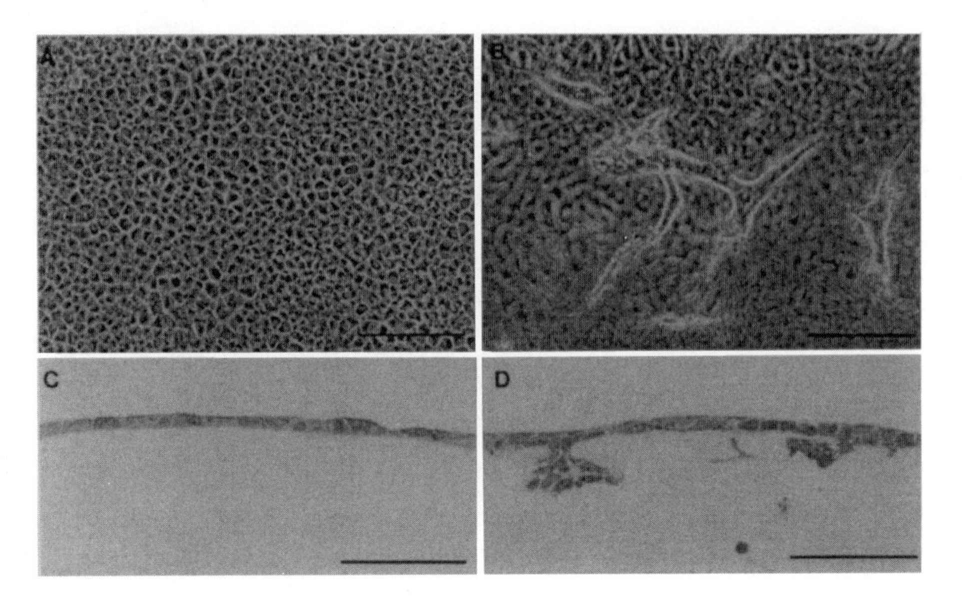


Fig. 4. Expression of Notch4 (int-3) confers an invasive phenotype to TAC-2 cells. TAC-2 cells were seeded onto the surface of a collagen gel and grown for 6 days in the presence of 20 ng/ml EGF (A,B) or 20 ng/ml EGF and 10μg/ml insulin (C,D). Whereas mock-transfected cells (A,C) form a monolayer exclusively confined to the surface of the gel, TAC-2 int-3 cells (B,D) invade the underlying matrix as thick branched cords. (A,B) Phase-contrast microscopy (focus is 20μm beneath the surface of the gel); bars, 150 μm. (C,D) Semi-thin sections perpendicular to the surface of the gel; bars, 100 μm.

We next investigated whether Notch4 (int-3) expression may alter contact-inhibition of cell proliferation. We found that whereas mock-transfected cells form cobblestone-like, contact-inhibited monolayers, TAC-2 int-3 cells grow in a disorganized criss-cross pattern and overlap each other in post-confluent cultures. These findings suggest that Notch4(int-3) blunt the responsiveness of TAC-2 cells to the normal regulatory mechanisms responsible for contact-inhibition of proliferation, thereby maintaining them in a hyperproliferative state. However, we found that TAC-2 int-3 cells are unable to form colonies when grown under anchorage-independent conditions. Thus, while exhibiting some properties of transformed cells (i.e., disorganized growth in collagen gels and cell multilayering in two-dimensional cultures), TAC-2 int-3 cells are still dependent on cell-substratum adhesion for proliferation.

Taken together, our results demonstrate that expression of Notch4 (int-3) in TAC-2 cells disrupts their ability to form organized lumen-containing structures, confers upon them an invasive phenotype, and suppresses contact-inhibition of cell proliferation. These alterations may represent important contributing factors in the development of anaplastic carcinomas observed in Notch4(int-3) transgenic mice[3,5,10].

# 4. CONCLUSION

The studies we have summarized above suggest that subversion of the normal morphogenetic properties of mammary epithelial cells represent an important mechanism by which loss of tumor suppressors or expression of oncogenes promotes malignant progression of epithelial tumors.

# 5. REFERENCES

1. Artavanis-Tsakonas S, Rand MD, Lake RJ. Notch signaling: cell fate control and signal integration in development. Science, 284, 770-776, 1999.
2. Fialka, I., Schwarz, H., Reichmann, E., Oft, M., Busslinger, M., and Beug, H. The estrogen-dependent c-JunER protein causes a reversible loss of mammary epithelial cell polarity involving a destabilization of adherens junctions. J Cell Biol 132, 1115-1132, 1996.
3. Gallahan D, Jhappan C, Robinson G, Hennighausen L, Sharp R, Kordon E, Callahan R, Merlino G, Smith GH. Expression of a truncated Int3 gene in developing secretory mammary epithelium specifically retards lobular differentiation resulting in tumorigenesis. Cancer Res 56, 1775-1785, 1996.
4. Irminger-Finger I, Soriano JV, Vaudan G, Montesano R, Sappino AP. In vitro repression of Brca1-associated RING domain gene, *Bard1*, induces phenotypic changes in mammary epithelial cells. J Cell Biol 143, 1329-1339, 1998.
5. Jhappan C, Gallahan D, Stahle C, Chu E, Smith GH, Merlino G, Callahan R. Expression of an activated Notch-related int-3 transgene interferes with cell differentiation and induces neoplastic transformation in mammary and salivary glands. Genes Dev 6, 345-355, 1992.
6. Petersen OW, Ronnov-Jessen L, Howlett A R, Bissell MJ. Interaction with basement membrane serves to rapidly distinguish growth and differentiation pattern of normal and malignant human breast epithelial. Proc Natl Acad Sci USA 89, 9064-9068, 1992.
7. Robbins J, Blondel BJ, Gallahan D, Callahan R. Mouse mammary tumor gene int-3: a member of the notch gene family transforms mammary epithelial cells. J Virol 66, 2594-2599, 1992.
8. Reichmann, E., Schwarz, H., Deiner, E. M., Leitner, I., Eilers, M., Berger, J., Busslinger, and M., Beug, H. Activation of an inducible c-FosER fusion protein causes loss of epithelial polarity and triggers epithelial-fibroblastoid cell conversion. Cell 71: 1103-1116, 1992.
9. Shearer M, Bartkova J, Bartek J, Berdichevsky F, Barnes D, Millis R, Taylor-Papadimitriou J. Studies of clonal cell lines developed from primary breast cancers indicate that the ability to undergo morphogenesis in vitro is lost early in malignancy. Int J Cancer 51, 602-612, 1992.

10. Smith GH, Gallahan D, Diella F, Jhappan C, Merlino G, Callahan R. Constitutive expression of a truncated INT3 gene in mouse mammary epithelium impairs differentiation and functional development. Cell Growth Differ 6, 563-577, 1995.
11. Soriano, J. V., Pepper, M. S., Nakamura, T., Orci, L., and Montesano, R. Hepatocyte growth factor stimulates extensive development of branching duct-like structures by cloned mammary gland epithelial cells. J Cell Sci 108, 413-430, 1995.
12. Thai TH, Du F, Tsan JT, Jin Y, Phung A, Spillmann MA, Massa HF, Muller CY, Ashfaq R, Mathis JM, Miller DS, Trask BJ, Baer R, Bowcock AM. Mutations in the Brca1-associated RING domain (BARD1) gene in primary breast, ovarian, and uterine cancers. Hum Mol Genet 7, 195-202, 1998
13. Uyttendaele H, Soriano JV, Montesano R, Kitajewski J. Notch4 and Wnt-1 proteins function to regulate branching morphogenesis of mammary epithelial cells in an opposing fashion. Dev Biol 196, 204-217, 199
14. Soriano JV, Uyttendaele H, Kitajewski J, Montesano R. Expression of an activated Notch4 (int-3) oncoprotein disrupts morphogenesis and induces an invasive phenotype in mammary epithelial cells in vitro. Manuscript submitted

23.

# Oncogene Mediated Signal Transduction in Transgenic Mouse Models of Human Breast Cancer

Peter M. Siegel, David L. Dankort and William J. Muller
*The Institute for Molecular Biology & Biotechnology, McMaster University, Hamilton, Ontario*

Key words: oncogenes, mouse

The ability of mammary epithelial cells to respond growth factor is dependent on specific growth factor receptors coupled to number of intracellular signaling pathways. Of relevance to this proposal, the development, maturation and differentiation of the mammary epithelial cell are dependent on the interplay of both hormones and growth factors. The development of the mammary gland is thought to involve a series of defined steps consisting of cell proliferation, differentiation and programmed cell death (apoptosis). After of the formation the primary mammary tree from its embryonic rudiment, there is a rapid expansion of ductal outgrowth through the mammary fat pad which is accompanied by the formation mammary terminal end buds (TEBs). By 10 weeks of age the mammary epithelium has reached the end of the fat pad and ceases further ductal outgrowth[1]. Following pregnancy, a further rapid expansion of the lobuloalveolar epithelium occurs leading to induction of terminal differentiation and lactation at birth[1,2]. After the pups have been weaned from the lactating mother, the mammary epithelium undergoes a rapid involution through the induction of programmed cell death (apoptosis)[3]. The balance of soluble growth factors, hormones and cell-substratum interactions controls the regulation of this cycle of proliferation, differentiation and apoptosis.

Mammary epithelial growth and differentiation is regulated by a number by growth factors and growth factor receptors. For example, implantation in the mammary fat pad of EGFR family ligands including EGF can dramatically accelerate lobuloalveolar development in virgin mammary glands[4]. Consistent with these observations, ectopic expression of another

EGFR ligand ,TGFα, in the mammary glands of transgenic mice results in dramatic impairment of normal mammary gland involution[5-8]. Similarly mammary epithelial specific expression of the ErbB-3 and ErbB-4 ligand, heregulin, leads to the induction of increased number of terminal end bud hyperplasias[9]. Conversely, germline inactivation of EGFR ligand amphiregulin results in a dramatic impairment of normal mammary ductal development[10].

The ability of growth factors to modulate proliferation, differentiation and apoptosis is thought to occur through activation of their cognate receptors and coupled signal transduction pathways. Indeed, there is evidence to suggest that alteration of growth factor receptor activity can have a dramatic effect on normal mammary gland development. For example, a naturally occurring mouse mutant (Waved-2 mice) carrying a catalytically inactive epidermal growth factor receptor (EGFR), exhibit a impaired mammary gland development and a severe lactation defect[11,12]. Ultimately, the activation of these growth factor receptors such as the EGFR family leads to the recruitment of a number of cytoplasmic signaling molecules. For example, stimulation of the intrinsic tyrosine activity of these growth factor receptors results in the tyrosine phosphorylation of specific phosphotyrosine residues within these receptors. Depending on the sequence context of these phosphotyrosine residues, specific src homology 2 (SH2) or phosphotyrosine binding/interaction domain (PTB) containing proteins can then be recruited to the activated growth factor receptor.

One class of SH2 containing proteins that are thought to play a critical role in RTK signal transduction is the phosphatidylinositol 3í (PI-3í) kinase. Genetic and biochemical analyses of PI-3í kinase has revealed that its recruitment to the inner face of the cytoplasmic membrane by its association with activated growth factor receptors results in the generation of phosinositide 3í lipids[13]. With respect to erbB-2, stimulation of the PI-3í kinase signaling pathway is dependent on its heterodimerization with erbB-3 which carries 6 consensus binding sites for the p85 subunit of the PI-3í kinase[14,15]. The lipid products generated by the PI-3í kinase are involved in the recruitment to the membrane of a number pleckstrin homology domain containing protein serine kinases including PDK1, ILK and Akt/PKB[16-21]. Following activation of PDK1 and ILK, Akt/PKB is phosphorylated on two distinct phosphorylation sites by these serine kinases and becomes enzymatically activated[18,20].

There is growing body of evidence that activation of the PI-3í/Akt kinase axis an have profound effects cell survival pathways. For example, activation of PI-3í kinase pathway has been implicated in preventing apoptotic cell death in a number of different tissue systems including the mammary epithelial cell[22-29]. Ultimately, activation of PI-3í kinase targets such as Akt result in inhibition of pro-apoptotic proteins such as Bad, Caspase-9 and the forkhead transcription factor family[30-32]. In addition, activation of the PI-3í kinase signaling pathway can activate transcription factors such as NF kappa beta that are involved in anti-apoptotic cellular response[33].

Growth factor receptor mediated activation of the PI-3í kinase can also have profound effects on epithelial cell morphology. Activation of either Met or erbB-2 receptor in MDCK cells gown in matrigel can induce branching morphogenesis that is sensitive to specific inhibitors of the PI-3í kinase. Moreover stimulation of either primary cells[34] or EpH4 cells[35] with hepatocyte growth factor (HGF), the ligand for the Met receptor results in branching morphogenesis without functional differentiation. The HGF branching morphogenesis signal was dependent on activation of the PI-3í kinase signaling axis[35]. Taken together these observations suggest that activation of the PI-3í signaling axis plays critical role in both mammary cell morphogenesis and survival responses.

Another signaling pathway that plays a critical role in modulating the epithelial proliferation is the activation of the Ras signaling pathway. In particular SH2 and PTB domain containing proteins such as Grb2 and Shc have been demonstrated to form specific complexes with these growth factor receptors and in turn activate the Ras signaling pathway[36-46]. Direct interaction of Grb2 adapter protein with activated growth factor receptor through it SH2 domain results in recruitment and stimulation of the guanine nucleotide exchange SOS to convert Ras from its inactive GDP bound state to its active GTP form[42, 46-48]. In addition to directly recruiting the Grb2/SOS complex, activated growth factor receptors can bind this complex indirectly through its association with the Shc adapter protein through specific tyrosine phosphorylation sites located within Shc[39,49-56]. More recently, several groups have suggested that in addition to recruitment of the Grb2/SOS complex, the interaction activated growth factor receptors with Shc can result in the stimulation of a Ras independent pathway involving the transcriptional upregulation of c-Myc transcription factor[54,56,57]. Activation of the Ras signaling pathway in turn can result in stimulation of a cascade of serine kinases including Raf, MEK and MAP kinase[58-61]. Ultimately, the downstream targets of these activated kinase cascades are nuclear transcription factors such as Fos, Jun and Ets proteins[62-65].

There is increasing body of evidence that receptor tyrosine kinases (RTKs) such as *erbB-2* can activate the Ras signaling pathway via multiple functionally redundant adapter proteins. Indeed, we have previously demonstrated that any one of four *erbB-2* tyrosine autophosphorylation sites (tyrosine 1144, 1201, 1227 and 1253) can independently mediate oncogenic transformation through activation of Ras signaling pathway[37]. We have further identified the tyrosine 1144 binding partner as Grb2 adapter protein and tyrosine 1227 as the Shc adapter protein[37]. In addition, to autophosphorylation sites that positively modulate *erbB-2* transformation, we identified at one tyrosine autophosphorylation site (1028) that negatively regulates *erbB-2* mediated transformation.

Another event that accompanies activation of activated growth factor receptors is stimulation of the Src family of tyrosine kinases. In fact RTKs such as Met and erbB-2 are known to associate with Src family kinases through specific phosphotyrosine residues within the receptor[66-71]. As a result of this physical interaction of RTKs with Src family kinase the specific activity of the src family kinases is increased[66,67,72]. Elevated Src family kinase activity can influence both cell cycle progression through its stimulation of cyclin D1[73] and cell substratum interaction via activation of the focal adhesion kinase[74-78]. Together these observations suggest that growth factor activation Src kinase family member play a critical role in mediating cell proliferation and migration.

Whereas it is clear that these signaling pathways are involved in the transduction of proliferative and differentiation signals from these cell surface receptors, the precise role of these signaling pathways play in mammary tumorigenesis is unclear. One of the important genetic changes tat are implicated in the genesis of human breast cancer is amplification and elevated expression of erbB-2/neu oncogene[79-81]. To more rigorously define the importance of *erbB-2/neu* expression in mammary tumorigenesis, transgenic mice have been generated which express the activated *neu* cDNA under the control of the mouse mammary tumor virus - long terminal repeat (MMTV-LTR) [82-84]. In the first report, all the female carriers from one transgenic line (TG.NF) demonstrated global mammary transformation involving all glands after 95 days[84]. The appearance of mammary gland abnormalities coincided with regions of transgene expression as determined by *in situ* hybridization. An independent transgenic line (TG.NK), characterized in the same study, demonstrated a non-uniform pattern of transgene expression that translated into a stochastic tumor phenotype in which mammary tumors arose next to morphologically normal epithelium. The phenotypes displayed by these two lines of transgenic mice suggested that expression of activated *neu* requires few, if any, additional events to transform the primary epithelial cell. In contrast, other transgenic strains carrying an MMTV/activated *neu* transgene develop focal mammary tumors that arose asynchronously in animals between 5 to 10 months of age[83]. The discrepancy in the phenotypes displayed by these different strains may reflect differences in the levels of activated Neu protein expressed in the mammary epithelium of the different strains. For example, transformation of the primary epithelial cell may require a critical threshold of activated Neu expression, which is not achieved in those animals exhibiting the stochastic tumor phenotype. Regardless of the explanation for such differences, these studies argue that activated *neu* can effectively transform the mammary epithelium of transgenic mice.

The strong transforming ability demonstrated by the *neu* oncogene in the mammary epithelium has prompted several groups to examine whether a similar point mutation within *erb*B-2 might occur in human breast tumors. Screens based on direct sequencing, sequence-specific oligonucleotide hybridization[85] or differential endonuclease digestion[86] have failed to detect the presence of the analogous mutation within the transmembrane domain of ErbB-2 in human breast cancers. The lack of activating mutations in the transmembrane domain, coupled with clinical observations suggesting that *erbB-2* is often amplified in human tumors, suggests that the primary mechanism operating in human breast cancer is overexpression of the normal, unaltered, receptor. To directly test the oncogenic potential of wild-type *neu* in the mammary epithelium, several strains of transgenic mice expressing the wild-type *neu* cDNA under the control of the MMTV-LTR have been established[87]. These animals develop focal mammary tumors with an average latency of approximately 7 months that frequently metastasize to the lung. Expression analysis of these mice revealed that in many cases the level of transgene RNA was significantly higher in the tumor than the surrounding epithelium, although in certain instances both the tumor and adjacent tissue expressed similar levels of the *neu* transgene. In animals belonging to the latter group, Neu protein was also detected in both tumor and adjacent mammary epithelium. However, analysis of the receptor's catalytic activity within these tissues revealed that the tumors alone exhibited elevated levels of Neu tyrosine kinase activity. Further characterization of these mice have demonstrated that mutations in the *neu* transgene are responsible for activation of the receptor in a majority of these mammary tumors[87-90].

Biochemical and genetic analyses of these alterations have revealed that these mutations involve a conserved set of cysteine residues located in the juxta transmembrane domain of erbB-2. Indeed, we have demonstrated that these alterations involve either a addition or deletion of cysteine residue within this region that result in the formation of disulfide bonded dimers of erbB-2 that are constitutively activated[88-90]. More recently, we have directly demonstrated that mammary-epithelial expression of these activated erbB-2 alleles leads to the rapid induction of metastatic mammary tumors[89]. In addition to the to the elevated expression of these activated *neu* alleles, we have observed a dramatic upregulation of the erbB-3 protein during tumorigenesis. Interestingly, the elevated expression of erbB-3 does not involve increases in erbB-3 transcript but rather appears to occur at either the level erbB-3 translation or stability[89]. Consistent with these transgenic studies, a survey of a number of erbB-2 expressing breast cancers revealed that 8/9 co-expressed erbB-3. Taken together, these observations argue that coexpression of erbB-3 and neu are critical events in mammary tumor progression.

## REFERENCES


1. Hennighausen, L. and G.W. Robinson, Think globally, act locally: the making of a mouse mammary gland. Genes Dev, 1998. **12**(4): 449-55.
2. Vonderhaar, B.K. and A.E. Greco, Lobulo-alveolar development of mouse mammary glands is regulated by thyroid hormones. Endocrinology, 1979. **104**(2): 409-18.
3. Humphreys, R.C., et al., Apoptosis in the terminal endbud of the murine mammary gland: a mechanism of ductal morphogenesis. Development, 1996. **122**(12): 4013-22.
4. Coleman, S., G.B. Silberstein, and C.W. Daniel, Ductal morphogenesis in the mouse mammary gland: evidence supporting a role for epidermal growth factor. Dev Biol, 1988. **127**(2): 304-15.
5. Sandgren, E.P., et al., Overexpression of TGF alpha in transgenic mice: induction of epithelial hyperplasia, pancreatic metaplasia, and carcinoma of the breast. Cell, 1990. **61**(6): 1121-35.
6. Sandgren, E.P., et al., Inhibition of mammary gland involution is associated with transforming growth factor alpha but not c-myc-induced tumorigenesis in transgenic mice. Cancer Res, 1995. **55**(17): 3915-27.
7. Matsui, Y., et al., Development of mammary hyperplasia and neoplasia in MMTV-TGF alpha transgenic mice. Cell, 1990. **61**(6): 1147-55.
8. Jhappan, C., et al., TGF alpha overexpression in transgenic mice induces liver neoplasia and abnormal development of the mammary gland and pancreas. Cell, 1990. **61**(6): 1137-46.
9. Krane, I.M. and P. Leder, NDF/heregulin induces persistence of terminal end buds and adenocarcinomas in the mammary glands of transgenic mice. Oncogene, 1996. **12**(8): 1781-8.
10. Luetteke, N.C., et al., Targeted inactivation of the EGF and amphiregulin genes reveals distinct roles for EGF receptor ligands in mouse mammary gland development. Development, 1999. **126**(12): 2739-50.
11. Fowler, K.J., et al., A mutation in the epidermal growth factor receptor in waved-2 mice has a profound effect on receptor biochemistry that results in impaired lactation. Proc Natl Acad Sci U S A, 1995. **92**(5): 1465-9.
12. Sebastian, J., et al., Activation and function of the epidermal growth factor receptor and erbB-2 during mammary gland morphogenesis. Cell Growth Differ, 1998. **9**(9): 777-85.
13. Whitman, M., et al., Type I phosphatidylinositol kinase makes a novel inositol phospholipid, phosphatidylinositol-3-phosphate. Nature, 1988. **332**(6165): 644-6.
14. Prigent, S.A. and W.J. Gullick, Identification of c-erbB-3 binding sites for phosphatidylinositol 3'- kinase and SHC using an EGF receptor/c-erbB-3 chimera. Embo J, 1994. **13**(12): 2831-41.
15. Soltoff, S.P., et al., ErbB3 is involved in activation of phosphatidylinositol 3-kinase by epidermal growth factor. Mol Cell Biol, 1994. **14**(6): 3550-8.

16. Alessi, D.R., et al., Characterization of a 3-phosphoinositide-dependent protein kinase which phosphorylates and activates protein kinase Balpha. Curr Biol, 1997. **7**(4): 261-9.
17. Alessi, D.R., et al., 3-Phosphoinositide-dependent protein kinase-1 (PDK1): structural and functional homology with the Drosophila DSTPK61 kinase. Curr Biol, 1997. **7**(10): 776-89.
18. Anderson, K.E., et al., Translocation of PDK-1 to the plasma membrane is important in allowing PDK-1 to activate protein kinase B. Curr Biol, 1998. **8**(12): 684-91.
19. Currie, R.A., et al., Role of phosphatidylinositol 3,4,5-trisphosphate in regulating the activity and localization of 3-phosphoinositide-dependent protein kinase-1. Biochem J, 1999. **337**(Pt 3): 575-83.
20. Delcommenne, M., et al., Phosphoinositide-3-OH kinase-dependent regulation of glycogen synthase kinase 3 and protein kinase B/AKT by the integrin-linked kinase. Proc Natl Acad Sci U S A, 1998. **95**(19): 11211-6.
21. Watton, S.J. and J. Downward, Akt/PKB localisation and 3' phosphoinositide generation at sites of epithelial cell-matrix and cell-cell interaction. Curr Biol, 1999. **9**(8): 433-6.
22. Gibson, S., et al., Epidermal growth factor protects epithelial cells against Fas-induced apoptosis. Requirement for Akt activation. J Biol Chem, 1999. **274**(25): 17612-8.
23. Kennedy, S.G., et al., Akt/Protein kinase B inhibits cell death by preventing the release of cytochrome c from mitochondria [In Process Citation]. Mol Cell Biol, 1999. **19**(8): 5800-10.
24. Khwaja, A., et al., Matrix adhesion and Ras transformation both activate a phosphoinositide 3-OH kinase and protein kinase B/Akt cellular survival pathway. Embo J, 1997. **16**(10): 2783-93.
25. Rodriguez-Viciana, P., et al., Role of phosphoinositide 3-OH kinase in cell transformation and control of the actin cytoskeleton by Ras. Cell, 1997. **89**(3): 457-67.
26. Stambolic, V., et al., Negative regulation of PKB/Akt-dependent cell survival by the tumor suppressor PTEN. Cell, 1998. **95**(1): 29-39.
27. Staveley, B.E., et al., Genetic analysis of protein kinase B (AKT) in Drosophila. Curr Biol, 1998. **8**(10): 599-602.
28. Sun, H., et al., PTEN modulates cell cycle progression and cell survival by regulating phosphatidylinositol 3,4,5,-trisphosphate and Akt/protein kinase B signaling pathway. Proc Natl Acad Sci U S A, 1999. **96**(11): 6199-204.
29. Webster, M.A., et al., Requirement for both Shc and phosphatidylinositol 3' kinase signaling pathways in polyomavirus middle T-mediated mammary tumorigenesis. Mol Cell Biol, 1998. **18**(4): 2344-59.
30. Datta, S.R., et al., Akt phosphorylation of BAD couples survival signals to the cell-intrinsic death machinery. Cell, 1997. **91**(2): 231-41.
31. Cardone, M.H., et al., Regulation of cell death protease caspase-9 by phosphorylation. Science, 1998. **282**(5392): 1318-21.
32. Brunet, A., et al., Akt promotes cell survival by phosphorylating and inhibiting a Forkhead transcription factor. Cell, 1999. **96**(6): 857-68.
33. Kane, L.P., et al., Induction of NF-kappaB by the Akt/PKB kinase. Curr Biol, 1999. **9**(11): 601-4.
34. Yang, Y., et al., Sequential requirement of hepatocyte growth factor and neuregulin in the morphogenesis and differentiation of the mammary gland. J Cell Biol, 1995. **131**(1): 215-26.
35. Niemann, C., et al., Reconstitution of mammary gland development in vitro: requirement of c- met and c-erbB2 signaling for branching and alveolar morphogenesis. J Cell Biol, 1998. **143**(2): 533-45.

36. Janes, P.W., et al., Activation of the Ras signalling pathway in human breast cancer cells overexpressing erbB-2. Oncogene, 1994. **9**(12): 3601-8.
37. Dankort, D.L., et al., Distinct tyrosine autophosphorylation sites negatively and positively modulate neu-mediated transformation. Mol Cell Biol, 1997. **17**(9): 5410-25.
38. Vijapurkar, U., K. Cheng, and J.G. Koland, Mutation of a Shc binding site tyrosine residue in ErbB3/HER3 blocks heregulin-dependent activation of mitogen-activated protein kinase. J Biol Chem, 1998. **273**(33): 20996-1002.
39. Pelicci, G., et al., Constitutive phosphorylation of Shc proteins in human tumors. Oncogene, 1995. **11**(5): 899-907.
40. Daly, R.J., M.D. Binder, and R.L. Sutherland, Overexpression of the Grb2 gene in human breast cancer cell lines. Oncogene, 1994. **9**(9): 2723-7.
41. Rozakis-Adcock, M., et al., Association of the Shc and Grb2/Sem5 SH2-containing proteins is implicated in activation of the Ras pathway by tyrosine kinases. Nature, 1992. **360**(6405): 689-92.
42. Lowenstein, E.J., et al., The SH2 and SH3 domain-containing protein GRB2 links receptor tyrosine kinases to ras signaling. Cell, 1992. **70**(3): 431-42.
43. Skolnik, E.Y., et al., The function of GRB2 in linking the insulin receptor to Ras signaling pathways. Science, 1993. **260**(5116): 1953-5.
44. Rozakis-Adcock, M., et al., The SH2 and SH3 domains of mammalian Grb2 couple the EGF receptor to the Ras activator mSos1 [see comments]. Nature, 1993. **363**(6424): 83-5.
45. Egan, S.E., et al., Association of Sos Ras exchange protein with Grb2 is implicated in tyrosine kinase signal transduction and transformation [see comments]. Nature, 1993. **363**(6424): 45-51.
46. Gale, N.W., et al., Grb2 mediates the EGF-dependent activation of guanine nucleotide exchange on Ras [see comments]. Nature, 1993. **363**(6424): 88-92.
47. Li, N., et al., Guanine-nucleotide-releasing factor hSos1 binds to Grb2 and links receptor tyrosine kinases to Ras signalling [see comments]. Nature, 1993. **363**(6424): 85-8.
48. Chardin, P., et al., Human Sos1: a guanine nucleotide exchange factor for Ras that binds to GRB2. Science, 1993. **260**(5112): 1338-43.
49. Skolnik, E.Y., et al., The SH2/SH3 domain-containing protein GRB2 interacts with tyrosine- phosphorylated IRS1 and Shc: implications for insulin control of ras signalling. Embo J, 1993. **12**(5): 1929-36.
50. Yamauchi, T., et al., Tyrosine phosphorylation of the EGF receptor by the kinase Jak2 is induced by growth hormone. Nature, 1997. **390**(6655): 91-6.
51. Pelicci, G., et al., A family of Shc related proteins with conserved PTB, CH1 and SH2 regions. Oncogene, 1996. **13**(3): 633-41.
52. Segatto, O., et al., Shc products are substrates of erbB-2 kinase. Oncogene, 1993. **8**(8): 2105-12.
53. Hashimoto, A., et al., Shc Regulates Epidermal Growth Factor-induced Activation of the JNK Signaling Pathway. J Biol Chem, 1999. **274**(29): 20139-20143.
54. Gotoh, N., M. Toyoda, and M. Shibuya, Tyrosine phosphorylation sites at amino acids 239 and 240 of Shc are involved in epidermal growth factor-induced mitogenic signaling that is distinct from Ras/mitogen-activated protein kinase activation. Mol Cell Biol, 1997. **17**(4): 1824-31.
55. Harmer, S.L. and A.L. DeFranco, Shc contains two Grb2 binding sites needed for efficient formation of complexes with SOS in B lymphocytes. Mol Cell Biol, 1997. **17**(7): 4087-95.

56. van der Geer, P., et al., The Shc adaptor protein is highly phosphorylated at conserved, twin tyrosine residues (Y239/240) that mediate protein-protein interactions. Curr Biol, 1996. **6**(11): 1435-44.
57. Gotoh, N., A. Tojo, and M. Shibuya, A novel pathway from phosphorylation of tyrosine residues 239/240 of Shc, contributing to suppress apoptosis by IL-3. Embo J, 1996. **15**(22): 6197-204.
58. Jelinek, T., et al., RAS and RAF-1 form a signalling complex with MEK-1 but not MEK-2. Mol Cell Biol, 1994. **14**(12): 8212-8.
59. Moodie, S.A., et al., Association of MEK1 with p21ras.GMPPNP is dependent on B-Raf. Mol Cell Biol, 1994. **14**(11): 7153-62.
60. Moodie, S.A. and A. Wolfman, The 3Rs of life: Ras, Raf and growth regulation. Trends Genet, 1994. **10**(2): 44-8.
61. Marshall, M.S., Ras target proteins in eukaryotic cells. Faseb J, 1995. **9**(13): 1311-8.
62. Whitmarsh, A.J., et al., Integration of MAP kinase signal transduction pathways at the serum response element. Science, 1995. **269**(5222): 403-7.
63. Deng, T. and M. Karin, c-Fos transcriptional activity stimulated by H-Ras-activated protein kinase distinct from JNK and ERK. Nature, 1994. **371**(6493): 171-5.
64. Hipskind, R.A., M. Baccarini, and A. Nordheim, Transient activation of RAF-1, MEK, and ERK2 coincides kinetically with ternary complex factor phosphorylation and immediate-early gene promoter activity in vivo. Mol Cell Biol, 1994. **14**(9): 6219-31.
65. Treisman, R., Ternary complex factors: growth factor regulated transcriptional activators. Curr Opin Genet Dev, 1994. **4**(1): 96-101.
66. Luttrell, D.K., et al., Involvement of pp60c-src with two major signaling pathways in human breast cancer. Proc Natl Acad Sci U S A, 1994. **91**(1): 83-7.
67. Muthuswamy, S.K., et al., Mammary tumors expressing the neu proto-oncogene possess elevated c-Src tyrosine kinase activity. Mol Cell Biol, 1994. **14**(1): 735-43.
68. Muthuswamy, S.K. and W.J. Muller, Activation of the Src family of tyrosine kinases in mammary tumorigenesis. Adv Cancer Res, 1994. **64**: 111-23.
69. Muthuswamy, S.K. and W.J. Muller, Activation of Src family kinases in Neu-induced mammary tumors correlates with their association with distinct sets of tyrosine phosphorylated proteins in vivo. Oncogene, 1995. **11**(9): 1801-10.
70. Muthuswamy, S.K. and W.J. Muller, Direct and specific interaction of c-Src with Neu is involved in signaling by the epidermal growth factor receptor. Oncogene, 1995. **11**(2): 271-9.
71. Ponzetto, C., et al., A multifunctional docking site mediates signaling and transformation by the hepatocyte growth factor/scatter factor receptor family. Cell, 1994. **77**(2): 261-71.
72. Courtneidge, S.A., et al., Activation of Src family kinases by colony stimulating factor-1, and their association with its receptor. Embo J, 1993. **12**(3): 943-50.
73. Lee, R.J., et al., pp60(v-src) induction of cyclin D1 requires collaborative interactions between the extracellular signal-regulated kinase, p38, and Jun kinase pathways. A role for cAMP response element-binding protein and activating transcription factor-2 in pp60(v-src) signaling in breast cancer cells. J Biol Chem, 1999. **274**(11): 7341-50.
74. Calalb, M.B., T.R. Polte, and S.K. Hanks, Tyrosine phosphorylation of focal adhesion kinase at sites in the catalytic domain regulates kinase activity: a role for Src family kinases. Mol Cell Biol, 1995. **15**(2): 954-63.
75. Klinghoffer, R.A., et al., Src family kinases are required for integrin but not PDGFR signal transduction. Embo J, 1999. **18**(9): 2459-71.
76. Lo, S.S., et al., Inhibition of focal contact formation in cells transformed by p185neu. Mol Carcinog, 1999. **25**(2): 150-4.

77. Oktay, M., et al., Integrin-mediated activation of focal adhesion kinase is required for signaling to Jun NH2-terminal kinase and progression through the G1 phase of the cell cycle. J Cell Biol, 1999. **145**(7): 1461-9.
78. Schaller, M.D., et al., Autophosphorylation of the focal adhesion kinase, pp125FAK, directs SH2- dependent binding of pp60src. Mol Cell Biol, 1994. **14**(3): 1680-8.
79. Andrulis, I.L., et al., neu/erbB-2 amplification identifies a poor-prognosis group of women with node-negative breast cancer. Toronto Breast Cancer Study Group. J Clin Oncol, 1998. **16**(4): 1340-9.
80. Slamon, D.J., Clark, G.M., Wong, S.G., Levin, W.J., Ullrich, A. and McGuire, W.L., Human breast cancer: correlation of relapse and survival with the amplification of the HER2/neu oncogene. Science, 1987. **235**: 177-182.
81. Slamon, D.J., Godolphin, W., Jones, L.A., Holt, J.A., Wong, S.G., Keith, D.E., Levin, W.J., Stuart, S.G., Udove, J., Ullrich, A., and Press, M.F., Studies of the HER-2/neu proto-oncogene in human breast and ovarian cancer. Science, 1989. **244**: 707-712.
82. Guy, C.T., R.D. Cardiff, and W.J. Muller, Activated neu induces rapid tumor progression. J Biol Chem, 1996. **271**(13): 7673-8.
83. Bouchard, L., et al., Stochastic appearance of mammary tumors in transgenic mice carrying the MMTV/c-neu oncogene. Cell, 1989. **57**(6): 931-6.
84. Muller, W.J., et al., Single-step induction of mammary adenocarcinoma in transgenic mice bearing the activated c-neu oncogene. Cell, 1988. **54**(1): 105-15.
85. Lemoine, N.R., et al., Absence of activating transmembrane mutations in the c-erbB-2 proto- oncogene in human breast cancer. Oncogene, 1990. **5**(2): 237-9.
86. Zoll, B., et al., Alterations of the c-erbB2 gene in human breast cancer. J Cancer Res Clin Oncol, 1992. **118**(6): 468-73.
87. Guy, C.T., et al., Expression of the neu protooncogene in the mammary epithelium of transgenic mice induces metastatic disease. Proc Natl Acad Sci U S A, 1992. **89**(22): 10578-82.
88. Siegel, P.M. and W.J. Muller, Mutations affecting conserved cysteine residues within the extracellular domain of Neu promote receptor dimerization and activation. Proc Natl Acad Sci U S A, 1996. **93**(17): 8878-83.
89. Siegel, P.M., et al., Elevated expression of activated forms of Neu/ErbB-2 and ErbB-3 are involved in the induction of mammary tumors in transgenic mice: implications for human breast cancer [In Process Citation]. Embo J, 1999. **18**(8): 2149-64.
90. Siegel, P.M., et al., Novel activating mutations in the neu proto-oncogene involved in induction of mammary tumors. Mol Cell Biol, 1994. **14**(11): 7068-77.

24.

# Caspases: Decoders of Apoptotic Signals During Mammary Involution

*Caspase activation during involution*

Andreas Marti[1], Hans Graber[1], Hedvika Lazar[1], Philipp M. Ritter[1], Anna Baltzer[1], Anu Srinivasan[2] and Rolf Jaggi[1]

*[1]Department for Clinical Research, University of Bern, Switzerland, [2]Idun Pharmaceuticals, La Jolla, USA*

Key words: apoptosis, caspase, Fas, involution, mammary gland

Abstract: At weaning most of the alveolar epithelial cells in the mammary gland die by apoptosis and are removed by phagocytosis. Caspases are a family of aspartate specific cysteine proteases. Activation of caspases is generally thought to represent a major and irreversible event in the apoptotic process. We analyzed caspase expression and activation during mammary gland involution. A quantitative RT-PCR based approach revealed that levels of mRNA expression of several caspases are induced during involution. Using an antibody that specifically recognizes activated caspases we measured a transient induction of caspase activity *in situ* and found a maximal activation at days two and three of involution. These data were corroborated by monitoring caspase-3 like activity in mammary extracts with a synthetic DEVD-afc peptide as caspase-3 substrate. Using Fas-deficient mice we present evidence that the Fas signaling pathway is not essential for caspase activation and apoptosis during mammary gland involution. In summary, signaling pathways during involution seem to involve activation of caspases as intraepithelial triggers of mammary epithelial cell apoptosis.

## 1. INTRODUCTION

Proliferation and programmed cell death (PCD) are two major principles contributing to many different biological processes during development and in adults. Apoptosis is a form of PCD which is morphologically characterized by membrane blebbing, condensation of chromatin and generation of apoptotic bodies consisting of membrane surrounded cellular fragments[1]. In many cases it is an energy dependent process requiring gene expression and a non-random fragmentation of genomic DNA is often observed during apoptosis[1, 2]. Induction of apoptosis involves the activation of specific death signaling pathways including the activation of proteases termed caspases. Caspases are a family of aspartate specific cysteine-proteases that are believed to be major effectors of apoptosis [3, 4]. They share typical structural and functional features such as an N-terminal pro-domain that is variable in length and sequence and two catalytic subunits. The prodomain mediates subcellular localization and protein/protein interaction; the catalytic subunits include a large ~17-20 kDa and a small ~10 kDa form. During apoptosis caspases are activated by a proteolytical separation of the large and small subunits that assemble to form the active enzyme complex. Many caspase substrates are cleaved by limited proteolysis, among them are poly(ADP-ribose) polymerase (PARP)[5], α-fodrin[6], MEKK-1[7], U1 associated 70 kDa protein[8], DNA fragmentation factor (DFF)[9, 10] and Bcl-2[11].

In the mammary gland apoptosis occurs in secretory epithelial cells during post-lactational involution[12,13]. Upon removing the pups from their mother, milk transiently accumulates in the glands resulting in a strong engorgement. At about three days of involution, milk is resorbed, alveolar structures collapse and epithelial cells undergo apoptosis. The lack of the suckling stimulus also leads to a drop of lactogenic hormone levels (e.g. prolactin and glucocorticoids) which is believed to contribute to the involution process in all glands[14,15]. Accumulation of milk and changes of hormone levels are possibly major triggers of apoptosis[15, 16, 17].

Several signal transduction pathways resulting in caspase activation have been described that involve death receptors such as Fas/APO-1[18]. Indeed, for induction of apoptosis in the prostate and the ovary it was proposed that activation of Fas may play an important role[19, 20]. These studies mainly relied on animals with defined genetic mutations in the Fas pathway[21]. Here we studied the involvement of the Fas pathway and caspases during involution of the mouse mammary gland. After weaning, elevated caspase-3 and caspase-9 mRNA expression and an induction of caspase activity were observed. In Fas-deficient mice (lpr mice) caspase activation and apoptosis as analyzed by terminal transferase assays (TUNEL) was indistinguishable from wildtype animals implicating that the Fas signaling pathway is not essential for mammary gland involution.

## 2. RESULTS AND DISCUSSION

For the mammary gland, little information is available about expression and activity of caspases during phases of epithelial cell apoptosis. For caspase-1 an induction at the mRNA level was shown during involution [22,23]; in addition, Krajewska and co-workers[24] demonstrated the presence of caspase-3 protein in ductal epithelial cells of human mammary biopsies. We investigated the expression of caspase-3 and caspase-9 in the mammary gland by quantitative RT-PCR (using Taqman technology). Total RNA of mammary tissue at lactation and 2, 3, 6, and 10 days of involution was prepared and Taqman-PCR was performed. As an internal reference 7S ribosomal RNA was measured. In addition, c-*fos*, a gene that was previously shown to be induced during involution[25], was also analyzed. After weaning, an induction of caspase-3 and caspase-9 mRNA expression was observed.

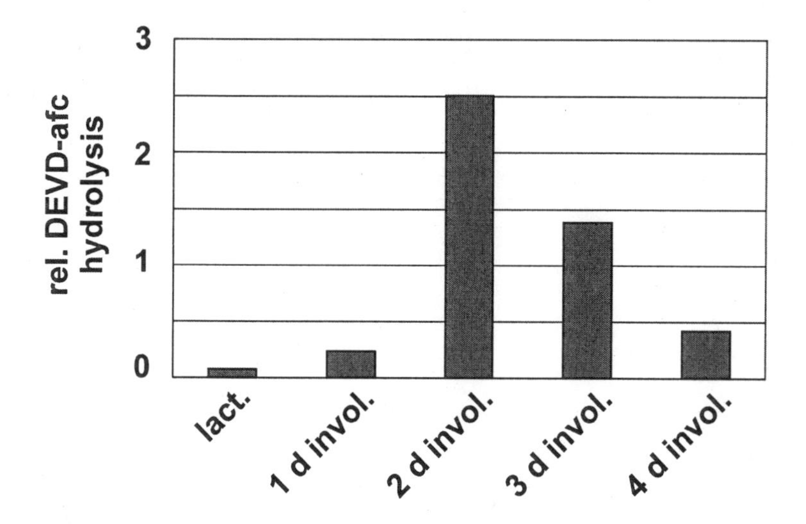


*Fig. 1.*Caspase cleavage activity during lactation and involution. Cytoplasmic extracts were prepared from mammary glands at lactation and 1, 2, 3 and 4 days of involution as described[25]. 25 μg protein were incubated in the presence of DEVD-afc in caspase buffer (Promega) and fluorescence was measured in a Fluorometer over a period of 1 h (200 cycles of measurement). Shown are the relative DEVD cleavage activities as arbitrary units of fluorescence.

Induction of c-*fos* and caspase-3 mRNA expression was maximal at day 2 (35 fold induction for c-*fos*, 3.3 fold induction for caspase-3) whereas caspase-9 mRNA levels were maximal at day 6 of involution (10.5 fold induction). These results indicate that, like caspase-1, caspase-3 and caspase-9 mRNA levels are induced during mammary gland involution.

To further investigate the involvement of caspases during involution, caspase activity was analyzed in mammary extracts derived of glands at lactation and 1, 2, 3, and 4 days of involution. Extracts were incubated with a synthetic caspase substrate (DEVD-afc) and cleavage of the substrate was analyzed in a fluorometer.

Fig. 1 shows that caspase activity was strongly and transiently induced, peaking at day 2 of involution. Caspase activation was further studied on mammary tissue sections using an antibody that specifically recognizes active caspase-3 but not the inactive zymogen (CM1 antibody)[26]. As shown in Fig. 2, a strong induction of active caspase-3 was observed in epithelial cells at 3 days of involution (panel B) whereas almost no active caspase was found in epithelial cells at lactation (panel A).

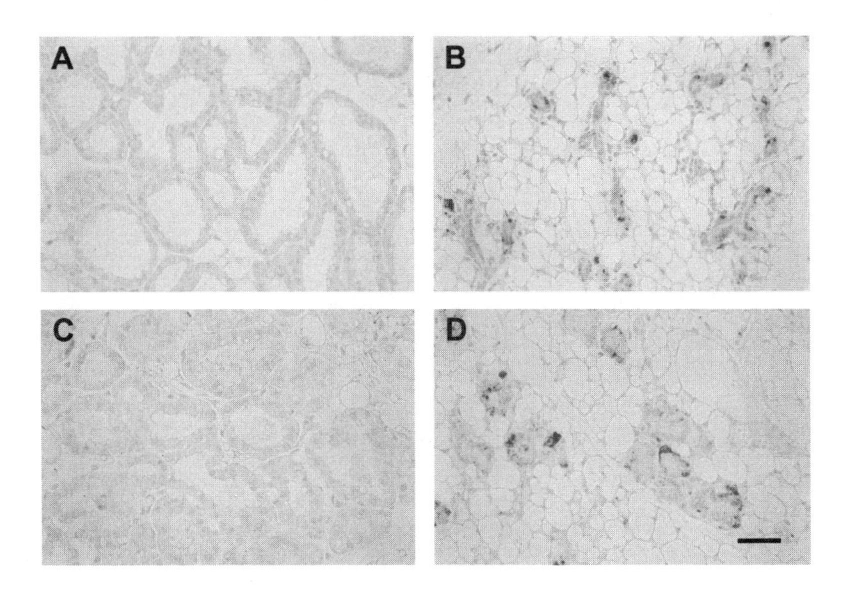


*Fig. 2.*Caspase activation during involution. Shown are mammary glands of wildtype (panels A, B) and Fas-deficient mice (panels C, D) at lactation (panel A, C) and at 3 days of involution (panel B, D). Paraffin embedded sections from paraformaldehyde fixed samples were incubated with an antibody (CM1) that recognizes active caspase-3[26]. Antigen-antibody complexes were detected with the EnVision System (DAKO, Glostrup, Denmark) using AEC as substrate. The bar represents 25 μm.

In order to investigate a potential involvement of Fas during mammary gland involution, Fas-deficient mice were analyzed immunohistochemically for caspase-3 like activity at lactation (Fig. 2, panel C) and at 3 days of involution (panel D). A similar caspase activation was observed in Fas-deficient animals as compared to wildtype animals.

These results indicate that caspases are activated during involution in the absence of functional Fas. To confirm these data, mammary sections derived of wildtype and Fas-deficient animals were analyzed for DNA fragmentation using a terminal transferase (TUNEL) assay (Fig. 3). Both, in wildtype and in Fas-deficient animals similar levels of DNA fragmentation were found at 2 days of involution (panels B, D) whereas almost no cells stained positive for fragmented DNA at lactation (panels A, C). In addition, total DNA was isolated from wildtype and Fas-deficient animals at lactation and 3 days of involution and analyzed by gel electrophoresis (data not shown). Again, the typical DNA ladder was observed in wildtype and in Fas-deficient animals at 2 and 3 days of involution but not at lactation. Taken together, these results indicate that mammary epithelial cell apoptosis occurs to a similar extent in Fas-deficient mice and in wildtype animals.

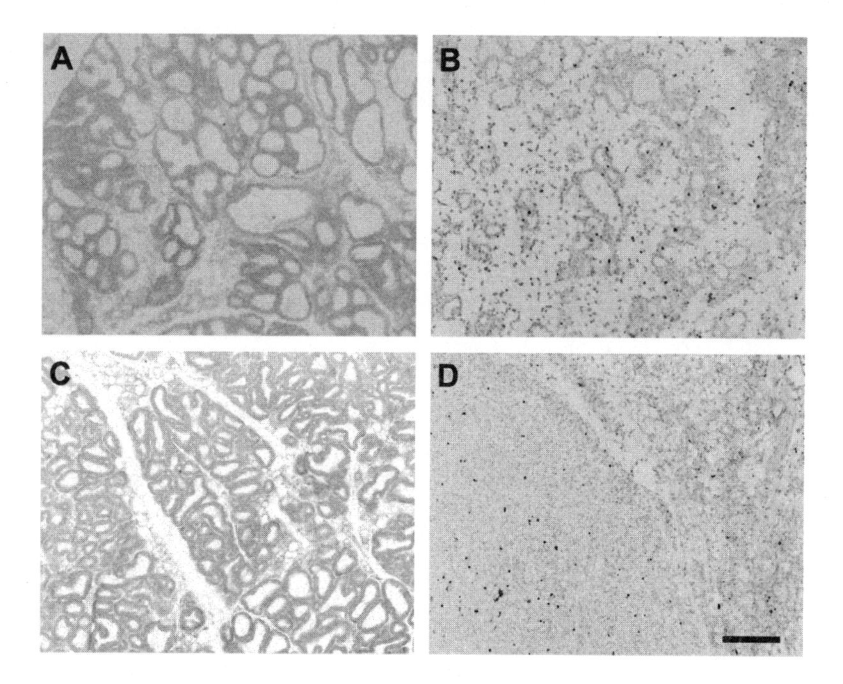


*Fig. 3.*DNA fragmentation in Fas-deficient animals during involution. Terminal transferase assays (TUNEL) were performed on mammary gland sections of wildtype (panels A and B) and Fas-deficient animals (panels C and D) at lactation (A and C) and at 2 days of involution (B and D). The bar represents 50 μm.

In summary, our data suggest that epithelial cell death that occurs during mammary gland involution involves an activation of caspases, both, at the mRNA and protein level. Cell death in the mammary gland seems to occur in a Fas independent manner.

**Acknowledgement**

This work was supported by the Swiss National Science Foundation, the Sandoz Foundation (Switzerland) and the Foundation for Clinical and Experimental Cancer Research (Switzerland).

## 3. REFERENCES

1. Wyllie, A. H., J. F. R. Kerr, and A. R. Currie: Cell death: The significance of apoptosis. Int. Rev. Cytol. 68, 251-306 (1980).
2. Schwartz, L. M., and B. A. Osborne: Programmed cell death, apoptosis and killer genes. *Immunol. Today* 14, 582-590 (1993).
3. Salvesen, G. S., and V. M. Dixit: Caspases: intracellular signaling by proteolysis. *Cell* 91, 443-446 (1997).
4. Thornberry, N. A., and Y. Lazebnik: Caspases: enemies within. *Science* 281, 1312-1316 (1998).
5. Tewari, M., L. T. Quan, K. O Rourke, S. Desnoyers, Z. Zeng, D. R. Beidler, G. G. Poirier, G. S. Salvesen, and V. M. Dixit: Yama/CPP32 beta, a mammalian homolog of CED-3, is a CrmA- inhibitable protease that cleaves the death substrate poly(ADP- ribose) polymerase. *Cell* 81, 801-809 (1995).
6. Martin, S. J., G. A. O'Brien, W. K. Nishioka, A. J. McGahon, A. Mahboubi, T. C. Saido, and D. R. Green: Proteolysis of fodrin (non-erythroid spectrin) during apoptosis. *J. Biol. Chem.* 270, 6425-6428 (1995).
7. Cardone, M. H., G. S. Salvesen, C. Widmann, G. Johnson, and S. M. Frisch: The regulation of anoikis: MEKK-1 activation requires cleavage by caspases. *Cell* 90, 315-323 (1997).
8. Casciola-Rosen, L., D. W. Nicholson, T. Chong, K. R. Rowan, N. A. Thornberry, D. K. Miller, and A. Rosen: Apopain/CPP32 cleaves proteins that are essential for cellular repair: a fundamental principle of apoptotic death. *J. Exp. Med.* 183, 1957-1964 (1996).
9. Enari, M., H. Sakahira, H. Yokoyama, K. Okawa, A. Iwamatsu, and S. Nagata: A caspase-activated DNase that degrades DNA during apoptosis, and its inhibitor ICAD. *Nature* 391, 43-50 (1998).
10. Liu, X., P. Li, P. Widlak, H. Zou, X. Luo, W. T. Garrard, and X. Wang: The 40-kDa subunit of DNA fragmentation factor induces DNA fragmentation and chromatin condensation during apoptosis. *Proc. Natl. Acad. Sci. USA* 95, 8461-8466 (1998).
11. Cheng, E. H., D. G. Kirsch, R. J. Clem, R. Ravi, M. B. Kastan, A. Bedi, K. Ueno, and J. M. Hardwick: Conversion of Bcl-2 to a Bax-like death effector by caspases. *Science* 278, 1966-1968 (1997).
12. Walker, N. I., R. E. Bennett, and J. F. Kerr: Cell death by apoptosis during involution of the lactating breast in mice and rats. *Am. J. Anat.* 185, 19-32 (1989).
13. Strange, R., F. Li, S. Saurer, A. Burkhardt, and R. R. Friis: Apoptotic cell death and tissue remodelling during mouse mammary gland involution. *Development* 115, 49-58 (1992).
14. Topper, Y. J., and C. S. Freeman: Multiple hormone interactions in the developmental biology of the mammary gland. *Physiol. Rev.* 60, 1049-1106 (1980).
15. Feng, Z., A. Marti, B. Jehn, H. J. Altermatt, G. Chicaiza, and R. Jaggi: Glucocorticoid and progesterone inhibit involution and programmed cell death in the mouse mammary gland. *J. Cell Biol.* 131, 1095-1103 (1995).

16. Marti, A., Z. Feng, H. J. Altermatt, and R. Jaggi: Milk accumulation triggers apoptosis of mammary epithelial cells. *Eur. J. Cell Biol.* 73, 158-165 (1997).
17. Li, M., X. Liu, G. Robinson, U. Bar-Peled, K. U. Wagner, W. S. Young, L. Hennighausen, and P. A. Furth: Mammary-derived signals activate programmed cell death during the first stage of mammary gland involution. *Proc. Natl. Acad. Sci. USA* 94, 3425-3430 (1997).
18. Ashkenazi, A., and V. M. Dixit: Death receptors: signaling and modulation. *Science* 281, 1305-1308 (1998).
19. Suzuki, A., A. Matsuzawa, and T. Iguchi: Down regulation of Bcl-2 is the first step on Fas-mediated apoptosis of male reproductive tract. *Oncogene* 13, 31-37 (1996).
20. Sakamaki, K., H. Yoshida, Y. Nishimura, S. Nishikawa, N. Manabe, and S. Yonehara: Involvement of Fas antigen in ovarian follicular atresia and luteolysis. *Mol. Reprod. Dev.* 47, 11-18 (1997).
21. Cohen, P. L., and R. A. Eisenberg: The lpr and gld genes in systemic autoimmunity: life and death in the Fas lane. *Immunol Today* 13, 427-428 (1992).
22. Boudreau, N., C. J. Sympson, Z. Werb, and M. J. Bissell: Suppression of ICE and apoptosis in mammary epithelial cells by extracellular matrix. *Science* 267, 891-893 (1995).
23. Lund, L. R., J. Romer, N. Thomasset, H. Solberg, C. Pyke, M. J. Bissell, K. Dano, and Z. Werb: Two distinct phases of apoptosis in mammary gland involution: proteinase-independent and -dependent pathways. *Development* 122, 181-193 (1996).
24. Krajewska, M., H. G. Wang, S. Krajewski, J. M. Zapata, A. Shabaik, R. Gascoyne, and J. C. Reed: Immunohistochemical analysis of in vivo patterns of expression of CPP32 (Caspase-3), a cell death protease. *Cancer Res.* 57, 1605-1613 (1997).
25. Marti, A., B. Jehn, E. Costello, N. Keon, G. Ke, F. Martin, and R. Jaggi: Protein kinase-A and AP-1 (c-Fos/JunD) are induced during apoptosis of mouse mammary epithelial cells. *Oncogene* 9, 1213-1223 (1994).
26. Srinivasan, A., K. A. Roth, R. O. Sayers, K. S. Shindler, A. M. Wong, L. C. Fritz, and K. J. Tomaselli: In situ immunodetection of activated caspase-3 in apoptotic neurons in the developing nervous system. *Cell Death Differ.* 5, 1004-1016 (1998).

25.

# The Role of Plasminogen Activator in the Bovine Mammary Gland

Ioannis Politis
*Department of Animal Production, Agricultural University of Athens, Athens, Greece*

Key words: plasminogen activator, bovine

## 1. INTRODUCTION

The major portion of mammary growth occurs during pregnancy. Following parturition, milk synthesis begins. Peak production in dairy cows occurs during the second month of lactation and then declines throughout the remaining months of lactation. The declining phase of lactation is known as gradual involution. When the lactation has ended, postlactation involution begins, characterized by a marked decrease in secretory tissue mass[1].

The plasmin-plasminogen system is thought to play an important role during cell proliferation in a variety of cellular systems. However, a relationship between bovine mammary cell growth and the plasminogen-activating system has not been established. Furthermore, while the physiologically important role of the plasminogen-activating system for mammary involution has been examined extensively, the role of the same system during mammary involution in dairy cows has received only limited attention.

In this paper, four main topics will be addressed:

a) Fundamental biochemical aspects of the plasminogen-activating system,
b) Regulation of mammary growth
c) The plasminogen-activating system during mammary involution
d) The plasminogen-activating system and mammary pathology.

## 2. BIOCHEMICAL ASPECTS

The process of plasminogen activation leads to the formation of a serine protease, plasmin, that in turn, is capable of degrading most extracellular proteins. This process is controlled by an interplay between plasminogen activators (PA) and PA inhibitors (PAI). There are two types of PA: urokinase-PA (u-PA) and tissue PA (t-PA). These two enzymes are products of different genes. Even though some exceptions exist, the general belief is that t-PA is mainly implicated in the maintenance of fluidity of the extracellular milieu (thrombolysis), while u-PA is involved in the proteolysis that accompany tissue remodelling events such as the involution of the mammary gland[2]. The activity of PA can be limited by the action of PAI. The most important, fast acting, PAI have been designated PAI-1 and PAI-2.

Recently, a specific receptor for u-PA (u-PAR) was discovered. The u-PAR is a protein that binds u-PA with high affinity. The primary function of u-PAR is to localize plasminogen-activation at the cell surface. The rate of plasminogen activation is enhanced when u-PA is bound to its receptor, leading to accelerated formation of plasmin[2].

## 3. REGULATION OF MAMMARY GROWTH

Several studies have shown that most factors regulating expression of PA-related genes were also among the most powerful regulators of growth of cultured cells. The concept that bovine mammary cell growth and PA expression are related has received limited attention.

A number of bovine mammary epithelial and myoepithelial cell lines were generated to allow appropriate experiments to be carried out[3]. Whether expression of PA-related genes are connected to the cell cycle was examined. Results indicated that expression of all components of the plasminogen-activating system (u-PA, PAI-1, u-PAR) occurs in both epithelial and myoepithelial cell lines but in a manner tightly connected to the cell cycle. As the cell cycle progressed, accumulated u-PA activity increased. Maximal activity occurred at the S-G2 border and decreased during the G2 and M phases[4].

Three experiments examined the effects of known regulators of bovine mammary cell growth on expression of PA-related genes. In the first experiment, the effect of insulin, insulin-like growth factor I (IGF-I) and IGF-II was examined. Results indicated that insulin, IGF-I and IGF-II increased cell proliferation. Furthermore, insulin, IGF-I and IGF-II increased by 2-4 fold expression of u-PA, PAI-1, and u-PAR mRNA indicating that there is a correlation between cell proliferation and expression of PA-related genes[3]. The second experiment examined the effect of phorbol myristate acetate (PMA) on mammary epithelial (BME-UV1) cell function. Against any preconceived notion, PMA caused a small inhibition of mammary epithelial cell proliferation. Furthermore, PMA caused 3-4 fold increases in expression of u-PA and u-PAR mRNA. Thus, induction of PA-related genes is not always associated with increased cell proliferation[3]. The third experiment examined the effects of extracellular calcium on mammary epithelial cell proliferation. Addition of extracellular calcium resulted in large increases in cell proliferation without any apparent effects on expression of u-PA, PAI-1 and u-PAR mRNA.

In conclusion, the three experiments described above allow two important conclusions to be made. First, increased expression of PA-related genes can occur without concomitant cell growth (PMA case). Second, accelerated cell growth can occur without increases in the expression of PA-related genes (calcium case). It is clear that expression of PA-related genes is likely to serve multiple functions in bovine mammary cells.

## 4. THE PLASMIN-PLASMINOGEN, IN VIVO STUDIES

A number of experiments were performed to evaluate the role of the plasminogen-activating system during gradual and postlactation involution. Both plasmin and plasminogen in milk increased with advancing lactation. However, the ratio of plasminogen to plasmin declined with advancing lactation, indicating enhanced conversion of plasminogen to plasmin during late lactation[3]. As expected, milk from late lactation cows contained more PA than did early lactation milk. Immunoreactive u-PA was present in high levels in mammary tissue obtained from late lactation cows (8 month in lactation) and involuting cows (2-6 days following cessation of milking) but not in mammary tissue obtained from early or middle lactation cows. In fact, the highest levels of immunoreactive u-PA were detected during late lactation. Thus, lactation and involution may occur concurrently during late lactation in dairy cows.

Whether administration of bST to dairy cows affected the plasmin-plasminogen system was examined. Administration of bST prevented the increase in plasmin activity that normally occurs during late lactation. Those results were interpreted to indicate that bST caused a retardation of gradual involution[3]. Very recently, using a co-culture of epithelial and myoepithelial cells, cultured on collagen, we were able to show that IGF-I (the factor that mediates the effects of bST) enhanced expression of PAI-1 mRNA, but not that of u-PA and u-PAR mRNA by both epithelial and myoepithelial cells. This provides further for the bST-plasmin hypothesis.

## 5. THE PLASMIN-PLASMINOGEN SYSTEM AND MAMMARY PATHOLOGY

Mastitis results in increased proteolytic activity that is related to increased concentrations of plasmin in milk. It appears that two mechanisms might be responsible for the increased amounts of plasmin in milk during mastitis. The first mechanism favors transport of both plasmin and plasminogen from blood to milk. The compromised permeability of the mammary epithelium allows transport of these two enzymes to milk[3]. The second mechanism favors increased conversion of plasminogen to plasmin by the action of both u-PA and t-PA. Milk from mastitic quarters contained 10- to 20-fold higher amounts of u-PA and t-PA than milk obtained from healthy quarters[5].

It has been suggested that some pathogenic bacteria including Staphylococcus aureus may use the plasminogen-plasmin system to penetrate and colonize mammalian tissue. S. aureus produces the bacterial PA-like protein staphylokinase, which converts the host plasminogen to plasmin. However, bovine strains of S. aureus are staphylokinase-negative. Thus, the plasmin-plasminogen system may be unimportant in bovine mammary infections or some other unknown mechanism compensates for the inability of S. aureus to produce staphylokinase. Using mammary epithelial and myoepithelial cell lines, it was found that S. aureus causes a large increase of u-PA production by mammary cells. Transcytosis of S. aureus across a mammary epithelial monolayer was enhanced by addition of bovine plasminogen[6]. Those data suggest that it might be possible that S. aureus causes an increase in expression of u-PA by mammary cells and this might compensate for the lack of intrinsic staphylokinase (PA-like) activity.

## 6. CONCLUSIONS

The plasmin-plasminogen system plays an important role in the bovine mammary gland. Increased conversion of plasminogen to plasmin occurs during late lactation and during postlactation involution. The highest levels of immunoreactive u-PA were detected during late lactation suggesting that lactation and involution may occur concurrently in the bovine mammary gland during late lactation. Other data indicated that expression of PA-related genes are not directly related to mitogenesis but are likely to serve multiple functions in the bovine mammary gland. Lastly, increased amounts of both u-PA and t-PA occur in milk obtained from mastitic quarters. In cultured mammary cells, bovine strains of Staphylococcus aureus upregulate expression of u-PA by mammary cells in an apparent attempt to compensate for their own inability to produce a PA-like protein that could facilitate tissue colonization. The latter needs to be confirmed in vivo.

## 7. REFERENCES

1. Knight, C.H., and Wilde, C. J., 1993, Mammary cell changes during pregnancy and lactation. Livest. Prod. Sci. 35:3-19.
2. Saksela, O., and Rifkin, D. B, 1988, Cell-associated plasminogen activation: regulation and physiological significance. Annu. Rev. Cell Biol. 4:93-118.
3. Politis, I., 1995, Plasminogen activator system: Implications for mammary cell growth and involution. J. Dairy Sci. 79:1097-1107.
4. Zavizion, B., White, J. H., and Bramley, A.J., 1998, Cell cycle-dependent fluctuation of urokinase-type plasminogen activator, its receptor, and inhibitors in cultured bovine mammary epithelial and myoepithelial cells. Bioch. Biophys. Acta 1403:141-150.
5. Heegard, C. W., Christensen, T., Rasmussen, M. D., Benfeldt, C., Jensen, N. E., Sjersen, K., Petersen, T. E., and Andreasen, P.A., 1994, Plasminogen activators in bovine milk during mastitis, an inflammatory disease. Fibrinolysis 8:22-27.
6. Zavizion, B., White, J. H., and Bramley, A. J., 1997, Staphylococcus aureus stimulates urokinase-plasminogen activator expression by bovine mammary cells. J. Infectious Dis. 176:1637-1640.

# 26.

# Regulation and Nutritional Manipulation of Milk Fat

## Low-Fat Milk Syndrome

[1]Dale.E. Bauman and [2]J. Mikko Griinari
*[1]Dept. of Animal Science, Cornell University, Ithaca, NY USA: [2]Dept. of Animal Science, University of Helsinki, Helsinki, FINLAND*

## 1. INTRODUCTION

Milk fat represents the major energy source in milk and the major energy cost in production of milk components. There has been a renewed interest in milk fat because changes in the economic value of milk fat favor increased production in some cases and reduced production in other cases. In addition, as more is learned about the biological roles of specific fatty acids, we have the opportunity to alter milk fatty acid composition to improve the healthfulness of milk as a food.

Milk fat is predominantly composed of triglycerides. In ruminant animals, milk fatty acids arise from *de novo* synthesis from acetate and β-hydroxybutyrate and from the uptake of circulating long chain fatty acids that originate from the diet or from body fat reserves. The fatty acids comprising milk triglycerides vary in chain length; milk fatty acids $C_4$ to $C_{14}$ and a portion of the $C_{16}$ arise from *de novo* synthesis and the remainder of $C_{16}$ and all of the longer chain fatty acids originate from the uptake of preformed fatty acids. Reviews have detailed the biosynthesis of milk fat[1,2] and summarized factors which affect milk fat yield and composition in dairy cows[3,4]. The high genetic correlation between fat yield and the yield of milk and other milk components makes it difficult to use selection to alter milk fat independently of other milk components. In contrast, nutrition and diet can be used to rapidly and extensively alter milk fat yield and composition. In particular, certain diets cause a dramatic reduction in milk fat synthesis and this is generally referred to as milk fat depression (MFD) or low-fat milk syndrome[5]. MFD is of obvious economic importance to producers but it is also of interest as a model to understand the regulation of the quantity and composition of milk fat.

## 2. GENERAL CHARACTERISTICS OF MILK FAT DEPRESSION

Diets that cause MFD on commercial dairy farms are generally divided into two broad groups. The first group involves diets that provide large amounts of readily digestible carbohydrate and reduced amounts of fibrous constituents. Of these, high grain/low forage diets (HG/LF) would be the most common. The second group represent dietary supplements of certain plant oils and fish oils. The presence of unsaturated fatty acids in the oils is essential and MFD occurs with dietary supplements of the oils themselves or with addition of full fat seeds containing the unsaturated fatty acids. In addition, the extent to which these two groups of diets induce MFD is related to other factors such as physical preparation of the diet (e.g. particle size and pelleting), management practices (e.g. feeding frequency), and animal factors (e.g. stage of lactation, level of intake)[3,4,5].

The classical diet-induced MFD occurs within a few days following dietary changes. Effects are specific and milk fat yield can be reduced by 50% or more with little or no change in milk yield or other milk components. In general, classical MFD results in a decrease yield of all fatty acids. However, the decline is greatest for fatty acids synthesized *de novo* so that the milk fat composition shifts toward lower proportions of short and medium chain fatty acids and greater concentrations of longer chain fatty acids. There is also a particularly noticeable increase in *trans* fatty acids with MFD. Buffering the rumen pH reduces or abolishes MFD which occurs with HG/LF diets and bypassing the rumen by abomasal infusion reduces or abolishes the MFD observed with oil supplements[4,5,6]. Thus, the induction of MFD must involve products produced by rumen bacteria that are related to the dietary-induced shifts in rumen fermentation.

The actual mechanisms involved in low-fat milk syndrome have not been fully elucidated but several theories have been proposed. These can be broadly summarized into two categories: theories which consider the depression to be a consequence of a shortage in the supply of lipid precursors to the mammary gland and theories that attribute MFD to a direct inhibition at the mammary gland of one or more steps in the synthesis of milk fat. The glucogenic-insulin theory is the main theory involving a lipid precursor shortage as the cause of MFD and it suggests that insulin plays a key role in the etiology of MFD[3,5]. The major theory that MFD is caused by a direct inhibition of mammary fatty acid synthesis involves products from incomplete or unusual biohydrogenation of polyunsaturated fatty acids in the rumen. *Trans* fatty acids have received special attention as a cause[5,6], and more recently this has been expanded to include *trans* fatty acids and related metabolites[7,8].

## 3. GLUCOGENIC-INSULIN THEORY AND MFD

Endocrine status can alter the partitioning of nutrients to specific tissues. Insulin has acute effects on adipose tissue rates of lipogenesis (stimulatory) and lipolysis (inhibitory), but the ruminant mammary gland is not responsive to changes in circulating insulin. This forms the basis for the glucogenic-insulin theory of MFD and it has been primarily applied to the MFD which occurs with diets high in readily digestible carbohydrate and low in fiber. High grain diets result in increased rumen production of propionate and increased hepatic rates of gluconeogenesis, which in turn increase pancreatic release of insulin. The elevated circulating concentrations of insulin, often two- to fivefold increases, enhance uptake of lipogenic precursors and decrease the release of fatty acids by adipose tissue. According to the glucogenic-insulin theory, this will result in an increase in adipose tissue use of acetate, β-hydroxybutyrate and diet-derived long chain fatty acids, and a reduction in mobilization of long chain fatty acids from body reserves. Thus, these overall changes preferentially channel nutrients to adipose tissue, resulting in a shortage of lipogenic precursors for the mammary gland and milk fat depression[3,5].

Evaluation of the involvement of insulin in MFD is complicated by its central role in glucose homeostasis. We have examined the role of insulin in milk fat synthesis by lactating cows by using a chronic hyperinsulinemic-euglycemic clamp to avoid hypoglycemia and counter-regulatory changes in glucose homeostasis. Milk yield was maintained during the 4-day insulin clamp and there was no evidence of insulin resistance based on circulating insulin concentrations, exogenous glucose required to maintain euglycemic (2 to 3 kg/d) and effects on circulating nonesterified fatty acids (antilipolytic effects). Relative to the glucogenic-insulin theory, we observed little or no effect on milk fat synthesis during studies with the insulin clamp (Table 1). Examination of milk fat composition revealed shifts had occurred, but they differed markedly from those observed during classical MFD (Fig. 1). Thus, our results with the insulin clamp technique offered no support for glucogenic-insulin theory as the cause of diet-induced MFD[9,10,11].

*Table 1.* Effect of hyperinsulinemic-euglycemic clamp on milk fat yield (kg/d)[1]

| | Standard Diet | | Supplemental Protein[2] | |
|---|---|---|---|---|
| Reference | Control | + Insulin | Control | + Insulin |
| McGuire et al.[9] | 1.26 | 1.30 | - | - |
| Griinari et al.[10] | 0.91 | 0.85 | 1.01 | 0.94 |
| Mackle et al.[11] | 0.86 | 0.77 | 0.86 | 0.83 |

[1] During the 4-day hyperinsulinemic-euglycemic clamps circulating insulin was approximately 4-fold over basal levels. Values represent milk fat yield on day 4 of the insulin clamp. [2]Supplemental amino acids were provided by abomasal infusion of casein or casein plus branched chain amino acids.

Others have infused propionate, glucose, or starch to examine the glucogenic-insulin theory (e.g. see[5,13,14,15]). In general, these infusions present a challenge to the lactating cow in terms of maintenance of glucose homeostasis. Nevertheless, effects on milk fat yield range from none to a modest reduction. Just as with the insulin clamp, the pattern of change in milk fat composition differs from classical MFD (Figure 1). The changes in the pattern of milk fat composition for glucose infusion and the insulin clamp are consistent with insulin inhibition of lipolysis and a reduction in mammary use of preformed fatty acids from body fat reserves. The extent of their effect on milk fat yield is undoubtedly related to the magnitude of the fatty acid contribution from body fat reserves. Overall, the results of infusion of glucogenic substrates and the hyperinsulinemic-euglycemic clamp do not support the glucogenic-insulin theory as the cause of classical diet-induced MFD.

## 4. *TRANS* FATTY ACIDS AND MFD

Direct inhibition of milk fat synthesis by *trans*-$C_{18:1}$ was first proposed more than two decades ago[5]. *Trans*-$C_{18:1}$ fatty acids arise as intermediates in the biohydrogenation of polyunsaturated fatty acids. Thus, their involvement in MFD was typically proposed for dietary supplements of plant and fish oils, but some proposed that *trans*-$C_{18:1}$ were also involved in MFD with high grain/low fiber diets[5,7]. Consistent with the theory, several groups demonstrated that feeding of partially hydrogenated vegetable oils to lactating dairy cows caused MFD (see review[6]) and Wonsil et al.[16] reported the outflow of *trans*-18:1 from the rumen correlated with the degree of MFD. Other studies summarized by Erdman[6] showed that the simple substitution of *trans*-C18:1 fatty acids for *cis*-18:1 fatty acids in abomasal infusions resulted in a dramatic reduction in milk fat yield.

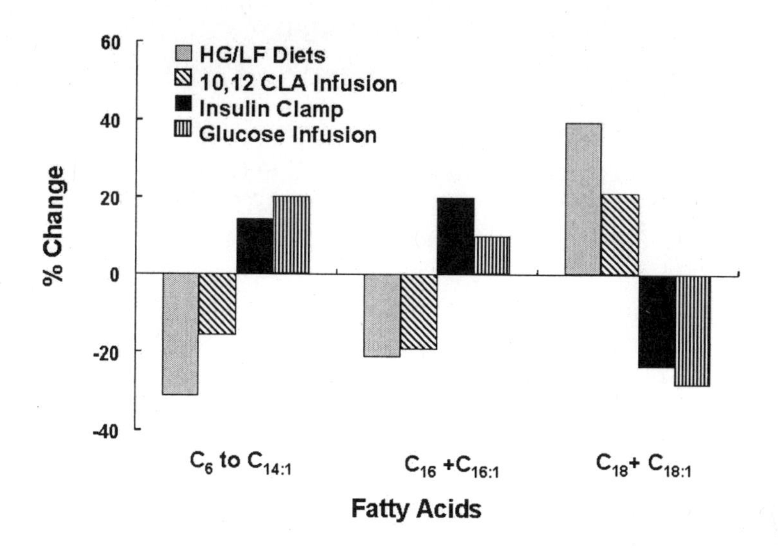


*Figure 1.* Changes in fatty acid composition of milk fat which occur with high grain/low fiber (HG/LF) diets[12], abomasal infusion of *trans*-10, *cis*-12 conjugated linoleic acid (10,12 CLA)[8], hyperinsulinemic-euglycemic clamp (insulin clamp)[10], and abomasal infusion of glucose[13].

Over a wide range of diets, there is a close relationship between the decrease in milk fat percentage and the increase in milk fat content of *trans*-$C_{18:1}$[6,7]. Occasionally when dietary oils are fed, the decrease in milk fat is less than predicted from the linear relationship with *trans*-C18:1. In examining this, we found that two conditions were required to get MFD - a dietary source of unsaturated fatty acids and an altered rumen fermentation which leads to an incomplete biohydrogenation[7]. We further demonstrated that MFD was related to a shift in the pattern of *trans* $C_{18:1}$ isomers to increased concentrations of *trans*-10 $C_{18:1}$[7]. In the biohydrogenation of linoleic acid, the first step is an isomerization to form *cis*-9, *trans*-11 conjugated linoleic acid (CLA). This is followed by two reductions to form *trans*-11 $C_{18:1}$ and then stearate. Consistent with this, *cis*-9, *trans*-11 CLA and *trans*-11 $C_{18:1}$ are the major fatty acid isomers of the biohydrogenation process typically found in milk fat. However, under some circumstances rumen biohydrogenation must be altered to give rise to the *trans*-10 $C_{18:1}$ observed during MFD. Putative biohydrogenation pathways for its formation have been proposed and in this case the initial isomerization step would form a *trans-10, cis*-12 CLA[17].

Based on the above discussion we postulated that *trans*-10 $C_{18:1}$ or related metabolites (e.g. *trans*-10, *cis*-12 CLA) might be the case of MFD[7]. In support of this we observed that increases in the milk fat content of *trans*-10

$C_{18:1}$ and *trans*-10-, *cis*-12 CLA were both inversely related to milk fat content[7,18]. To directly examine their role in MFD, we compared effects of abomasal infusion of *cis*-9, *trans*-11 and *trans*-10, *cis*-12 isomers of CLA[8]. The 9,11 CLA isomer had no impact on milk fat whereas a 4-day abomasal infusion of less than 10 g/d of *trans*-10, *cis*-12 CLA resulted in over a 40% reduction in milk fat content and yield (Figure 2). Furthermore, the shifts in milk fat composition observed with *trans*-10, *cis*-12 CLA parallel those which occur with classical dietary-induced MFD (Fig. 1).

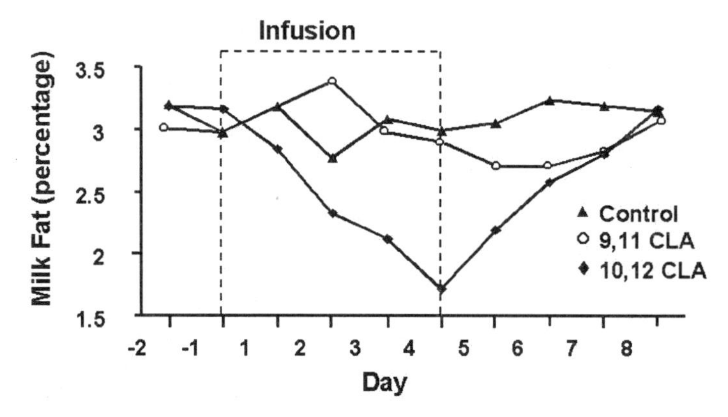


*Figure 2.* Temporal pattern of milk fat content during abomasal infusion of conjugated linoleic acid (CLA) isomers. Infusions were of 10 g/d of *cis*-9, *trans*-11 CLA or *trans*-10, *cis*-12 CLA[8].

The mechanisms whereby *trans*-10, *cis*-12 CLA inhibits milk fat synthesis represents an active area of research. Specific mechanisms have not been established but based on the changes in milk fat composition they must involve *de novo* fatty acid synthesis and the desaturase system[8]. The regulation of key enzymes such as acetyl CoA carboxylase, fatty acid synthase, and $\Delta^9$-desaturase are logical possibilities.

## 5. CONCLUSIONS

Certain diets induce MFD in lactating dairy cows and several theories have been proposed as the basis. One of these is the glucogenic-insulin theory, but recent investigations offer no support for this theory as the cause of MFD. Another theory is that intermediates formed in rumen biohydrogenation of unsaturated fatty acids inhibit fat synthesis at the mammary gland. Recent studies suggest that pathways of rumen biohydrogenation are altered under certain dietary conditions and this results in a shift in the CLA and *trans* $C_{18:1}$ isomers that are produced. Specifically, diet-induced MFD corresponds to increases in milk fat content of *trans*-10, *cis*-12 CLA and *trans*-10 $C_{18:1}$, and the former has been shown to be a potent inhibitor of milk fat synthesis.

## 6. REFERENCES


1. Barber, M.C., Clegg, R.A., Travers, M.T., and Vernon, R.G., 1997, Lipid metabolism in the lactating mammary gland. *Biochim. Biophys. Acta.* 1347:101-126.
2. German, J.B., Morand, L., Dillard, C.J., and Xu, R., 1997, Milk Fat Composition: Targets for Alteration of Function and Nutrition. In *Milk Composition and Biotechnology* (R.A.S. Welch, D.J.W. Burns, S.R. Davis, A.I. Popay, and C.G. Prosser, eds.), CAB International, New York, NY, pp. 35-72.
3. Sutton, J.D., 1989, Altering milk composition by feeding. *J. Dairy Sci.* 72:2801-2814.
4. Palmquist, D.L., Beaulieu, A.D., and Barbano, D.M., 1993, Feed and animal factors influencing milk fat composition. *J. Dairy Sci.* 76:1753-1771.
5. Davis, C.L., and Brown, R.E., 1970, Low-fat Milk Syndrome. In *Physiology of Digestion and Metabolism in the Ruminant* (A.T. Phillipson, ed.), Oriel Press Limited, Newcastle upon Tyne, UK. pp. 545-565.
6. Erdman, R., 1996, Milk fat depression: some new insights. Tri-State Dairy Nutrition Conference, Fort Wayne, IN, pp. 1-16.
7. Griinari, J.M., Dwyer, D.A., McGuire, M.A., Bauman, D.E., Palmquist, D.L., and Nurmela, K.V.V., 1998, *Trans*-octadecenoic acids and milk fat depression in lactating dairy cows. *J. Dairy Sci.* 81:1251-1261.
8. Baumgard, L.H., Corl, B.A., Dwyer, D.A., SΦbρ, A., and Bauman, D.E., 1999, Identification of the conjugated linoleic acid isomer which inhibits milk fat synthesis. *Am. J. Physiol.* (in press).
9. McGuire, M.A., Griinari, J.M., Dwyer, D.A., and Bauman, D.E., 1995, Role of insulin in the regulation of mammary synthesis of fat and protein. *J. Dairy Sci.* 78:816-824.
10. Griinari, J.M., McGuire, M.A., Dwyer, D.A., Bauman, D.E., and Palmquist, D.L., 1997, Role of insulin in the regulation of milk fat synthesis in dairy cows. *J. Dairy Sci.* 80:1076-1084.
11. Mackle, T.R., Dwyer, D.A., Ingvartsen, K.L., Chouinard, P.Y., Lynch, J.M., Barbano, D.M., and Bauman, D.E., 1999, Effects of insulin and amino acids on milk protein concentration and yield from dairy cows. *J. Dairy Sci.* 82:1512-1524.
12. Storry, J.E., and Rook, J.A.F., 1965, The effects of a diet low in hay and high in flaked maize on milk-fat secretion and on the concentrations of certain constituents in the blood plasma of the cow. *Brit. J. Nutr.* 19:101-109.
13. Hurtaud, C., Rulquin, H., and Verite, R., 1998, Effects of graded duodenal infusions of glucose on yield and composition of milk from dairy cows. 1. Diets based on corn silage. *J. Dairy Sci.* 81:3239-3247.

14. Reynolds, C.K., Sutton, J.D., and Beever, D.E., 1997, Effects of feeding starch to dairy cattle on nutrient availability and production. In *Recent Advances in Animal Nutrition* (P.C. Garnsworthy and J. Wiseman, eds.), Butterworths, London, UK, pp. 105-134.
15. Lemosquet, S., Rideau, N., Rulquin, H., Faverdin, P., Simon, J., and Verite, R., 1997, Effects of duodenal glucose infusion on the relationship between plasma concentrations of glucose and insulin in dairy cows. *J. Dairy Sci.* 80:2854-2865.
16. Wonsil, B.J., Herbein, J.H., and Watkins, B.A., 1994, Dietary and ruminal deprived *trans*-18:1 fatty acids alter bovine milk lipids. *J. Nutr.* 124:556-565.
17. Griinari, J.M., and Bauman, D.E., 1999, Biosynthesis of CLA and Incorporation into Milk Fat. In: *Advances in Conjugated Linoleic Acid Research, Vol. 1* (M.P. Yurawecz, M.M. Mossoba, J.K.G. Kramer, M.W. Pariza, and G. J. Nelson, eds.), AOCS Press, Champaign, IL, pp. 180-200.
18. Griinari, J.M., Nurmela, K., Dwyer, D.A., Barbano, D.M. and Bauman, D.E., 1999, Variation of milk fat concentration of conjugated linoleic acid and milk fat percentage is associated with a change in ruminal biohydrogenation. *J. Anim. Sci.* 77(Suppl. 1):117-118 (abstr.).

27.

# Effect of Feeding Pattern and Behaviour on Hormonal Changes and Milk Composition

Giuseppe Bertoni
Istituto di Zootecnica, Facoltà di Agraria, Università Cattolica del Sacro Cuore, Piacenza, Italy

Key words: Dairy cows, feeding pattern, hormones, milk

Abstract: There is no doubt that feeding and nutrient availability modify the metabolism through hormonal changes. It is otherwise interesting to better define the influence of the imposed or freely chosen meal distribution during the day; in fact the daily feed intake in a relatively short time, namely during the daylight, seems to induce a clear anabolic phase in afternoon-night (high insulin and low urea) and a catabolic one in the morning (low insulin and high urea); among the possible consequences, the lower fat content in the morning milking and the higher one in the afternoon, are of great importance. The factors that can influence these daily changes are many, namely the meal size, the day or night, the stage of lactation and perhaps a genetic effect. These results can be useful for a better interpretation of blood parameters and of the relationships between feeding and milk composition.

## 1. INTRODUCTION

It is well known that feeding influences the availability of nutrients which in turn represent the precursors of milk constituents. It is also known that mammary gland uptake and utilisation of these precursors are strongly influenced by the hormonal status[1] as well as by the mammary gland itself[2]. These facts could explain many differences in milk yield and composition that the availability of nutrients alone cannot fully justify, namely the different behaviour of milk fat content in early or late lactation in cows fed different concentrate: forage ratios[3]; the same is true for the protein content (our Institute unpublished results).

The obvious consequence is that many different factors are involved in the milk components synthesis and not all are well known as suggested by Maas et al.[4] for the protein.

## 2. FEEDING, ENDOCRINE CHANGES AND MAMMARY RESPONSE

One of the reasons is surely the effects of hormones that in turn can be modified in their release, blood levels, peripheral activity etc.; feeding is one of the causes of these variations, but our interest has been attracted to the feeding pattern. Quite old works have shown the influence of concentrate partitioning on the rumen pH and on milk yield[5]; even the frequency of feeding and methods of feeding (e.g., total mixed rations) are commonly known to influence milk composition[3]. Also Yang and Varga[6] did show that feeding concentrate four times daily increased milk fat and proteins. Similar results were obtained by Sutton et al.[7] and Sutton et al.[8], but later[9] these authors reconsidered their results and concluded that the partitioning of nutrients may be more critically related to the postprandial changes in circulating volatile fatty acids and insulin, and that the postprandial increase in insulin measured after feeding a 70% concentrate ration twice daily may have been sufficient to stimulate milk-fat depression.

Nevertheless, the concept of a tight relationship between VFA and hormone levels that Bassett[10] emphasized for insulin and glucagon, does not seem so sure; in fact the last paper of Bergman[11], speaking about the indirect effects of volatile fatty acids on metabolism, stated: "despite the above clear-cut effects of VFA on insulin and glucagon secretion in ruminants, there is still uncertainty about the physiological importance of propionate and butyrate as regulators of hormone release". Furthermore, it is longer known that in ruminants the postprandial insulin raise is mainly consequence of nervous-endocrine (digestive tract) effects and this explains the contemporary glucose decline[12].

## 3. FEED DISTRIBUTION

Since the late '70s we tried to understand the effects of the meals on the blood metabolites[13], but only in the early '80s some hormones were also considered; namely, after the meal there is a rise of insulin and thyroid hormones, while cortisol is suddenly reduced[14]. Nevertheless, it was only in the early '90s [15] that we paid attention to the metabolites and hormones as influenced not only by the meals, but also by the type and frequency of them during the day and how such influence can be affected by the lactation stage as well as by the genetic merit. The most interesting results of this last research work can be summarized: [16]:

a) in the early lactation the thyroid hormone raise after the meal is very small respect to the advanced stages of lactation;
b) also in the early stages, insulin and GH show an opposite behaviour respect to the late lactation, namely insulin is unchanged or reduced after the meal and GH is increased.

Nevertheless we were also able to show that small concentrate meals do not have relevant influence respect to the big forage-concentrate meals which effects, in turn, depend on the interval from the previous meal[17,18], when the meals are not symmetric. The effects on milk yield cannot be reliably quantified in this kind of experiments of short period; on the contrary, the milk composition is modified when 3 vs. 2 meals are administered: i.e. milk fat and protein contents are increased[17].

A similar experiment with not yet published data has been carried out on 5 cows in their early lactation, switched from 12-12 to 8-16 h intervals immediately after a daily blood sampling from 7.00 a.m. (1st meal) to 11 p.m.; the same blood sampling was repeated after 4 days. The treatment was repeated 15-30, 40-60 and 80-100 days after calving. The very short period did not allow any important difference at the milk level, while blood variations of urea (fig. 1), insulin (fig. 2) and $T_3$ confirm the data of previous

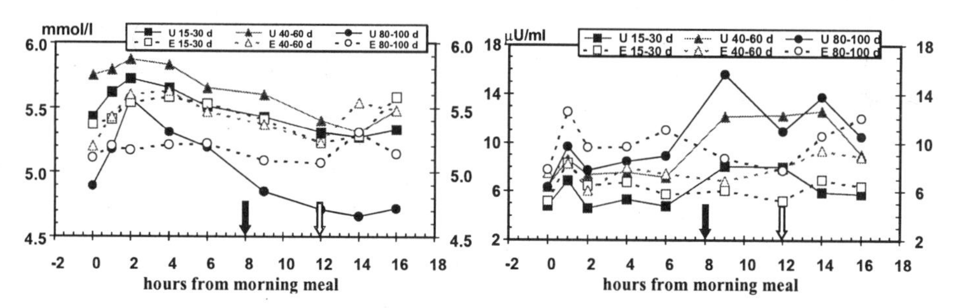


Fig.1-2 Daily pattern of changes of urea (mmol/l) and insulin (µU/ml) in dairy cows fed forages at equal (12-12 h: E - - ) or unequal (8-16 h: U — ) intervals and checked at 15-30, 40-60 and 80-100 days of lactation.

experiment, but furthermore it has been observed that:

a) urea is increased after 1st and not after 2nd meal in 8-16 h intervals, therefore its level fluctuate much more and particularly at the 3rd month of lactation;

b) the rise of insulin is more evident after the 2nd meal in the 8-16 h intervals and it becomes more remarkable as much as the lactation stage is advanced.

These data seemed promising for a better understanding of well known relationships between feeding, digestive activity, endocrine-metabolic response and milk yield and composition. Two new experiments were therefore carried out:

*a*) the first to see blood changes, every 1-2 h for a whole day, and milk differences between morning and afternoon milkings of 2 groups of dairy cows fed at 8-16 h intervals but with the 8 h interval alternatively during the day (7.00 a.m. and 3.00 p.m.) or the night (7 p.m. and 3 a.m.). The main results as presented by Trevisi et al[19] can be summarized:

1. the nightly fed animals show a slower intake, particularly after the 2nd meal;
2. this seems to allow a more spread nutrient availability during the 24 h because the insulin (fig. 3) and urea (fig. 4) pattern of changes are index of a relatively better constancy, particularly if compared with the daylight fed animals that show a long and deep anabolic phase (high insulin and low urea) from late afternoon to early morning while a shorter catabolic phase appears in the early morning (12-15 hours after the last meal start);
3. a small positive effect could be seen on the milk yield (and lactose synthesis), but it is particularly evident for the fat content that tends to be similar between the 2 daily milkings for the nightly fed animals and otherwise tends to be lower in the morning (fig. 5) after the long anabolic phase appearing in the daylight fed animals with 8-16 intervals;

*b*) within the dairy cows bred in our tied experimental stable, and with a perfectly symmetric distribution of meals, we had the opportunity to choose quick and slow eaters, the first being characterized by the tendency to eat the diet in few hours after the meal start. The animals were bled

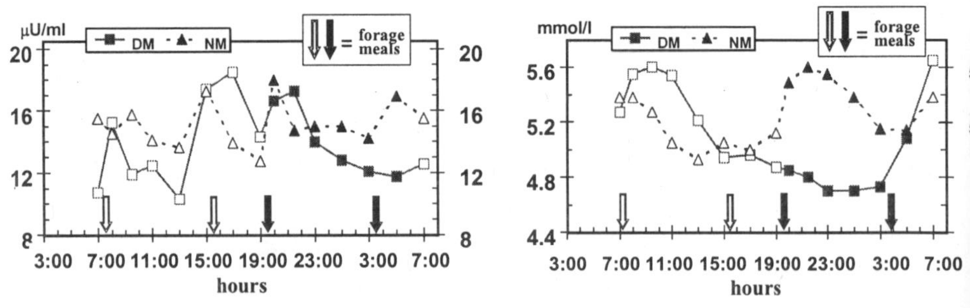


Fig.3-4 Daily pattern of changes of insulin (µU/ml) and urea (mmol/l) in dairy cows fed diurnal or nocturnal meals at 8-16 h intervals.

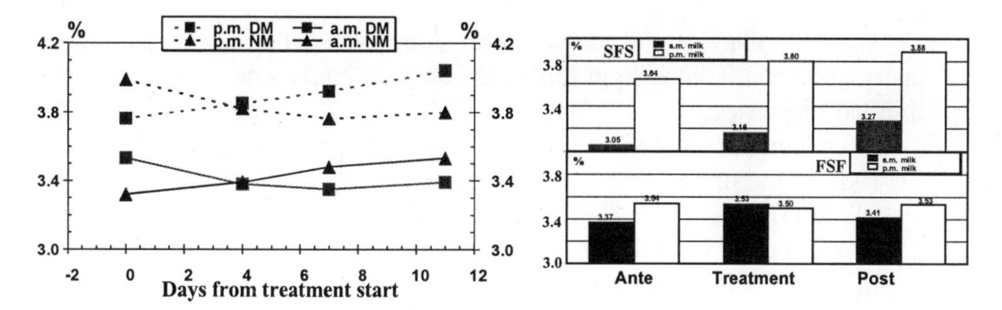


Fig. 5 Milk fat content in a.m. and p.m. milkings with diurnal and nocturnal meals.
Fig. 6 Average milk fat content of morning and afternoon milkings in dairy cows with a different ingestion rate (slow = S or fast = F) before, during and after an attempt to modify their feeding behaviour.

during the day before and after the quick eaters were obliged to a slower eating rate dividing the diet in 4 x 6 h meals. The results cannot be considered definitive, however the fat content of the morning milking was much lower in the slow eaters, while it was only slightly lower in the fast eaters (fig. 6) and it became similar or slightly higher (3,53 vs. 3,50%) compared to the afternoon milking when the fast eaters were obliged to slow down the eating rate. These results suggest that the slow eaters are in a situation similar to the 8-16 h intervals distribution of meals. In fact the insulin behaviour during the day shows that in fast eaters it suddenly increases after each meal but remains generally lower than in the slow eaters, in which insulin level is characterized by a gradual increase during the day and by a supposed higher value during the night. Moreover, the quick eaters obliged to slow down are not comparable to the "natural" slow eaters because they rapidly eat the available feeds and the post-prandium peak of insulin occurs 4 times instead of 2.

## 4. FINAL REMARKS

As recently underlined by Dawson[20], the way of feed intake during the day (light and dark) has an effect on the nutrient availability and, in turn, on the levels of some hormones, namely insulin. As suggested by Dawson[20], well distributed meals maintain a "constant" concentration of insulin that may suppress the tissue protein degradation with an improved protein gain. The interest in the dairy cows may be different, because insulin may be more important for milk composition (fat and protein); it is however interesting to consider the possibility that this could be due to the way of feed intake and that other hormones may be involved.

Anyhow, our results suggest that some different factors, like meal distribution restricted to light time and slow feed intake rate, can cause some peculiar responses:

1. at endocrine-metabolic level with a well differentiated anabolic phase (late afternoon till night) from a shorter catabolic phase (early morning till noon);
2. at milk composition level, particularly for fat content, much lower in the milking following the anabolic phase.

Besides a confirmation of an insulin role on the milk fat content, these interesting results suggest some stimulating remarks, particularly for future experiments:

1. the behaviour of metabolites and hormones can be different according to the feeding strategy, many times unknown because the animals can choose it; this impose some caution to interpreter blood parameters if measured;
2. small or big meals can influence the animal metabolism and the lactation stage can contribute to modify such influence;
3. perhaps there is a genetic effect on the feed intake rate and this could influence the milk fat contents of 2 consecutive milkings, the question is whether this could be used as a diagnostic tools for feeding habits evaluation.

## 5. REFERENCES


1. Bertoni G. Feeding and bovine milk quality: endocrine and metabolic factors. Zoot. Nutr. Anim. 22, 205-214, 1996.
2. Bequette BJ, Backwell FRC. Amino acid supply and metabolism by the ruminant mammary gland. Proc. of the Nutr. Soc. 56, 593-605, 1997.
3. Kennelly JJ, Glimm DR. The biological potential to alter the composition of milk. Can. J. Anim. Sci. 78(supp.), 23-26, 1998.
4. Maas JA, France J, McBride BW. Application of a mechanistic model of bovine milk protein synthesis to examine the use of isotope labeling methods. J. Dairy Sci. 81, 2440-2450, 1998.
5. Kaufmann W. Influence of the composition of the ration and the feeding frequency on pH-regulation in the rumen and on feed intake in ruminants. Livestock. Prod. Sci. 3, 103-114, 1976.
6. Yang C-M J, Varga GA. Effect of three concentrate feeding frequencies on rumen protozoa, rumen digesta kinetics, and milk yield in dairy cows. J. Dairy Sci. 72, 950-957, 1989.
7. Sutton JD, Broster WH, Napper DJ, Siviter JW. Feeding frequency for lactating cows: effects on digestion, milk production and energy utilization. British Journal of Nutrition 53, 117-130, 1985.

8. Sutton JD, Hart IC, Broster WH, Elliot RJ, Schuller E. Feeding frequency for lactating cows: effects on rumen fermentation and blood metabolites and hormones. British Journal of Nutrition 56, 181-192, 1986.
9. Sutton JD, Hart IC, Morant SV, Schuller E, Simmonds AD. Feeding frequency for lactating cows: diurnal patterns of hormones and metabolites in peripheral blood in relation to milk-fat concentration. British Journal of Nutrition 60, 265-274, 1988.
10. Bassett JM. Dietary and gastro-intestinal control of hormones regulating carbohydrate metabolism in ruminants. *In*: Digestion and Metabolism in the Ruminant. Proceedings of the IV International Symposium on Ruminant Physiology. Sydney, Australia, 383-398, 1975.
11. Bergman EN. Energy contributions of volatile fatty acids from the gastrointestinal tract in various species. Physiol. Rev. 70(2), 567-590, 1990.
12. Riis PM. Adaptation of metabolism to various conditions. Nutritional and other environmental conditions. *In*: Dinamic biochemistry of animal production, Amsterdam: Elsevier, 319-357, 1983.
13. Bertoni G, Galimberti A. Il Profilo Metabolico nelle lattifere: esecuzione ed interpretazione. Atti Soc. Ital. Buiatria IX, 241-251, 1977.
14. Bertoni G, Lombardelli R, Piccioli Cappelli F. Variazioni ormonali post-prandiali in lattifere con alimentazione ed in fase fisiologica diverse. Atti 2° Congr. Naz. S.I.B.C.A. 139-148, 1986.
15. Bertoni G, Lombardelli R. Effetti dell'alimentazione sul quadro endocrino e metabolico degli animali da latte. Atti IX Congr. Naz. A.S.P.A. Roma 2, 1161-1169, 1991.
16. Bertoni G, Lombardelli R, Cappa V. Variazioni di alcuni metaboliti ed ormoni nel corso della giornata e in fasi fisiologiche diverse in bovine da latte. Zoot. Nutr. Anim. 21, 271-283, 1995.
17. Lombardelli R, Bertoni G, Piccioli Cappelli F. Frequenza dei pasti nelle bovine da latte: effetti a livello metabolico, endocrino e produttivo. Zoot. Nutr. Anim. 19, 123-130, 1993.
18. Bertoni G, Trevisi E, Bani P. Metabolic effects of two different lapses without concentrate in early lactating dairy cows. Livestock Prod. Sci. 39, 139-140, 1994.
19. Trevisi E, Lombardelli R, Piccioli Cappelli F, Bertoni G. Effect of different daily distribution of feeds on blood insulin and milk composition. Cost 825 – Mammary Gland Biology, Tours, 16-18 Sept., 1999.
20. Dawson JM. Variation in nutrient supply and effects on whole body anabolism. *In*: Proc. VIIIth Protein metabolism and nutrition Symp.-EAAP Publ. n. 96 Wageningen Pers., 101-126, 1999.

# 28.

# Genetic Polymorphisms in Milk Protein Genes and their Impact on Milk Composition

Peter Dovc
*Dept. of Animal Science, Biotechnical Faculty, University of Ljubljana, Domzale, Slovenia*

Key words: lactoprotein genes, genetic variants, casein, β-lactoglobulin, alternative splicing, gene expression, pseudogene.

Abstract: About 40 different genetic variants are present at bovine milk protein loci and considerable differences in allele frequencies were observed among breeds. Advanced electrophoretic methods and recently, the DNA technology have been applied for identification and characterisation of milk protein genetic variants. Mutations with impact on gene expression have been identified. The influence of lactoprotein genetic variants on milk composition and cheese making ability has been studied extensively. The most prominent effects have been found for β-LG A/B and κ-CN A/B in cattle and $\alpha_{s1}$-CN E in goat.

## 1. INTRODUCTION

Six major milk proteins in ruminants are encoded by four casein genes, beta lactoglobulin- (β-LG) and alpha lactalbumin- (α-LA) gene. In cattle four casein genes ($\alpha_{s1}$-CN, β-CN, $\alpha_{s2}$-CN, κ-CN) are clustered on chromosome 6 occupying a genomic region of about 250 kb[1]. The gene order and transcriptional orientation are highly conserved among ruminant species. While casein genes are present in ruminants as single copy loci, pseudogenes for β-LG were found in cattle[2] , goat and sheep[3]. In addition, pseudogene for α-LA, flanked by two directly repeated LINE sequences, was found in cattle[4]. The length of the transcription unit and number and length of exons vary among milk protein genes considerably[5]. Due to the large number of short exons and presence of retroposon elements in the bovine $\alpha_{s2}$-CN gene, it has been proposed[6] that $\alpha_{s2}$- and $\beta$-CN gene evolved from the common ancestor by a couple of internal duplications. The availability of genomic sequences for all lactoprotein genes and analysis of allelic differences at molecular level enabled identification of new alleles and shed new light onto the regulatory mechanisms involved in the regulation of milk protein gene expression.

## 2. GENETIC POLYMORPHISMS AT LACTOPROTEIN LOCI

The genetic typing of milk was traditionally performed using electrophoretic separation of milk proteins in starch and polyacrylamide gels. These methods allowed only identification of deletions and mutations, which caused the change of the net charge of the protein. The application of more sensitive electrophoretic methods (isoelectric focusing, IEF) allowed faster routine typing of lactoproteins and led to the identification of new genetic variants at $\alpha_{s1}$-CN-, $\kappa$-CN- and $\beta$-LG locus[7,8]. In cattle, goat and sheep the sequence and position of amino acid differences in almost all the known variants has been established using biochemical methods. Besides the single amino acid substitutions, which are the most common type of mutations, some deletions[9] as well as changes in the number of phosphate groups have been observed. In cattle, which is compared with goat and sheep more polymorphic species, some 40 genetic variants at six major lactoprotein loci have been identified (Table 1).

The introduction of DNA technology enabled typing of lactoprotein genes in both sexes independent on lactation status. Using this technology, new allelic variants have been characterised[10] and identification of mutations within the non coding region became feasible. It has been shown that the deletion of 13 aa in bovine $\alpha_{s1}$-CN A is caused by exon 4 skipping due to the point mutation within the splice donor site and not by deletion of genomic DNA[11]. Similarly, shorter variants of ovine and caprine $\alpha_{s1}$-CN are associated with exon 16 skipping[12] and differential splicing of exons 11, 13 and 16 of pre-mRNA[13], respectively. In addition, mutations within the non-coding regions might have important quantitative effects as demonstrated by the occurrence of the LINE element in the exon 19 (last untranslated exon) of the goat $\alpha_{s1}$-CN E allele. The presence of the LINE element could be responsible for the reduced mRNA stability and consequently for the three-fold reduction of $\alpha_{s1}$-CN E expression in goat[14]. Allele A and B specific polymorphisms were found in the proximal promoter region of the $\beta$-LG gene[15], which might cause differential expression of both alleles due to different AP-2 binding affinity. The bovine $\kappa$-CN A and B alleles are also expressed differentially[16], but no allele specific polymorphisms could be found in the proximal promoter region. However, allele specific polymorphism within the 3'-UTR of the κ-CN gene might be related with differential stability of the allele specific mRNA transcripts[17].

Table 1: The most common alleles of six major bovine lactoprotein genes

| Locus | Alleles | |
|---|---|---|
| **$\alpha_{s1}$-CN** | A: deletion: aa 14-26<br>B: reference allele<br>C: $_{192}$Glu→Gly<br>D: $_{53}$Ala→ThrP | E: $_{59}$Gln→Lys; $_{192}$Glu→Gly<br>F: $_{66}$SerP→Leu<br>G: 371 bp insertion in the exon 19 |
| **$\alpha_{s2}$-CN** | A: reference allele<br>B: ND | C: SerP→Gly<br>D: delet. aa 51 → 59, exon VIII skipping |
| **β-CN** | A1: $_{67}$Pro→His<br>A2: reference allele<br>A3: $_{106}$His→Gln<br>B: $_{67}$Pro→His; $_{122}$Ser→Arg | C: $_{35}$SerP→Ser; $_{37}$Glu→Lys $_{67}$Pro→-His<br>D: $_{18}$SerP→Lys<br>E: $_{36}$Glu→Lys |
| **κ-CN** | A: reference allele<br>B: $_{136}$Thr→Ile; $_{148}$AspAla<br>C:$_{81}$Asn→Asp; $_{97}$Arg→His; $_{136}$Thr→Ile; $_{148}$Asp→Ala | E: $_{155}$Ser→Gly<br>F:$_{10}$Arg→His<br>G:$_{97}$Arg→Cys |
| **β-LG** | A:$_{64}$Gly→Asp;$_{118}$Ala→Val<br>B: reference allele<br>C: $_{59}$Glu→His<br>D:$_{45}$Glu→Gln<br>E:$_{158}$Glu→Gly<br>F:$_{50}$Pro→Ser; $_{129}$Asp→Tyr $_{158}$Glu→Gly | G: $_{78}$Ile→Met; $_{158}$Glu→Gly<br>H: ND<br>I: $_{106}$Glu→Gly<br>J: $_{126}$Pro→Leu<br>W: $_{56}$Ile→Leu<br>X: ND |
| **α-LA** | A: $_{10}$Arg→Glu<br>B: reference allele<br>C: ND | |

ND: not determined

## 3. ALLELE FREQUENCIES IN DIFFERENT BREEDS

The distribution of lactoprotein allele frequencies among different cattle breeds varies considerably. For the majority of breeds $\alpha_{s2}$-CN and α-LA locus are considered to be monomorphic. In most breeds $\alpha_{s1}$-CN B and β-CN A1 and A2 are the most common alleles. High frequencies of κ-CN A allele were found in Ayershire, Holstein Friesian, Fleckvieh, Guernsey and Red Danish Dairy Cattle, whereas κ-CN B allele is present at intermediate frequencies in Jersey, Normande and Brown Swiss. However, in Jersey population in California a very high frequency of κ-CN B was found[18]. Distribution of β-LG A and B alleles is close to intermediate frequencies in the majority of the breeds with exception of Canadian Shorthorn, Red

Danish Dairy Cattle and Ayershire[19], where the frequency of B allele is around 0.8. The most common haplotype for the $\alpha_{s1}$-CN, β-CN and κ-CN locus in most breeds is B/A2/A with the frequencies of 0.30 - 0.40. In Europe β-CN A1 and $\alpha_{s1}$-CN C are almost absent in the Northern breeds and $\alpha_{s2}$-CN D and β-LG D are mainly found in and around the Alpine area.

# 4. LACTOPROTEIN GENETIC VARIANTS, MILK COMPOSITION AND CHEESE MAKING PROPERTIES

Several studies examined the effects of milk protein genetic variants on milk production traits, milk composition and technological properties of milk. However, results from those studies are sometimes conflicting with respect to the significance and the size of genotype effects[20]. The age of the animals, stage of lactation, different genetic background in different breeds and statistical model can bias the results considerably. The association with higher milk, fat and protein yields has been shown for $\alpha_{s1}$-CN B and β-CN A alleles as well as positive effect of κ-CN BB and β-LG AA genotypes on protein yield[21]. Significant positive effect on fat percentage, overall solids and casein content was found for κ-CN B and β-LG B alleles[22]. In addition, positive additive effect of β-LG allele B on casein content and casein number has been found[23]. Quantitative effects of β-LG alleles A and B and $\alpha_{s1}$-CN alleles B and C were demonstrated in the Simmental and Brown Alpine breeds[24]. Comparison of statistical models analysing single locus and multi-locus effects revealed only small number of statistically significant effects[20]. Considering the technological properties of milk it has been shown that milk containing κ-CN B variant has better rennetability, better curd firmness, better cheese yield and better recovery of fat[25]. In the Finnish Ayershire population, where 8% of non coagulating milk occurs, a positive effect of k-CN B allele on curd firmness compared with k-CN E was found[26]. In addition it has been shown that quantitative differences in Parmesan cheese production between k-CN AA and k-CN BB milk were greater than expected from differences in milk composition[27].

In conclusion, the most striking effects of lactoprotein genetic variants in cattle were found for κ-CN allele B, which is associated with shorter coagulation time, faster rate of firming and production of firmer curd, and for β-LG B allele which has major effect on β-LG content (A>B) and positive effect on casein number. In goat The major effect on casein content in goat has $\alpha_{s1}$-CN, where the E allele is expressed at three-fold lower level[14].

# 5. REFERENCES

1. Rijnkels, M., Kooiman, P. M., de Boer, H.A., Pieper, F.R. 1997. Organization of the bovine casein gene locus. *Mammalian Genome* **8**: 148-152.
2. Passey, R.J. and Mackinlay, A.G. 1995. Characterization of a second, apparently inactive, copy of the bovine β-lactoglobuline gene. *Eur. J. Biochem.* **233**(3): 736-743.
3. Folch, J. M., Coll, A., Hayes, H. C., Sanchez, A. 1996. Characterization of a caprine beta-lactoglobulin pseudogene, identification and chromosomal localization by in situ hybridization in goat, sheep and cow. *Gene* **177**(1-2): 87-91.
4. Vilotte, J. L., Soulier, S., Mercier, J.C. 1993. Complete sequence of a bovine alpha-lactalbumin pseudogene: the region homologous to the gene flanked by two directly repeated LINE sequences. *Genomics* **16**(2): 529-32.
5. Mercier,J.C. and Vilotte, J.L. 1993. Structure and Function of Milk Protein Genes. *J.Dairy Sci.* **76**: 3079-3098.
6. Groenen, M. A. M., Dijkhof, R. J. M., Verstege, A. J. M. and van der Poel, J. J. 1993. The complete sequence of the gene encoding bovine $\alpha_{s2}$-casein. *Gene***123**: 187-193
7. Krause, I., Buchberger, J., Weiss, G., Pflügler, M., Klostermeyer, H. 1989. Isoelectric focusing in immobilized pH gradients with carrier ampholytes added for high-resolution phenotyping of bovine β-lactoglobulins: characterization of a new genetic variant. *Electrophoresis* **9**: 609-613
8. Erhard, G. 1993. Allele frequencies of milk proteins in German cattle breeds and demonstration of $\alpha_{s2}$- casein variants by isoelectric focusing. *Arch.Tierz.,Dummerstorf* **36**: 145-152
9. Grosclaude, F., Joudrier, P., Mahe, M.F. 1978. Polymorphisme de la caséine $\alpha_{s2}$ bovine: étroite liasons du locus $\alpha_{s2}$-Cn avec les loci $\alpha_{s1}$-Cn, β-Cn et κ-Cn; mise en évidence d'une délétion dans le variant $\alpha_{s2}$-CN D. *Ann. Génét. Sél. Anim.* **10**: 313-327.
10. Rando, A., Di Gregorio, P., Ramunno, L., Mariani, P., Fiorella, A., Senese, C., Marletta, D., Masina, P. 1998. Characterization of the CSN1A G Allele of the Bovine $\alpha_{s1}$-Casein Locus by the Insertion of a Relict of a Long Interspersed Element. *J. Dairy Sci.* **81**: 1735-1742.
11. Mohr, U., Koczan, D. Linder, D., Hobom, G., Erhardt, G. 1994. A single point mutation results in A allele-specific exon skipping in the bovine $\alpha_{s1}$-casein mRNA. *Gene* **143**: 187-192.
12. Passey, R., Glenn, W., Mackinlay, A. 1996. Exon skipping in the ovine $\alpha_{s1}$-casein gene. *Comp. Biochem. Physiol.* **114B**: 389-394.
13. Ferranti,P., Addeo, F., Malorni, A., Chianese, L., Leroux, C., Martin, P. 1997. Differential splicing of pre-messenger RNA produces multiple forms of mature caprine $\alpha_{s1}$-casein. *Eur. J. of Biochem.* **249**: 1-7.
14. Perez, M. J., Leroux, C., Bonastre, A. S., Martin, P. 1994. Occurrence of a LINE sequence in the 3'UTR of the goat alpha s1-casein E-encoding allele associated with reduced protein synthesis level. *Gene* **147**: 179-187.
15. Lum, L. S., Dovc, P., Medrano, J. F. 1997. Polymorphisms of Bovine β-Lactoglobulin Promoter and Differences in the Binding Affinity of Activator Protein-2 Transcription Factor. *J. Dairy Sci.* **80**: 1398-1397.
16. VanEenennaam, A., Medrano, J.F. 1991. Differences in allelic protein expression in the milk of heterozygous κ-casein cows. *J. Dairy Sci.* **74**: 1491-1496.
17. Debeljak, M., Susnik, S., Marinsek-Logar, R., Medrano J. F., Dovc, P. Allelic differences in bovine kappa-CN gene related to the regulation of gene expression. *Eur. J. Physiol. Pflügers Archiv* (in press).

18. VanEenennaam, A., Medrano, J. F. 1991. Milk protein polymorphisms in California dairy cattle. *J. Dairy Sci* **74**: 1730-1742.
19. Buchberger, J. 1995. Genetic polymorphisms of milk proteins: Differences between breeds. *Bulletin of the IDF* **304**: 5-6.
20. Bovenhuis, H., van Arendonk, J. A. M., Korver S. 1992. Associations between Milk Protein Polymorphism and Milk Production Traits. *J.Dairy Sci.* **75**: 2549-2559.
21. Ng-Kwai-Hang, K. F., Hajes, J. F., Moxley, J. E., Monardes,H.G. 1986. Relationships between Milk Protein Polymorphism and Mayor Milk Constituents in Holstein-Friesian Cows. *J.Dairy Sci.* **69**: 22-26.
22. Ron, M., Yoffe, O., Ezra, E., Medrano, J. F., Weller, I. J. 1994. Determination of effects of milk protein genotype on production traits of Israeli Holsteins. *J. Dairy Sci.* **77**: 1106-1113.
23. Lunden, A., Nilson, M., Janson, L. 1997. Marked Effect of β-Lactoglobulin Polymorphism on the Ratio of Casein to Total Protein in Milk. *J. Dairy Sci.***80**: 2996-3005.
24. Graml,R., Weiss, G., Buchberger, J., Pirchner, F. 1989. Different rates of synthesis of whey protein and casein by alleles of the β-lactoglobulin and $\alpha_{s1}$-casein locus in cattle. *Genet. Sel. Evol.* **21**: 547-554.
25. Jakob, E. and Puhan, Z. 1992. Technological Properties of Milk as Influenced by Genetic Polymorphism of Milk Proteins. *Int. Dairy Journal* **2**: 157-178.
26. Ikonen, T., Ahlfors, K., Kempe, R., Ojala, M., Ruottinen, O. 1999. Genetic parameters for the milk coagulation properties and prevalence of noncoagulating milk in Finnish Dairy Cows. *J. Dairy Sci.* **82**: 205-214.
27. Aleandri, R., Buttazzoni, G., Schneider, J.C., Caroli, A., Davoli, R. 1990. The effects of milk protein polymorphisms on milk components and cheese-producing ability. *J. Dairy Sci.* **73**: 241-255.

# 29.

# Mammary Gland Immunology Around Parturition

## *Influence of stress, nutrition and genetics*

Karin Persson Waller

*Department of Obstetrics and Gynaecology, Centre for Reproductive Biology, Swedish University of Agricultural Sciences, Uppsala, Sweden*

Key words: Bovine, mammary, immunology, stress, nutrition, genetics

Abstract: Adequate immune functions are essential for the defence against udder infections. Detailed knowledge about the immune response and important defence factors is essential in order to find new ways for the prevention and treatment of udder infections leading to mastitis. Work should be concentrated on ways of minimising the negative influence on immune functions and/or ways of stimulating these functions, especially during periods of immune suppression. A depression of important immune functions has been reported around parturition and there is a higher prevalence of clinical mastitis and other diseases during this period. Immunosuppression is often associated with high levels of glucocorticoids in blood, a common finding around parturition and during stressful conditions. A number of stressors are present around calving, e.g. parturition, onset of lactation and changes in feeding and management regimes. Adequate management including feeding strategies and routines are important for the immune functions. Metabolic stress as well as deficiencies in vitamins and minerals around parturition and during the first month of lactation can have a negative influence on the immune functions and thereby increase the risks for udder infections and mastitis. There seem to be a genetic variability in certain immune functions among periparturient cows. This might indicate a possibility to find markers for genetic selection of individuals with a well-developed immune system without negative effects on milk productivity.

## 1. INTRODUCTION

Udder infections and mastitis are major problems for the dairy industry throughout the world, and the yearly costs are substantial. The risks for udder infections are highest around drying-off and parturition, and the prevalence of clinical mastitis is higher during the first month after calving. Other diseases prevalent in the periparturient period, such as retained placenta, parturient paresis and ketosis, increase the risk for clinical mastitis.

Adequate immune functions are essential for the defence against udder infections. However, increasing evidence show that several important immune functions are depressed during certain periods, in particular around calving. This immune suppression renders the animals more susceptible to infections and diseases like mastitis. The causes for the immune suppression are not fully understood but several factors are involved such as stress, changes in hormonal patterns as well as in feeding and management, onset of lactation and high milk production, metabolic disturbances and genetics.

Detailed knowledge about the immune responses and important host defence factors is essential in order to find better ways for the prevention and treatment of udder infections and mastitis. Work should be concentrated on ways of minimising the negative influence on immune functions and/or ways of stimulating these functions, especially during periods of immune suppression, to increase the natural ability of the cow to resist, or defend herself, against udder infections.

This paper will consist of a review of the knowledge in bovine mammary gland immunology around parturition with special emphasis on the influence of stress, nutrition and genetics.

## 2. IMMUNE FUNCTIONS AROUND CALVING

An increasing number of reports describe changes in leukocyte trafficking patterns and decreased functions of a number of immunological parameters in dairy cows around parturition. Alterations in immune responsiveness during this period has lately been reviewed[47,67].

The numbers and proportions of white blood cells (WBC) change around parturition. The highest total leukocyte counts occur at parturition, then return to normal levels within a few weeks[22,52]. In a recent study, we detected significantly higher numbers of total WBC, polymorphonuclear neutrophils (PMN) and monocytes in blood at parturition compared with 1 month before and 1 month after parturition, but the numbers of lymphocytes and eosinophils did not change[76]. This is somewhat different from a study by Moreira da Silva et al[52] who reported an increase of lymphocytes, but not of monocytes, at parturition.

The greatest numbers of PMN in blood are found on the day of parturition[22]. At parturition and onset of lactation there is an increasing demand for PMN which are mobilised from the marginating pool, due to high levels of cortisol at parturition, and from the bone marrow. Only small numbers of mature PMN are stored in the bone marrow, therefore the numbers of immature neutrophils in circulation increase, with the highest proportion occurring during the first week after parturition[22,35,52]. In immature PMN, several important functions are reduced such as phagocytosis, intracellular killing and chemotaxis[13,14,32]. The numbers of circulating mature neutrophils was negatively correlated with severity of mastitis.

The lymphocyte patterns in both blood and milk also change around parturition. The proportion of T-cells declines and a number of T-cell subsets (CD4, CD8, WC1) vary significantly during this period[69]. The proportions of several cell populations (CD2, CD4, CD8, MHCII) are lower in milk than in blood following parturition, while the proportion of WC1+ cells is higher in milk[47]. The proportions of CD4+ cells decrease around parturition, both in blood and mammary tissue, and this is correlated with a decrease in interferon-γ (IFN-γ) secretion by lymphocytes and in IFN-γ concentrations in colostrum[31,61]. The ratio of CD4+/CD8+ T cells is lower in milk than in blood, indicating a higher proportion of CD8+ cells in the mammary gland. Moreover, Shafer-Weaver and Sordillo[61] observed preferential trafficking of CD8+ T-cells of the suppressor type rather than of the cytotoxic type into the mammary gland around calving. This was supported by the finding that interleukin-4 (IL-4) is the main cytokine in mononuclear cells from cows at parturition, while IFN-γ is the main cytokine in monocytes from cows in middle to later stages of lactation[61]. Moreover, mononuclear cells from periparturient cows produce more tumor necrosis factor-α than cows in mid-lactation[66].

Several important functions of blood and mammary neutrophils and lymphocytes decrease during the weeks immediately before calving. Most commonly they are at their lowest at, or just after, calving and return to normal levels within three weeks after calving[10,35,36,44,62]. Reduced functions of the leukocytes diminish their ability to kill bacteria and to stimulate the immune defence making the animals more susceptible to infections. Impaired PMN function is clearly associated with a higher incidence and severity of periparturient mastitis[6,63].

Impaired phagocytosis, respiratory burst, superoxide anion production, random migration, chemotaxis, and myeloperoxidase activity of PMN has been observed around parturition[6,19,35,59]. However, other studies report increased, or unchanged, phagocytosis by blood PMN after calving[15]. Moreover, the activity in PMN of the enzyme acyloxyacyl hydrolase (AOAH), important for degradation of endotoxin, decreases after calving[14]. The ability of neutrophils to migrate from blood to tissues is impaired at parturition due to a decreased proportion of cells expressing adhesion receptors[44,76]. Lymphocyte functions, like mitogen-induced proliferation, are also depressed around calving[36,47,59]. Moreover, mammary gland lymphocytes have a lower capacity than blood cells, possibly due to less efficient presentation of antigen by antigen-presenting cells of the mammary gland.

Cytokines are essential for adequate immune functions. The levels of interleukin-2 (IL-2) and IFN-γ in mammary gland secretions decrease around calving[65] which might contribute to the increased susceptibility to infections. Moreover, the capacity of B-cells from periparturient cows to secrete IgM is impaired which might be caused by the impaired production of IFN-γ and IL-2 [11,39]. In addition, at parturition the corticosteroid levels are high which can block cellular production of IL-1, IL-8 and tumor necrosis factor- (TNF-a) as a response to endotoxin stimulation[39].

What is the reason behind the depressed functional capacity of leukocytes? Metabolic and hormonal changes after parturition are probably important factors. Increased levels of beta-hydroxybutyrate, not uncommon during early lactation, can for example reduce the respiratory burst activity of PMN, probably due to impaired production of hydrogen peroxide, and inhibit lymphocyte functions. There is also a significant relationship between negative energy balance, increased blood ketone levels and impaired PMN function. This will be further elucidated in Section 4.

Large changes in many hormones occur around parturition. Such changes in progesterone, estradiol 17ß, cortisol, IGF-1 and growth hormone around parturition can modulate PMN and lymphocyte functions[47,52]. Recently, bovine pregnancy-associated glycoprotein (BPAG), originating in bovine fetal cotyledons, has been associated with immune suppression[15]. The plasma concentration of BPAG increases around parturition and may be associated with inhibition of PMN function. Plasma levels of the endogenous opioids ß-endorphin and met-enkephalin also increase during the periparturient period with the peak level at parturition which also might influence the immune functions[2].

Plasma cortisol levels are increased shortly after calving. A number of potentially stressful factors, like parturition, onset of lactation and changes in management, influence the cow around calving probably through additional release of cortisol. The changes in immune functions around calving are similar to those observed *in vivo* and *in vitro* after treatment with corticosteroids.

## 3. INFLUENCE OF STRESS

Stress, occurring for example around calving or due to adverse environmental factors, can have negative effects on the immune response and make both adults and young animals less resistant to infectious diseases. This is probably mainly due to increased levels of corticosteroids which can affect both the functions and numbers of leukocytes, and thus increase the host's susceptibility to infections[22]. The effects of stress on leukocyte trafficking and immune response were reviewed by Kehrli et al[39].

Treatment of cows with synthetic glucocorticoids, like dexamethasone, negatively influences the numbers and proportions of leukocytes in blood by down-regulation of adhesion molecules on blood neutrophils[3,5]. Such changes are correlated with leukocytosis and increased shedding of *Staphylococcus aureus* in subclinically infected mammary glands, which reinforces the potential risk of treating infected animals with potent synthetic glucocorticoids.

Administration of cortisol and dexamethasone induce neutrophilia in cattle without significant increase in immature neutrophils suggesting that glucocorticoids decrease the marginal pool of neutrophils and reduce the efficiency of their egress from blood to tissues[5]. As mentioned, glucocorticoids decrease the adherence of PMN to the endothelium by altering the expression of leukocyte adhesion molecules CD62L and CD18, possibly by inhibiting production of regulating cytokines. This will negatively affect the ability of neutrophils to migrate from blood to sites of infections. Moreover, dexamethasone reduces the expression of intercellular adhesion molecule-1 (ICAM-1) on endothelial cells *in vitro*[8]. Glucocorticoids do not influence PMN phagocytosis of *E. coli in vitro,* and results from a study in periparturient cows suggest that phagocytosis of blood PMN was not affected by increased plasma cortisol levels[14]. Moreover, increased levels of cortisol did not have any effect on PMN chemiluminescence[26]. However, corticosteroids can interfere with adherence of IgG and complement to leukocytes *in vitro*[60].

An increase in plasma cortisol decreases the numbers of circulating lymphocytes[30]. The numbers of γδT-cells in blood decline after glucocorticoid administration, but the expression of CD62L on the cells is not affected. This suggests that glucocorticoids can enhance migration of γδT-cells from blood into tissues[4]. However, the ability of lymphocytes and monocytes to produce INF-γ and IgM in vitro is negatively affected by corticosteroids[12,54]. Glucocorticoids can also decrease the expression of MHC class I and II molecules on mononuclear cells[4,54], and inhibit endotoxin-stimulated cytokine production and gene expression of the cytokines IL-1, IL-8 and TNF-α [39].

# 4. INFLUENCE OF NUTRITION

## 4.1 General aspects

Adequate nutrition is essential for a good immune response and thus for overall health and the ability of the animal to resist diseases. The balance between protein and energy, the feeding routines and eating time are factors influencing animal health by potentially increasing the risk for metabolic disturbances. Differences in feeding strategy and/or routines can be factors causing high or low incidence of clinical mastitis[24]. Of course, good hygienic quality of the feed is also essential.

Metabolic stress is a common state in high-producing dairy cows at the beginning of lactation due to negative energy balance leading to the mobilisation of considerable amounts of tissue reserves. Negative energy balance and elevated blood ketone levels could play a role in the decreased neutrophil function observed after parturition and increased incidence of disease. Increased levels of ß-hydroxybutyrate (BHB) and ketone bodies can have a negative effect on neutrophil functions such as chemotaxis, phagocytosis and killing[27,41,68]. BHB can also inhibit the responses of bovine lymphocytes[41]. Kremer et al[42] reported that cases of coliform mastitis were more severe and of longer duration in ketotic cows compared with non-ketotic cows. *In vitro* studies by Hoeben et al[28] suggested that BHB and acetoacetic acid can alter the numbers of circulating neutrophils due to their effects on bone marrow progenitor cells.

Balanced feeding during the dry period is a prerequisite for good feed consumption after calving thus avoiding metabolic stress and increased risk for disease during the early lactation period. Overfeeding results in a larger demand on tissue reserves and higher levels of free fatty acids in the blood after calving, which can be negative for both humoral and cellular immune responses[58,73]. Underfeeding during the dry period can also cause metabolic diseases and immunosuppression[18,34]. A high fat content in the liver after calving can increase the risk for metabolic disturbances, and is associated with a decreased function of blood and uterine granulocytes[75].

The amino acid glutamine has several important metabolic functions. The demand of the dairy cow for glutamine increases in association with metabolic stress[23,43], such as at the beginning of lactation. During this period, the plasma content of glutamine decreases substantially[49]. Substantial amounts of glutamine are metabolised in the udder where it is used for synthesis of milk protein, as a source of energy and as substrate for cell proliferation[48]. Glutamine is a key substance also in the immune defence, such as for proliferation of lymphocytes and macrophages[57]. The glutamine

glutathione peroxidase, which is essential for the protection of cells and tissues from auto-oxidative damage from production of oxygen radicals[56]. Deficiency in Se resulted in reduced neutrophil migration into the udder, impaired neutrophil intracellular killing and extracellular hydrogen peroxide production, but did not affect neutrophil phagocytosis and superoxide anion production[21,29] Se supplementation can improve udder health by reducing the severity and duration of cases of clinical mastitis[29]. Moreover, *in vitro* supplementation of Se can enhance production of neutrophil chemotaxins by bovine mammary macrophages and enhance proliferation of peripheral blood lymphocytes[53]. Plasma Se decreases at calving according to Weiss et al (1990). However, in our study, plasma and erythrocyte Se were higher at calving than at one month before or after calving, indicating that Se deficiency was not a problem during this period[76].

Vitamin E is important for both cellular and humoral immune functions. This may be elicited through its effect on cell membrane stability and regulatory role in biosynthesis of various inflammatory mediators[56,64]. Dietary supplementation of vitamin E is of value, especially as the concentration of this vitamin in fodder decreases with length of storage. Moreover, the serum concentration of vitamin E drops during the weeks around calving, a period of immune depression and increased susceptibility to disease, reaching lowest levels 1 day after calving[20,72]. Supplementation with vitamin E before and after calving increased neutrophil intracellular killing of bacteria, and reduced the numbers of new intramammary infections[29]. Politis et al[55] observed that the suppression in blood neutrophil and monocyte functions postpartum was abrogated when cows were supplemented with vitamin E before and after parturition. Vitamin E supplementation did not affect lymphocyte proliferation and mammary macrophage production of superoxide, chemotactic factors and IL-1, or MHC class II expression[55]. However, vitamin E enhanced production of chemotaxins by bovine mammary macrophages and stimulated blood lymphocyte proliferation *in vitro*[53]. Dietary supplementation with vitamin E should be considered as a preventive measure in the control of mastitis especially during the late stage of the dry period and around calving.

Vitamin A and ß-carotene are also important for mucosal integrity and stability. ß-carotene can act independently as an oxygen radical scavenger. Both substances have stimulatory effects on immune cell populations and have been correlated with an increased resistance to disease[67]. Deficiencies were related with severity of mastitis[7]. Vitamin A and ß-carotene plasma concentrations decrease at parturition, reaching the lowest concentrations of vitamin A at, or just after, calving and the lowest concentrations of ß-carotene occur somewhat later[20,33]. The concentrations remained low for the first weeks after parturition. Concentrations of both vitamin A and ß-carotene were higher in colostrum than in milk[33]. Mastitic cows had a lower concentration of vitamin A in blood plasma during the first weeks of lactation than non-mastitic cows, but the opposite was the case for ß-carotene[33]. However, in milk, there were no significant differences in the concentrations of either substance between mastitic and non-mastitic cows. The concentration of vitamin A was negatively correlated with the SCC in milk during lactation while the opposite was the case for ß-carotene[33]. Vitamin A supplementation around calving was shown to stimulate the functions of blood lymphocytes and phagocytes[51].

Limited information is available about the importance of Cu and Zn in relation to the risk for diseases in dairy cows. However, both substances have important biological functions. For example they have importance for certain anti-oxidative enzyme systems and can thus protect cells and tissues from detrimental effects of oxidative substances which are released during phagocytosis and killing by leukocytes[56]. Deficiency in Cu can decrease the antibacterial effect of the immune defence[25,56]. Limited information is available regarding Cu levels in blood around calving. In our own study, the Cu levels were significantly lower one month before calving than at calving and one month after calving[76] which is in line with findings by Xin et al[74]. Supplementation of Cu resulted in a considerable decrease in the number of infected udder quarters at calving compared with untreated controls[25]. Moreover, the cell count in milk tended to be lower in the group that received extra Cu.

Zn is important for the integrity of the skin, which is the first barrier against infections, and has also importance for immune functions. Zn deficiency can cause degeneration of lymphoid tissues and have a negative influence on immune cells[25,56]. Few studies have been made to study the relationship between Zn and mastitis. However, deficiency in Zn can predispose cows to secondary infections, which can be reversed by supplementation of Zn[25,56]. Problems in connection with deficiency in Zn can be exacerbated by a feeding regime containing high amounts of calcium during early lactation. Moreover, plasma Zn concentrations decrease markedly at parturition, but returns to normal values within a few days[20,76]. This might be connected with the immune suppression observed during this period. During the acute phase of inflammation, Zn is removed from the blood circulation and stored in the liver where it is bound to the acute phase protein methallothionein[46]. A similar release of acute phase proteins at parturition may explain the decrease in Zn levels.

## 5. INFLUENCE OF GENETICS ON THE IMMUNE RESPONSE

The genetic influence on immunological parameters and disease resistance is another area of interest. Studies are aimed towards the possibility of selecting cows with improved immune response, without negative effects on milk productivity. It is also likely that genetic factors can have an effect on how an individual cow can handle stress in different situations.

Heritabilities for clinical mastitis has been estimated to be low[9,16]. Somatic cell counts (SCC) have a higher heritability, and might be a better selective tool[9,16]. Moderate genetic correlation has been estimated between clinical mastitis and SCC. In Scandinavia, a combination of these parameters is included in the breeding values of bulls with the aim of improving disease resistance. High milk production and various udder traits (such as udder suspensory ligament, fore udder attachment and udder depth) are associated with increased incidence of mastitis[1,16]. However, Detilleux et al[11] showed that cows selected for high milk production had significantly higher numbers of circulating neutrophils and mononuclear cells and higher neutrophil chemiluminescence and directed migration, than cows with average production.

Variations in leukocyte sub-populations may have a role in the probability of an individual cow developing mastitis, as the proportions of neutrophils, macrophages, and CD4+ or CD8+ T-lymphocytes differ significantly between cows[45]. Kelm et al[40] used different measures of mastitis, a number of immunological assays and molecular markers. The molecular markers accounted for up to 40 % of the variation in estimated breeding value (EBV) for measures of mastitis. In cows with a low EBV for somatic cell score, the neutrophils functioned more effectively at maximal immunosuppression, the serum immunoglobulin concentration was lower, and the numbers of circulating mononuclear cells were higher. Significant differences between sire progeny groups within lines for lymphocyte blastogenes, neutrophil function and serum conglutinin activity have also been observed[37]. Moreover, results by Fitzpatrick et al[17] indicated that immunological assays may be useful in identifying bulls whose progeny would be associated with a higher resistance to intramammary infections with *S. aureus*.

A selection program for animals resistant to immunosuppression and disease around parturition would be especially desirable. Genetic variability in certain immune functions, i.e. total numbers of neutrophils, neutrophil chemokinesis, assays of neutrophil respiratory burst associated with phagocytosis, and in serum concentrations of immunoglobulins and hemolytic complement activity was observed in periparturient dairy cows[10]. Mallard et al[47] showed that cows at peripartum may be categorised as high or low responders to ovalbumin and *E. coli* J5 immunisations, and that the heritability for specific antibody responses was moderate to high. These findings may be used to identify animal with high resistance to disease.

Several groups have investigated associations between alleles of the bovine MHC (BoLA) and disease resistance or immune response parameters[50,71]. Such associations were found, but the relationships varied among breeds and it is not clear which gene, or genes, are responsible for these associations.

The importance of adhesion molecules, essential for leukocyte migration, has recently been highlighted. Genetic leukocyte adhesion deficiencies leading to chronic, or even fatal, infections have been observed in several animal species including cattle bovine leukocyte adhesion deficiency (BLAD)[38]. All species affected show signs of chronic and recurrent infections due to a deficiency in the chemotactic and phagocytic properties of leukocytes, particularly neutrophils. However, there does not seem to be any differences in neutrophil and macrophage function or in the risk for udder infections at calving between BLAD carriers and noncarriers[70].

## 6. REFERENCES

1. Alexandersson G. Genetiska samband mellan exteriöra juveregenskaper och klinisk mastit hos förstakalvande SRB-kor. Publication 195, Department of Animal Breeding and Genetics, Swedish University of Agricultural Sciences, Uppsala, Sweden, 1998.
2. Aurich JE, Dobrinski I, Hoppen H-O, Grunert E. ß-endorphin and met-enkephalin in plasma of cattle during pregnancy, parturition, and the neonatal period. J Reprod Fertil, 89, 605-612, 1990.
3. Burton JL, Kehrli ME Jr. Regulation of neutrophil adhesion molecules and shedding of Staphylococcus aureus in milk of cortisol- and dexamethasone-treated cows. Am J Vet Res 56, 997-1006, 1995.
4. Burton JL, Kehrli ME Jr. Effects of dexamethasone on bovine circulating T lymphocyte populations. J Leukocyte Biology, 59, 90-99, 1996.
5. Burton JL, Kehrli ME Jr, Kapil S, Horst RL. Regulation of L-selectin and CD18 on bovine neutrophils by glucocorticoids: effects of cortisol and dexamethasone. J Leukocyte Biol 57, 317-325, 1995.
6. Cai T-Q, Weston PG, Lund LA, Brodie B, McKenna DJ, Wagner WC. Association between neutrophil functions and periparturient disorders in cows. Am J Vet Res, 55, 934-943, 1994.
7. Chew BP, Hollen LL, Hillers JK, Herlugson ML. Relationship between vitamin A and ß-carotene in blood plasma and milk and mastitis in Holsteins. J Dairy Sci 65, 2111-2118, 1982.
8. Cronstein BN, Kimmel SC, Levin RI, Martiniuk F, Weissmann G. A mechanism for the anti-inflammatory effects of corticosteroids: The glucocorticoid receptor regulates leukocyte adhesion to endothelial cells and expression of endothelial-leukocyte adhesion

molecule 1 and intercellular adhesion molecule 1. Proc Natl Acad Sci USA, 89, 9991-9995, 1992.

9. De Haas Y. Genetic parameters of mastitis traits and protein production in the first two lactations of Swedish red and White cows. Publication no 17. Department of Animal Breeding and Genetics, Swedish University of Agricultural Sciences, Uppsala, Sweden, 1998.
10. Detilleux JC, Koehler KJ, Freeman AE, Kehrli ME Jr, Kelley DH. Immunological parameters of periparturient Holstein cattle: Genetic variation. J Dairy Sci 77, 2640-2650, 1994.
11. Detilleux JC, Kehrli ME Jr, Stabel JR, Freeman AE, Kelley DH. Study of immunological dysfunction in periparturient Holstein cattle selected for high and average milk production. Vet Immunol Immunopathol, 44, 251-267, 1995.
12. Doherty ML, Bassett HF, Quinn PJ, Davis WC, Monaghan ML. Effects of dexamethasone on cell-mediated immune responses in cattle sensitized to *Mycobacterium bovis*. Am J Vet Res, 56, 1300-1306, 1995.
13. Dosogne H, Burvenich C, van Werven T, Roets E, Noordhuizen-Stassen EN, Goddeeris B. Increased surface expression of CD11b receptors on polymorphonuclear leukocytes is not sufficient to sustain phagocytosis during Escherichia coli mastitis in early postpartum dairy cows. Vet Immunol Immunopath 60, 47-59, 1997a.
14. Dosogne H, Burvenich C, Paape MJ, Capuco A, Fenwick B. Defense against *Escherichia coli*: Phagocytosis and detoxification of endotoxins by neutrophils. Flem Vet J, 66, Suppl., 243-268, 1997b.
15. Dosogne H, Burvenich C, Freeman AE, Kehrli ME Jr, Detilleux JC, Sulon J, Beckers JF, Hoeben D. Pregnancy-associated glycoprotein and decreased polymorphonuclear leukocyte function in early post-partum dairy cows. Vet Immunol Immunopathol, 67, 47-54, 1999.
16. Emanuelson U, Danell B, Philipsson J. Genetic parameters for clinical mastitis, somatic cell counts, and milk production estimated by multiple-trait restricted maximum likelihood. J Dairy Sci 71, 467-476, 1988.
17. Fitzpatrick JL, Logan KE, Stear MJ, McGuirk B, Platt DJ. Immunological responses of Holstein/Friesian cattle to Staphylococcus aureus. In: Proceedings of the 5th International Veterinary Immunology Symposium, November 8-13, Punjab Agricultural University, Ludhiana, India, 117, 1998.
18. Franklin S, Young J, Nonnecke B. Effects of ketones, acetate, butyrate and glucose on bovine lymphocyte proliferation. J Dairy Sci, 80, 2507-2516, 1991.
19. Gilbert RO, Gröhn YT, Miller PM, Hoffman DJ. Effect of parity on periparturient neutrophil function in dairy cows. Vet Immunol Immunopatol, 36, 75-82, 1993.
20. Goff JP, Stabel JR. Decreased plasma retinol, α-tocopherol, and zinc concentration during the periparturient period: Effect of milk fever. J Dairy Sci 73, 3195-3199, 1990.
21. Grasso P, Scholz RW, Erskine RJ, Eberhart RJ. Phagocytosis, bactericidal activity, and oxidative metabolism of mammary neutrophils from dairy cows fed selenium-adequate and selenium-deficient diets. Am J Vet Res, 51, 269-274, 1990.
22. Guidry AJ, Paape MJ, Pearson RE. Effects of parturition and lactation on blood and milk cell concentrations, corticosteroids, and neutrophil phagocytosis in the cow. Am J Vet Res, 37, 1195-1200, 1976.
23. Hall JC, Heel K, McCauley R. Glutamine: A review. Brit J Surg 83, 305-312, 1996.
24. Hallén Sandgren C. Field experiences of mastitis risk factors with special emphasis on heifer management and replacement. In: Proceedings of the Nordic Seminar Regarding Future Aspects of Mastitis Prevention, January 21-22, Aarhus, Denmark, 1998.
25. Harmon RJ, Torre RM. Copper and zinc: do they influence mastitis? In: Proceedings of the 33rd Annual Meeting of the National Mastitis Council, Orlando, FL., National Mastitis Council, Inc., Arlington, VA, USA, 54, 1994.

26. Hoeben D, Burvenich C, Heyneman R, Hamann J. Respiratory burst of bovine granulocytes around parturition: Modulation by antibiotics. Flem Vet J, 66, Suppl., 269-295, 1997a.
27. Hoeben D, Heyneman R, Burvenich C. Elevated levels of beta-hydroxybutyric acid in periparturient cows and in vitro effect on respiratory burst activity of bovine neutrophils. Vet Immunol Immunopathol, 58, 165-170, 1997.
28. Hoeben D, Burvenich C, Massart-Leen AM, Lenjou M, Nijs G, Van Bockstaele D, Beckers JF. In vitro effect of ketone bodies, glucocorticoids and bovine pregnancy-associated glycoprotein on cultures of bone marrow pregenitor cells of cows and calves. Vet Immunol Immunopathol, 68, 229-240, 1999.
29. Hogan JS, Weiss WP, Smith KL. Role of vitamin E and selenium in host defense against mastitis. J Dairy Sci 76, 2795-2803, 1993.
30. Hopster H, van der Werf JT, Blokhuis HJ. Stress enhanced reduction in peripheral blood lymphocyte numbers in dairy cows during endotoxin-induced mastitis. Vet Immunol Immunopathol, 66, 83-97, 1998.
31. Ishikawa H, Shirahata T, Hasegawa K. Interferon-γ production of mitogen stimulated peripheral lymphocytes in perinatal cows. J Vet Med Sci, 56, 735-738, 1994.
32. Jain NC, Paape MJ, Berning LM, Salgar SK, Worku M. Functional competence and monoclonal antibody reactivity of neutrophils from cows injected with *Escherichia coli* endotoxin. Comp Haematol Int, 1, 10-20, 1991.
33. Johnston LA, Chew BP. Peripartum changes of plasma and milk vitamin A and ß-carotene among dairy cows with or without mastitis. J Dairy Sci 67, 1832-1840, 1984.
34. Kaneene J, Miller R, Herdt T, Gardiner J. The association of serum nonesterified fatty acids and cholesterol, management and feeding practices with peripartum disease in dairy cows. Prev Vet Med, 31, 59-72, 1997.
35. Kehrli ME Jr, Nonnecke BJ, Roth JA. Alterations in bovine neutrophil function during the periparturient period. Am J Vet Res, 50, 207-214, 1989.
36. Kehrli ME Jr, Nonnecke BJ, Roth JA. Alterations in bovine lymphocyte function during the periparturient period. Am J Vet Res, 50, 215-220, 1989
37. Kehrli ME Jr, Weigel KA, Freeman AE, Thurston JR, Kelley DH. Bovine sire effects on daughters' in vitro blood neutrophil functions, lymphocyte blastogenesis, serum complement and conglutinin levels. Vet Immunol Immunopath 27, 303-319, 1991.
38. Kehrli ME Jr, Shuster DE, Ackermann MR. Leukocyte adhesion deficiency among Holstein cattle. Cornell Vet 82, 103-109, 1992.
39. Kehrli ME, Burton JL, Nonnecke BJ, Lee EK. Effects of stress on leukocyte trafficking and immune responses: Implications for vaccination. Adv Vet Med, 41, 61-81, 1999.
40. Kelm SC, Detilleux JC, Freeman AE, Kehrli ME Jr, Dietz AB, Fox LK, Butler JE, Kasckovics I, Kelley DH. Genetic association between parameters of innate immunity and measures of mastitis in periparturient Holstein cattle. J Dairy Sci 80, 1767-1775, 1997.
41. Klucinski W, Degorski A, Miernik-Degorska E, Targowski S, Winnicka A. Effect of ketone bodies on the phagocytic activity of bovine milk macrophages and polymorphonuclear leukocytes. Zentralbl Veterinarmed A, 35, 632-639, 1988.
42. Kremer WDJ, Burvenich C, Noordhuizen-Stassen EN, Grommers FJ, Schukken YH, Brand A. Severity of experimental E. coli mastitis in ketonemic and nonketonemic cows. J Dairy Sci, 76, 348-3436, 1993.
43. Lacey JM, Wilmore DW. Is glutamine a conditionally essential amino acid? Nutrition Rev 48, 297-309, 1990.
44. Lee E-K, Kehrli ME Jr. Expression of adhesion molecules on neutrophils of periparturient cows and neonatal calves. Am J Vet Res, 59, 37-43, 1998.
45. Leitner G, Chaffer M, Winkler M, Glickman A, Weisblit L, Krifucks O, Ezra E, Saran A. The determination of milk leukocyte population in heifers free of udder infection as a tool for the genetic improvement of dairy cows. In: Proc. of the 5th Int. Vet. Immunol.

Symposium, November 8-13, Punjab Agricultural University, Ludhiana, India, 119, 1998.
46. Lappalainen R, Kaartinen L, Veijalainen K, Kuosa PL, Sankari S, Pyörälä S, Sandholm M. Sequential changes of mineral and trace elements in milk during endotoxin-induced mastitis as analyzed by particle induced X-rays (PIXE), gamma-ray emission (PIGE) and ion selective electrodes. J Vet Med B, 35, 664-676, 1988.
47. Mallard BA, Dekkers JC, Ireland MJ, Leslie KE, Sharif S, Van Kampen CL, Wagter L, Wilkie BN. Alteration in immune responsiveness during the peripartum period and its ramification on dairy cow and calf health. J Dairy Sci, 81, 585-595, 1998.
48. Meijer GA, Van der Meulen J, van Vuuren AM. Glutamine is a potentially limiting amino acid for milk production in dairy cows: A hypothesis. Metabolism 42, 358-364, 1993.
49. Meijer GA, Van der Meulen J, Bakker JG, van der Koelen CJ, Van Vuuren AM. Free amino acids in plasma and muscle of high yielding dairy cows in early lactation. J Dairy Sci, 78, 1131-1141, 1995.
50. Mejdell CM, Lie O, Solbu H, Arnet EF, Spooner RL. Association of major histocompatibility complex antigens with AI bull progeny test results for mastitis, ketosis and fertility in Norwegian cattle. Anim Genet, 25, 99-104, 1994.
51. Michal JJ, Heirman LR, Wong TS, Chew BP, Frigg M, Volker L. Modulatory effects of dietary beta-carotene on blood and mammary leukocyte function in periparturient dairy cows. J Dairy Sci, 77, 1408-1421, 1994.
52. Moreira da Silva F, Burvenich C, Massart-Leen AM, Russell-Pinto FF. Effect of some reproduction hormones on the oxidative burst of bovine neutrophils. Flem Vet J, 66, 297-314, 1997.
53. Ndiweny N, Finch JM. Effects of in vitro supplementation of bovine mammary gland macrophages and peripheral blood lymphocytes with alpha-tocopherol and sodium selenite: implications for udder defences. Vet Immunol Immunopathol, 47, 111-121, 1995.
54. Nonnecke BJ, Burton JL, Kehrli Jr ME. Associations between function and composition of blood mononuclear leukocyte populations from dexamethasone-treated Holstein bulls. J Dairy Sci, 80, 2403-2410, 1997.
55. Politis I, Hidiroglou M, Batra TR, Gilmore JA, Gorewit RC, Scherf H. Effects of vitamin E on immune function of dairy cows. Am J Vet Res, 56, 179-184, 1995.
56. Reddy PG, Frey RA. Nutritional modulation of immunity in domestic food animals. Adv Vet Sci Comp Med, 35, 255-281, 1990.
57. Rohde T, McLean DA, Klarlund-Petersen B. Glutamine, lymphocyte proliferation and cytokine production. Scand J Immun, 44, 648-650, 1996.
58. Rukkwamsuk T, Wensing T, Geelen M. Effect of overfeeding during the dry period on regulation of adipose tissue metabolism in dairy cows during the periparturient period. J Dairy Sci, 81, 2904-2911, 1998.
59. Saad AM, Concha C, Astrom G. Alterations in neutrophil phagocytosis and lymphocyte blastogenesis in dairy cows around parturition. Zentr Vet Med B, 36, 337-345, 1989.
60. Schreiber AD, Parsons J, McDermett P. Effect of corticosteroids on the human monocyte and IgG and complement receptors. J Clin Invest, 56, 1189-1197, 1975.
61. Shafer-Weaver KA, Sordillo LM. Bovine CD8+ suppressor lymphocytes alter immune responsiveness during the postpartum period. Vet Immunol Immunopath 56, 53-64, 1997.
62. Shafer-Weaver KA, Pighetti GM, Sordillo LM. Diminished mammary gland lymphocyte functions parallel shifts in trafficking patterns during the postpartum period. Proc Soc Exp Biol Med 212, 271-279, 1996.
63. Shuster DE, Lee EK, Kehrli ME. Bacterial growth, inflammatory cytokine production and neutrophil recruitment during coliform mastitis in cows within ten days after calving. Am J Vet Res, 57, 1569-1575, 1996.

64. Smith KL, Hogan JS, Weiss WP. Dietary vitamin E and selenium affect mastitis and milk quality. J Animal Sci 75, 1659-1665, 1997.
65. Sordillo LM, Redmond MJ, Campos M, Warren L, Babiuk LA. Cytokine activity in bovine mammary gland secretions during the periparturient period. Can J Vet Res 55, 298-301, 1991.
66. Sordillo LM, Pighetti GM, Davis MR. Enhanced production of bovine tumor necrosis factor-alpha during the periparturient period. Vet Immunol Immunopath 49, 263-270, 1995.
67. Sordillo LM, Shafer-Weaver K, DeRosa D. Immunobiology of the mammary gland. J Dairy Sci 80, 1851-1865, 1997.
68. Suriyasathaporn W, Daemen AJ, Noordhuizen-Stassen EN, Dieleman SJ, Nielen M, Schukken YH. Beta-hydroxybutyrate levels in peripheral blood and ketone bodies supplemented in culture media affect the in vitro chemotaxis of bovine leukocytes. Vet Immunol Immunopathol, 68, 177-186, 1999.
69. Van Kampen C, Mallard BA. Effects of peripartum stress and health on circulating bovine lymphocyte subsets. Vet Immunol Immunopathol 59, 79-91, 1997.
70. Wanner JM, Rogers GW, Kehrli ME, Cooper JB. Intramammary infections in primiparous Holsteins: Heritabilities and comparisons of bovine leukocyte adhesion deficiency carriers and noncarriers. J Dairy Sci 81, 3293-3299, 1998.
71. Weigel KA, Kehrli ME, Freeman AE, Thurston JR, Stear MJ, Kelley DH. Associations of class I bovine lymphocyte antigen complex alleles with in vitro blood neutrophil functions, lymphocyte blastogenesis, serum complement, and conglutinin levels in dairy cattle. Vet Immunol Immunopathol, 27, 321-335, 1991.
72. Weiss WP, Todhunter DA, Hogan JS, Smith KL. Effect of duration of supplementation of selenium and vitamin E on periparturient cows. J Dairy Sci, 73, 3187-3194, 1990.
73. Wentink GH, Rutten VP, van den Ingh TS, Hoek A, Muller KE, Wensing T. Impaired specific immunoreactivity in cows with hepatic lipidosis. Vet Immunol Immunopatol, 56, 77-83, 1997.
74. Xin Z, Waterman DF, Hemken RW, Harmon RJ. Copper status and requirement during the dry period and early lactation in multiparous Holstein cows. J Dairy Sci, 76, 2711-2716, 1993
75. Zerbe H, Schneider N, Obadnik C, Schuberth H-J, Wensing T, Kruip T, Grunert E, Leibold W. The negative energy balance reduces the neutrophil defense capacity in cows. In: Proceedings of the 5th International Veterinary Immunology Symposium, November 8-13, Punjab Agricultural University, Ludhiana, India, 185, 1998
76. Meglia G and K. Persson Waller, unpublished data

# 30.

# Cells and Cytokines in Inflammatory Secretions of Bovine Mammary Gland

Céline Riollet, Pascal Rainard, Bernard Poutrel
*Laboratoire de Pathologie Infectieuse et Immunologie, INRA, 37380 Nouzilly, France*

Key words: Neutrophils, T cells, cytokines, mastitis, cattle

Abstract: In response to invading bacteria, the mammary gland is protected by a variety of defence mechanisms, which can be separated into two distinct categories: innate immunity and specific immunity. Milk somatic cells consist of several cell types, including neutrophils, macrophages, lymphocytes and a smaller percentage of epithelial cells. In the healthy lactating mammary gland, macrophages are the predominant cell type whereas neutrophils are the major cell population during early inflammation. Following a bacteria invasion, neutrophil recruitment is elicited by inflammatory mediators that are produced in the infected gland by cells, possibly macrophages, activated by bacteria phagocytosis or responding to bacterial toxins or metabolites. Several cytokines, including interleukin- (IL-) 1β, IL-6, IL-8, tumour necrosis factor- (TNF-) α and interferon- (IFN-) γ are known to be important to elicit the acute phase response and allow the accumulation of leukocytes at the site of infection. In addition to their role in early non-specific defences, macrophages also play a key role in the specific immune system, as antigen processing and presenting cells for the T cells. Few lymphocytes are found in milk of healthy glands where the predominant phenotype is CD8+ T cells. During the inflammatory reaction, T cells are recruited in milk and CD4+ cells become the predominant phenotype. The understanding of the specific and non-specific immune mechanisms involved in the mammary gland defence against invading bacteria may lead to the development of new vaccines and to the use of cytokines to design immunomodulatory strategies for the control of bovine mastitis.

# 1. INTRODUCTION

Inflammation of the mammary gland caused by invading pathogens is common among lactating dairy cows and is a major cause of economic losses. The mammary gland is protected by a variety of defence mechanisms, which can be separated into two categories: innate immunity and specific immunity. During the early stages of infection, non-specific responses are the predominant defences. They are mediated by the physical barrier of the teat end, macrophages, neutrophils, natural killer like cells, and by certain soluble factors such as lactoferrin, complement, lysozyme and enzyme lactoperoxydase. The specific immune system recognizes specific determinants of a pathogen that facilitate selective elimination. Recognition of pathogenic factors is mediated by antibody molecules, macrophages, and several lymphoid populations. Both innate and specific immunity are coordinated to provide optimal protection of the mammary gland from infection[65].

The various strategies utilized by the immunoinflammatory system to deal with bacterial pathogens rely heavily on the family of intercellular signalling proteins known as cytokines. These molecules are essential local and systemic hormones and major homeostatic mediators which form part of the system of innate immunity. Some of these cytokines (interleukin- (IL-) 1, IL-6, IL-8, tumour necrosis factor- (TNF-) $\alpha$) are involved in the induction of inflammation[17]. Other cytokines (for example: IL-1 receptor antagonist (IL-1ra), IL-10) play a role in downregulating inflammatory processes[27,28]. Many of them are involved in the activation of T and B lymphocytes, their clonal expansion and interactions with non-lymphoid cells. It is the balance between these cytokines, along with other factors (other mediators such as eicosanoids, nitric oxide (NO), soluble cytokine receptors, etc ) that controls the induction, perpetuation and cessation of inflammation[30].

The investigation and characterization of cattle leukocyte differentiation antigens has resulted in the identification of a large number of surface molecules. Differential expression of these molecules has allowed the identification of functionally distinct subpopulations of cells, enabling *in vitro* and *in vivo* studies of their roles to be made in immune responses following infection or injection of antigens. Step by step, mechanisms of the defence of the bovine mammary gland are better understood and appear as a complex network of interactions between bacteria and host, involving innate and specific immunity, orchestrated by cytokines.

## 2. THE INNATE IMMUNITY: THE FIRST LINE OF DEFENSE OF THE MAMMARY GLAND

Bacterial pathogens that are able to traverse the teat end opening must then escape the antibacterial activities of the mammary gland microenvironment in order to establish disease. The activities of resident and newly recruited leukocytes during the early stages of pathogenesis play a pivotal role in the establishment of intramammary infection. Differential cellular counts reveal that milk somatic cells consist of several cell types, including neutrophils, macrophages, lymphocytes, and a smaller percentage of epithelial cells. Depending on the healthy or infectious status of the gland, SCCs vary in number and composition (Table 1).

Table. 1 SCC in healthy and infected bovine mammary gland[12,37,45,65].

| | SCC | SCC composition (%) | | | |
|---|---|---|---|---|---|
| | (cells/ml of milk) | Neutrophils | Macrophages | lymphocytes | Epithelial cells |
| Healthy quarter | <200,000 often <100,000 | 0 - 11 | 66 – 88 | 10 - 27 | 0 – 7 |
| Infected quarter | >200,000 | 50 - 95 | 9 – 32 | 14 - 24 | 0 – 9 |

Macrophages are the predominant cell type in milk and tissues of healthy mammary gland and their primary function is phagocytosis and intracellular killing of invading microorganisms and removal of milk fat from involuting glands[47]. Mammary macrophages may act as initiators of infection and, on stimulation by phagocytosis, release chemotactic activity for neutrophils, thereby amplifying the inflammatory response[13]. Few hours after infusion of *Escherichia coli* lipopolysaccharide into a healthy lactating gland, large amounts of the pro-inflammatory cytokines IL-1 and IL-6 are found in milk, while high concentrations of IL-1, IL-6 and TNF-α are found after *E. coli* infusion[60,61]. These inflammatory cytokines have local and systemic effects. At the systemic level, they mediate the acute phase reaction (fever, high serum cortisol concentration, release of acute phase proteins, leukopenia, etc) [18,33,39]. At the local level, they trigger the neutrophil migration from the blood into the infected mammary gland, through several actions. They activate the endothelial cells to express more E-selectin and P-selectin. These molecules allow neutrophils to bind more tightly in these areas. This binding, as well as stimulation by inflammatory mediators, enhances expression and adhesiveness of another neutrophil adhesion molecule, Mac-1 (known also as CD11b/CD18), which is a member of the $\beta_2$-integrin family of leukocyte adhesion molecules. At the same time that Mac-1 expression increases, L-selectin is proteolytically shed from the neutrophil surface[35]. The Mac-1 molecule allows neutrophils to bind tightly to activated endothelium via another endothelial adhesion molecule, ICAM-1. This second adhesive interaction allows neutrophils to migrate along the endothelial surface and into tissues along a concentration gradient of chemoattractants, one of the most potent being IL-8 [31]. Studies showed that IL-1 and TNF-α stimulate IL-8 secretion[36,50], and IL-8 was suggested to be an important mediator of IL-1-induced neutrophil migration[11]. This neutrophil recruitment from the circulation to the focus of infection is essential in the defence of the mammary gland against invading bacteria and the promptitude of the recruitment and the amount of recruited neutrophils are determining for the outcome of the infection.

Once at the site of infection, neutrophils phagocyte and kill bacteria through a respiratory burst that produces hydroxyl and oxygen radicals, which are key components of the oxygen-dependent killing mechanism. Inflammatory cytokines such as IL-1, TNF-α or interferon- (IFN-) γ can enhance neutrophil phagocytosis and/or their bactericidal activity[55,64,66]. In addition to their phagocytic capabilities, neutrophils are a source of small antibacterial peptides, the defensins, which are able to kill a number of pathogens that cause mastitis[57].

If the acute phase response allows the elimination of the bacteria which represent the inflammatory stimuli, then neutrophil recruitment ceases and SCC returns to healthful levels. This is what it is observed during clinical mastitis, mainly induced by *E. coli*, and accompanied by a spontaneous cure of the infected gland within a few days after the beginning of the infection[60,62]. If bacteria are able to survive this immediate host response, then the inflammation and the leukocyte migration continue. Prolonged diapedesis of leukocytes causes damage to mammary parenchyma tissue, resulting in decreased production of milk[25]. This is the case of *S. aureus* mastitis which are chronic and most of the time subclinical. Several months after the beginning of the infection, neutrophils are still the major cells present in infected gland (Riollet et al, submitted).

## 3. THE SPECIFIC IMMUNE DEFENSES OF THE MAMMARY GLAND

Most of the studies undertook to understand the mechanisms of the defence of the mammary gland against pathogens have focused on the innate immunity mediated by macrophages, neutrophils, natural killer-like cells and soluble factors. Recently, attention has been paid to the role of the specific immunity, particularly to the role of the lymphocyte population, in the protection of the mammary gland against bacterial infections[48,67,68]. In the lactating healthy gland, the lymphocyte population consists predominantly of T cells, with few B cells. The T lymphocytes can be subdivided into αβ T cells, which include CD4+ (T-helper) lymphocytes and CD8+ (T-cytotoxic or T-suppressor) lymphocytes, and γδ T lymphocytes[65]. In normal mammary gland secretions, T cells mainly express the αβ T cell receptor and are predominantly CD8+[48,67]. As in human milk, bovine milk T lymphocytes display the phenotype of memory cells[3,67,70]. Any change in one particular subset of lymphocytes can have effects upon the activity of other cell subsets[10,49].

The functional significance of the various lymphocyte subpopulations found in the bovine mammary gland remains unclear. Even in healthy glands, variations in the composition of the lymphocyte population are observed during a lactation cycle but the consequences of these changes on the defence of the mammary gland are still hypothetical[48,67]. Recently, it has been described shifts in cell functions of trafficking CD4+ and CD8+ cells during the postpartum period and these shifts may be one reason for increased incidence of mastitis during this lactation stage[58,59].

In infected mammary glands, specific immune response to bacteria is mediated by lymphocytes that recognize antigens through membrane receptors that are specific for invading pathogens. In contrast to healthy gland, CD4+ cells are the predominant phenotype of the T lymphocyte population[68]. The CD4+ lymphocytes produce cytokines in response to recognition of an antigen-major histocompatibility complex (MHC) class II complex on B lymphocytes or macrophages. Through the ability to secrete certain cytokines, CD4+ cells play an important role in activating B lymphocytes, T lymphocytes, macrophages and various other cells that participate in the immune response. Regulatory cytokines have been shown to polarize the T helper subsets and the concept prevails that immune responses against specific pathogens can be categorized as either Th1 or Th2 responses[46]. IL-12 has been shown to be produced by macrophages and B cells to favour the development of Th1 cells[69] and IL-4 is the critical cytokine which favours the development of Th2 cells[56]. IL-2 and IFN-γ are the major cytokines associated with the Th1 cells which promote cellular responses against intracellular pathogens and virus. The Th2 lymphocytes mainly secrete IL-4, IL-5 and IL-10 and promote humoral immunity. However, in cattle and humans, in contrast to mice, IL-10, which exerts anti-inflammatory effects, was found to be produced by and to downregulate all subtypes of Th cells[7,15] . It has been shown that once activated, each type of Th cells may be able to regulate the proliferation and/or function of the other[20,24] and it is now admitted that the paradigm of Th1 or Th2 responses is oversimplified and that it is rather one type of response that is predominant compared to the other type[6,34].

It is now well established that CD8+ cells can exert either cytotoxic or suppressor function[32]. Cytotoxic T cells recognize and eliminate altered self cells via antigen presentation in conjunction with MHC class I molecules. Therefore, cytotoxic cells may act as scavengers, removing old or damaged secretory cells, the presence of which could increase the susceptibility of the mammary gland to infections[67]. Suppressor T lymphocytes are thought to control or modulate the immune response. Park et al[49] demonstrated that in bovine mammary glands infected with *S. aureus*, the CD4+ cell proliferation was suppressed by activated CD8+ cells, which could therefore enhance persistent intramammary infection by *S. aureus*. As for CD4+ cells, different cytokines have been found to be expressed by the two subpopulations of CD8+ cells: IL-4 has been associated with the suppressor phenotype, and IFN-γ with the cytotoxic phenotype[32,59].

Recently, authors emphasized on the possible role of γδ T cells in the bovine immune system[26] but this role is still unclear, especially in response to infectious pathogens in the mammary gland. The WC1+ subpopulation represents a minor portion of lymphocyte population in the normal mammary secretions[54,67,71], and WC1+ cells are not recruited in chronic *S. aureus* infected mammary secretions[54], suggesting that if these cells play a role in the defence of the mammary gland against bacterial infection, it is rather in the mammary tissue than in milk. This would be in accordance with the fact that they preferentially migrate to the epithelial surfaces and do not recirculate extensively[2,40]. There are indications that γδ T cells can mediate cytotoxicity with variable involvement of MHC[41], suggesting they may be able to destroy altered epithelial cells[40].

In addition to the T cells, there is a second subset of lymphocytes which is the B cells. Their primary role is to produce antibodies against invading pathogens. B lymphocytes utilize their cell surface receptors to recognize specific pathogens. They can internalize, process and present antigen in the context of MHC class II molecules to T-helper cells. Upon presentation of the processed antigen to T-helper cells, cytokines, such as IL-2, IL-4, IL-6 and IL-10, are secreted by the T cells, which in turn induces proliferation and differentiation of the B lymphocytes into either plasmocytes that produce antibodies or memory cells. Their proportion among the total lymphocyte population in milk does not increase when infection occurs[54].

## 4. COORDINATION OF INNATE AND SPECIFIC IMMUNITY IN THE MAMMARY GLAND

First of all, generation of effective immunity involves both antigen-presenting cells and lymphocytes. Macrophages play a key role in antigen processing and presentation[23,51]. Antigens from bacteria are processed within macrophages and appear on the membrane in association with MHC class I or II.

Moreover, macrophages secrete cytokines that are known to regulate T cell differentiation. For example, it has been shown that IL-12 enhances development of cytotoxic and IFN-γ producing CD8+ cells[44], and polarizes the CD4+ T cells toward a Th1 type response[69]. On the contrary, cytokines expressed by lymphocytes can act on macrophage cytokine production and functions. As an example, IL-10, secreted among other cells by Th2 cells, has been shown to inhibit the ability of macrophages to stimulate Th1-T cell clones to produce IFN-γ[22] and to have an inhibitory effect on lipopolysaccharide- (LPS) induced cytokine production by macrophage cell lines[21].

In addition, traditionally, mature neutrophils have been considered to be terminally differentiated cells lacking the ability to synthesize proteins. In recent years, however, it has become increasingly clear that this view is outdated. There now exists compelling evidence that neutrophils can release a number of cytokines, chemokines, growth factors, and interferons, both *in vitro* and *in vivo*[8,38]. In view of the broad spectrum of biological activities exerted by cytokines and of the fact that neutrophils represent the first cell type to migrate toward inflamed tissue, and given that most neutrophil responses are stimulus specific, it can be reasonably inferred that neutrophils not only play an important role in eliciting and sustaining inflammation, but may also significantly contribute to determine or at least to influence the evolution of subsequent host responses[9]. No studies have been reported concerning the cytokine secretion of neutrophils in milk of infected mammary glands. Nevertheless, as in the other *in vivo* animal models[9,43], neutrophils recruited in the bovine mammary gland may likely synthesize pro- and/or anti-inflammatory cytokines which may play a crucial role in the communication between innate cells and adaptive cells.

## 5. CONCLUSION

Therefore, a variety of signals from both cells of innate immune system and specific immune system is produced during an immune response and this molecule network mediates cross-regulation of the different cellular actors in order to eliminate invading bacteria. However, it is certainly more complicated ! Indeed, currently, it is accepted that the interaction between microorganisms and host cells is a linear causal chain of events in which the exogenous organism induces the host to synthesize cytokines which warn and protect the host or induce tissue pathology or both. Recently, authors[29] advanced the hypothesis that bacteria have the ability to modulate host cytokine networks and are, in turn, able to respond to such networks in order to minimize host inflammatory responses. Of significance to the hypothesis are recent reports of the existence of bacterial proteins which have the ability to downregulate pro-inflammatory cytokine networks. For example, YopB protein from *Yersinia enterocolitica* has been found to inhibit TNF-α production by host cells[4]. In addition, host cytokines have been reported to act on bacteria. For example, IL-1, IL-2 and growth macrophage-colony stimulating factor (GM-CSF) have been claimed to stimulate the growth of virulent strains of *E. coli*[16,52]. Of course, this leaves the key question to be answered - how do bacteria control host cytokine networks?

The understanding of specific and non-specific immune mechanisms and the understanding of the bacterial-host interaction mechanisms will open up new avenues for the development of new vaccines and for the design of new immunomodulatory strategies to control bovine mastitis. It has already been reported promising results for use of cytokines, such as IL-1, IL-2 and IFN-γ, in the prevention or treatment of bovine mastitis[14,53,63].

The immunoinflammatory system, however, contains a paradox. Cytokines, which are central to the protective role of this system, are also the major mediators of the pathology which accompanies infections. The most striking example of this is Gram-negative septic shock which can be prevented by neutralizing a number of pro-inflammatory cytokines including IL-1, IL-6 and TNF-α or by administering IL-10 [42]. This illustrates the importance of the balance of the cytokine network in the control of tissue pathology and the pivotal nature of this balance to host survival.

At last, the role of the epithelial cells has to be considered in the defence of the bovine mammary gland against bacterial infections. Indeed, several studies have demonstrated that epithelial cells can generate a variety of inflammatory mediators upon interaction with the bacteria[1,5,19]. The contribution of epithelial cells to leukocyte recruitment has still not been deeply studied but it could be helpful to understand and control the interaction network that occurs between pathogens and host.

## 6. REFERENCES

1. Agace WW, Hedges SR, Ceska M, Svanborg C. Interleukin-8 and the neutrophil response to mucosal gram-negative infection. J Clin Invest 2, 780-785, 1993.
2. Allison JP, Havran WL. The immunobiology of T cells with invariant gamma delta antigen receptors. Annu Rev Immunol , 679-705, 1991.
3. Bertotto A, Gerli R, Fabietti G, Crupi S, Arcangeli C, Scalise F, Vaccaro R. Human breast milk T lymphocytes display the phenotype and functional characteristics of memory T cells. Eur J Immunol 8, 1877-1880, 1990.

4. Beuscher HU, Rodel F, Forsberg A, Rollinghoff M. Bacterial evasion of host immune defense: *Yersinia enterocolitica* encodes a suppressor for tumor necrosis factor alpha expression. Infect Immun 4, 1270-1277, 1995.
5. Boudjellab N, Chang-Tang HS, Li X, Zhao X. Interleukin-8 response by bovine mammary epithelial cells to lipopolysaccharide stimulation. Am J Vet Res , 1563-1567, 1998.
6. Brown WC, Rice-Ficht AC, Estes DM. Bovine type 1 and type 2 responses. Vet Immunol Immunopathol 1-2, 45-55, 1998.
7. Brown WC, Woods VM, Chitko-McKown CG, Hash SM, Rice-Ficht AC. Interleukin-10 is expressed by bovine type 1 helper, type 2 helper, and unrestricted parasite-specific T-cell clones and inhibits proliferation of all three subsets in an accessory-cell-dependent manner. Infect Immun 11, 4697-4708, 1994.
8. Cassatella MA. The production of cytokines by polymorphonuclear neutrophils. Immunol Today 1, 21-26, 1995.
9. Cassatella MA. Neutrophil-derived proteins: selling cytokines by the pound. Adv Immunol , 369-509, 1999.
10. Chiodini RJ, Davis WC. The cellular immunology of bovine paratuberculosis: immunity may be regulated by CD4+ helper and CD8+ immunoregulatory T lymphocytes which down-regulate gamma/delta+ T-cell cytotoxicity. Microb Pathog 5, 355-367, 1993.
11. Colditz IG, Zwahlen RD, Baggiolini M. Neutrophil accumulation and plasma leakage induced in vivo by neutrophil-activating peptide-1. J Leukoc Biol 2, 129-137, 1990.
12. Concha C, Holmberg O, Astrom G. Cells found in non-infected and staphylococcus-infected bovine mammary quarters and their ability to phagocytose fluorescent microspheres. Zentralbl Veterinarmed [B] 5, 371-378, 1986.
13. Craven N. Generation of neutrophil chemoattractants by phagocytosing bovine mammary macrophages. Res Vet Sci 3, 310-317, 1983.
14. Daley MJ, Coyle PA, Williams TJ, Furda G, Dougherty R, Hayes PW. *Staphylococcus aureus* mastitis: pathogenesis and treatment with bovine interleukin-1 beta and interleukin-2. J Dairy Sci 12, 4413-4424, 1991.
15. Del Prete G, De Carli M, Almerigogna F, Giudizi MG, Biagiotti R, Romagnani S. Human IL-10 is produced by both type 1 helper (Th1) and type 2 helper (Th2) T cell clones and inhibits their antigen-specific proliferation and cytokine production. J Immunol 2, 353-360, 1993.
16. Denis M, Campbell D, Gregg EO. Interleukin-2 and granulocyte-macrophage colony-stimulating factor stimulate growth of a virulent strain of *Escherichia coli.* Infect Immun 5, 1853-1856, 1991.
17. Dinarello CA. Interleukin-1. Adv Pharmacol , 21-51, 1994.
18. Eberhart RJ, Natzke RP, Newbould FHS. Coliform mastitis - a review. J Dairy Sci , 1-22, 1979.
19. Eckmann L, Kagnoff MF, Fierer J. Epithelial cells secrete the chemokine interleukin-8 in response to bacterial entry. Infect Immun 11, 4569-4574, 1993.
20. Fiorentino DF, Bond MW, Mosmann TR. Two types of mouse T helper cell. IV. Th2 clones secrete a factor that inhibits cytokine production by Th1 clones. J Exp Med 6, 2081-2095, 1989.
21. Fiorentino DF, Zlotnik A, Mosmann TR, Howard M, O'Garra A. IL-10 inhibits cytokine production by activated macrophages. J Immunol 11, 3815-3822, 1991a.

22. Fiorentino DF, Zlotnik A, Vieira P, Mosmann TR, Howard M, Moore KW, O'Garra A. IL-10 acts on the antigen-presenting cell to inhibit cytokine production by Th1 cells. J Immunol 10, 3444-3451, 1991.
23. Fitzpatrick JL, Cripps PJ, Hill AW, Bland PW, Stokes CR. MHC class II expression in the bovine mammary gland. Vet Immunol Immunopathol 1-2, 13-23, 1992.
24. Gajewski TF, Fitch FW. Anti-proliferative effect of IFN-gamma in immune regulation. I. IFN- gamma inhibits the proliferation of Th2 but not Th1 murine helper T lymphocyte clones. J Immunol 12, 4245-4252, 1988.
25. Harmon RJ, Heald CW. Migration of polymorphonuclear leukocytes into the bovine mammary gland during experimentally induced *Staphylococcus aureus* mastitis. Am J Vet Res 6, 992-998, 1982.
26. Hein WR, Mackay CR. Prominence of gamma delta T cells in the ruminant immune system. Immunol Today 1, 30-34, 1991.
27. Henderson, B. and Bodmer, M. *In* Therapeutic modulation of cytokines. Boca Raton: CRC Press 1996.
28. Henderson B, Poole S. Modulation of cytokine function: therapeutic applications. Adv Pharmacol , 53-115, 1994.
29. Henderson B, Poole S, Wilson M. Bacterial modulins: a novel class of virulence factors which cause host tissue pathology by inducing cytokine synthesis. Microbiol Rev 2, 316-341, 1996a.
30. Henderson B, Poole S, Wilson M. Microbial/host interactions in health and disease: who controls the cytokine network? Immunopharmacology 1, 1-21, 1996b.
31. Huber AR, Kunkel SL, Todd RF, Weiss SJ. Regulation of transendothelial neutrophil migration by endogenous interleukin-8 [published errata appear in Science 1991 Nov 1;254(5032):631 and 1991 Dec 6;254(5037):1435]. Science 5028, 99-102, 1991.
32. Inoue T, Asano Y, Matsuoka S, Furutani-Seiki M, Aizawa S, Nishimura H, Shirai T, Tada T. Distinction of mouse CD8+ suppressor effector T cell clones from cytotoxic T cell clones by cytokine production and CD45 isoforms. J Immunol 6, 2121-2128, 1993.
33. Jackson JA, Shuster DE, Silvia WJ, Harmon RJ. Physiological responses to intramammary or intravenous treatment with endotoxin in lactating dairy cows. J Dairy Sci 3, 627-632, 1990.
34. Kelso A. Th1 and Th2 subsets: paradigms lost? Immunol Today 8, 374-379, 1995.
35. Kishimoto TK, Jutila MA, Berg EL, Butcher EC. Neutrophil Mac-1 and MEL-14 adhesion proteins inversely regulated by chemotactic factors. Science 4923, 1238-1241, 1989.
36. Larsen CG, Anderson AO, Oppenheim JJ, Matsushima K. Production of interleukin-8 by human dermal fibroblasts and keratinocytes in response to interleukin-1 or tumour necrosis factor. Immunology 1, 31-36, 1989.
37. Lee CS, Wooding FB, Kemp P. Identification, properties, and differential counts of cell populations using electron microscopy of dry cows secretions, colostrum and milk from normal cows. J Dairy Res 1, 39-50, 1980.
38. Lloyd AR, Oppenheim JJ. Poly's lament: the neglected role of the polymorphonuclear neutrophil in the afferent limb of the immune response. Immunol Today 5, 169-172, 1992.
39. Lohuis JA, Verheijden JH, Burvenich C, van Miert AS. Pathophysiological effects of endotoxins in ruminants. 1. Changes in body temperature and reticulo-rumen motility, and the effect of repeated administration. Vet Q 2, 109-116, 1988.
40. Mackay CR, Hein WR. Marked variations in gamma delta T cell numbers and distribution throughout the life of sheep. Curr Top Microbiol Immunol , 107-111, 1991.

41. Mackay CR, Hein WR, Brown MH, Matzinger P. Unusual expression of CD2 in sheep: implications for T cell interactions. Eur J Immunol 11, 1681-1688, 1988.
42. Marchant A, JL Vincent, M Goldman. *In* Novel therapeutic strategies in the treatment of sepsis. New York: Marcel Dekker 1996.
43. Matsukawa A, Yoshinaga M. Neutrophils as a source of cytokines in inflammation. Histol Histopathol 2, 511-516, 1999.
44. Mehrotra PT, Wu D, Crim JA, Mostowski HS, Siegel JP. Effects of IL-12 on the generation of cytotoxic activity in human CD8+ T lymphocytes. J Immunol 5, 2444-2452, 1993.
45. Miller RH, Paape MJ, Fulton LA. Variation in milk somatic cells of heifers at first calving. J Dairy Sci 11, 3782-3790, 1991.
46. Mosmann TR, Coffman RL. TH1 and TH2 cells: different patterns of lymphokine secretion lead to different functional properties. Annu Rev Immunol , 145-173, 1989.
47. Outteridge PM, Lee CS. Cellular immunity in the mammary gland with particular reference to T, B lymphocytes and macrophages. Adv Exp Med Biol , 513-534, 1981.
48. Park YH, Fox LK, Hamilton MJ, Davis WC. Bovine mononuclear leukocyte subpopulations in peripheral blood and mammary gland secretions during lactation. J Dairy Sci 4, 998-1006, 1992.
49. Park YH, Fox LK, Hamilton MJ, Davis WC. Suppression of proliferative response of BoCD4+ T lymphocytes by activated BoCD8+ T lymphocytes in the mammary gland of cows with *Staphylococcus aureus* mastitis. Vet Immunol Immunopathol 2, 137-151, 1993.
50. Persson K, Larsson I, Hallen SC. Effects of certain inflammatory mediators on bovine neutrophil migration in vivo and in vitro. Vet Immunol Immunopathol 2, 99-112, 1993.
51. Politis I, Zhao X, McBride BW, Burton JH. Function of bovine mammary macrophages as antigen-presenting cells. Vet Immunol Immunopathol 4, 399-410, 1992.
52. Porat R, Clark BD, Wolff SM, Dinarello CA. Enhancement of growth of virulent strains of *Escherichia coli* by interleukin-1 [see comments]. Science 5030, 430-432, 1991.
53. Reddy PG, Reddy DN, Pruiett SE, Daley MJ, Shirley JE, Chengappa MM, Blecha F. Interleukin 2 treatment of *Staphylococcus aureus* mastitis. Cytokine 3, 227-231, 1992.
54. Riollet C, Rainard P, Poutrel B. Cell subpopulations and cytokine expression in bovine milk in response to chronic *Staphylococcus aureus* infection. submitted for publication, 1999.
55. Sample AK, Czuprynski CJ. Priming and stimulation of bovine neutrophils by recombinant human interleukin-1 alpha and tumor necrosis factor alpha. J Leukoc Biol 2, 107-115, 1991.
56. Seder RA, Paul WE, Davis MM, Fazekas dS. The presence of interleukin 4 during in vitro priming determines the lymphokine-producing potential of CD4+ T cells from T cell receptor transgenic mice. J Exp Med 4, 1091-1098, 1992.
57. Selsted ME, Tang YQ, Morris WL, McGuire PA, Novotny MJ, Smith W, Henschen AH, Cullor JS. Purification, primary structures, and antibacterial activities of beta-defensins, a new family of antimicrobial peptides from bovine neutrophils. J Biol Chem 9, 6641-6648, 1993.
58. Shafer-Weaver KA, Corl CM, Sordillo LM. Shifts in bovine CD4+ subpopulations increase T-helper-2 compared with T-helper-1 effector cells during the postpartum period. J Dairy Sci , 1696-1706, 1999.
59. Shafer-Weaver KA, Sordillo LM. Bovine CD8+ suppressor lymphocytes alter immune responsiveness during the postpartum period. Vet Immunol Immunopathol 1-2, 53-64, 1997.

60. Shuster DE, Kehrli MEJ, Rainard P, Paape M. Complement fragment C5a and inflammatory cytokines in neutrophil recruitment during intramammary infection with *Escherichia coli*. Infect Immun 8, 3286-3292, 1997.
61. Shuster DE, Kehrli MEJ, Stevens MG. Cytokine production during endotoxin-induced mastitis in lactating dairy cows. Am J Vet Res 1, 80-85, 1993.
62. Shuster DE, Lee EK, Kehrli MEJ. Bacterial growth, inflammatory cytokine production, and neutrophil recruitment during coliform mastitis in cows within ten days after calving, compared with cows at midlactation. Am J Vet Res 11, 1569-1575, 1996.
63. Sordillo LM, Babiuk LA. Controlling acute *Escherichia coli* mastitis during the periparturient period with recombinant bovine interferon gamma. Vet Microbiol 2, 189-198, 1991a.
64. Sordillo LM, Babiuk LA. Modulation of bovine mammary neutrophil function during the periparturient period following in vitro exposure to recombinant bovine interferon gamma. Vet Immunol Immunopathol 4, 393-402, 1991b.
65. Sordillo LM, Shafer-Weaver K, DeRosa D. Immunobiology of the mammary gland. J Dairy Sci 8, 1851-1865, 1997.
66. Steinbeck MJ, Roth JA. Neutrophil activation by recombinant cytokines. Rev Infect Dis 4, 549-568, 1989.
67. Taylor BC, Dellinger JD, Cullor JS, Stott JL. Bovine milk lymphocytes display the phenotype of memory T cells and are predominantly CD8+. Cell Immunol 1, 245-253, 1994.
68. Taylor BC, Keefe RG, Dellinger JD, Nakamura Y, Cullor JS, Stott JL. T cell populations and cytokine expression in milk derived from normal and bacteria-infected bovine mammary glands. Cell Immunol 1, 68-76, 1997.
69. Trinchieri G. Interleukin-12: a proinflammatory cytokine with immunoregulatory functions that bridge innate resistance and antigen-specific adaptive immunity. Annu Rev Immunol , 251-276, 1995.
70. Wirt DP, Adkins LT, Palkowetz KH, Schmalstieg FC, Goldman AS. Activated and memory T lymphocytes in human milk. Cytometry 3, 282-290, 1992.
71. Yang TJ, Ayoub IA, Rewinski MJ. Lactation stage-dependent changes of lymphocyte subpopulations in mammary secretions: inversion of CD4+/CD8+ T cell ratios at parturition. Am J Reprod Immunol 5, 378-383, 1997.
72. Yssel H, De Waal M, Roncarolo MG, Abrams JS, Lahesmaa R, Spits H, de Vries JE. IL-10 is produced by subsets of human CD4+ T cell clones and peripheral blood T cells. J Immunol 7, 2378-2384, 1992.

31.

# Immune Surveillance of Mammary Tissue by Phagocytic Cells

[1]Max J. Paape, [1]Kimberly Shafer-Weaver, [2]Anthony V. Capuco, [3]Kaat Van Oostveldt and [3]Christian Burvenich
*[1]Immunology and Disease Resistance Laboratory, [2]Gene Evaluation and Mapping Laboratory, United States Department of Agriculture, Beltsville, MD and [3]University of Ghent, Faculty of Veterinary Medicine, Merelbeke, Belgium*

Key words: Immune surveillance, mastitis

Abstract: The leukocytes in milk consist of lymphocytes, neutrophil polymorphonuclear leukocytes (PMN) and macrophages. Lymphocytes together with antigen-presenting cells function in the generation of an effective immune response. Lymphocytes can be divided into two distinct subsets, T- and B- lymphocytes, that differ in function and protein products. The professional phagocytic cells of the bovine mammary gland are PMN and macrophages. In the normal mammary gland macrophages are the predominate cells which act as sentinels to invading mastitis causing pathogens. Once the invaders are detected, macrophages release chemical messengers called chemoattractants that cause the directed migration of PMN into the infection. Migration of neutrophils into mammary tissue provides the first immunological line of defense against bacteria that penetrate the physical barrier of the teat canal. However, their presence is like a double-edged sword. While the PMN are phagocytosing and destroying the invading pathogens, they inadvertently release chemicals which induces swelling of secretory epithelium cytoplasm, sloughing of secretory cells, and decreased secretory activity. Permanent scarring will result in a loss of milk production. Resident and newly migrated macrophages help reduce the damage to the epithelium by phagocytosing PMN that undergo programmed cell death through a process called apoptosis. Specific ligands on the neutrophil surface are required for directed migration and phagocytosis. In response to infection, freshly migrated leukocytes express greater numbers of cell surface receptors for immunoglobulins and complement and are more phagocytic than their counterparts in blood. However, phagocytic activity rapidly decreases with continued exposure to inhibitory factors such as milk fat globules and casein in mammary secretions. Compensatory hypertrophy in non-mastitic quarters partially compensates for lost milk production in diseased quarters. Advances in molecular biology are making available the tools, techniques, and products to study and modulate host-parasite interactions. For example the cloning and expression of proteins that bind endotoxin may provide ways of reducing damaging effects of endotoxin during acute coliform mastitis. The successful formation of bifunctional monoclonal antibodies for the targeted lysis of mastitis causing bacteria represents a new line of therapeutics for the control of mastitis in dairy cows.

## 1. INTRODUCTION

The first line of defense against mammary infection is the teat canal. Bacteria that pass this barrier and enter the teat cistern meet the second line of defense: phagocytic leukocytes. Phagocytes, consisting of neutrophil polymorphonuclear leukocytes (PMN) and macrophages, ingest and kill mastitis pathogens. Bacterial invasion and growth within the mammary gland is the main cause of mastitis. Invading bacteria settle next to the epithelial cells lining the mammary ducts, absorbing nutrients from milk while expelling harmful toxins that attack and destroy the epithelium. While helpless against the invading horde of bacteria, the epithelium and white blood cells called macrophages soon release chemical messengers called cytokines that signal the body for help. These chemicals increase blood flow to the udder and open spaces between the endothelial cells lining the capillary bed of the udder allowing for the release of blood plasma into the milk. If this bacterium has been encountered before, specific antibodies will pass into the milk along with the plasma. Soon PMN, a specialized form of white blood cells, migrate directly from blood into the bacterial hoard. The PMN release potent oxidants that destroy not only some of the bacteria but also some of the epithelial cells lining the ducts and alveoli within the udder. The PMN also combats the bacteria directly by ingestion or phagocytosis, aided by antibodies that attach to the bacteria and allow the PMN to recognize them as foreign. For the PMN it is a dead end mission. After ingestion and release of their chemicals most of the PMN perish. Next macrophages migrate in through the pores of the capillaries. Damage to the epithelium is limited by induction of programmed cell death (apoptosis) in PMN and their engulfment by macrophages. Through this process damaging chemicals are walled off within dying PMN that are then ingested by macrophages to minimize damage to the epithelium of the udder. Within several hours lymphocytes arrive at the site of the infection and take the battle to another level of immunological defense, as they recognize antigens through membrane receptors that are specific for invading pathogens. Soon the balance of the struggle is tipped away from the invading bacteria. However, vast areas of the epithelium have been destroyed. Extensive scarring will result in a loss of secretory epithelium and milk production will never again reach normal levels. If all of the bacteria are not completely destroyed, the cellular drama becomes chronic and a subclinical form of the disease may continue for the remainder of lactation.

## 2. CELLS IN MILK

The cells in normal milk consist of lymphocytes, PMN, macrophages and epithelial cells. Because of the presence of epithelial cells, the term milk somatic cells was coined[54]. In mammary quarters free from bacterial infection macrophages are the predominate cell type (35-79%), followed by PMN (3-26%), lymphocytes (10-24%) and epithelial cells (2-15%) (37,46,50). During early and late lactation the percentage of PMN tends to increase while the percentage of lymphocytes decreases[46,69]. In infected mammary glands the percentage of neutrophils can approach 100%[63]. Total milk somatic cell count (SCC) from non-infected mammary quarters are low averaging 50,000/ml, and cows with subclinical intramammary infections will exceed this level[50,52]. Recent examination of Dairy Herd Improvement records during 1996-97 from 3,349,186 lactations obtained from 22 states in the USA, average SCC ranged from 254,000 to 485,000/ml[45]. Means by month indicated lowest SCC in November - January and peak counts in July - August. The percentage of herds exceeding 750,000, the cell count limit for normal milk in the USA, was as high as 14% in some states. However, 15 to 25% of the herds that exceeded 750,000 also were above 750,000 on the next test day. Thus, it would appear that some of the herds are harboring cows with subclinical mastitis and would have problems meeting the European Union cell count limit of 400,000 cells/ml of milk.

## 3. LYMPHOCYTES

Generation of an effective specific immune response involves both antigen-presenting cells and lymphocytes. Lymphocytes are the only cells of the immune system that have specificity as they recognize antigens through membrane receptors that are specific for invading pathogens. Upon antigen recognition, lymphocytes can become effector or memory cells, and these functions can be augmented by repeated exposure to the same antigen. Lymphocytes can be divided into two distinct subsets, T- and B-lymphocytes, that differ in function and protein products. T-lymphocytes can be further subdivided, by their T-cell receptor (TCR), into $\alpha\beta$ and $\gamma\delta$ T-lymphocytes. Lymphocytes that express the $\alpha\beta$ TCR include CD4+ (T-helper) and CD8+ (T-cytotoxic/suppressor) effector cells. Although natural killer (NK) cells do not specifically recognize antigen, they are classified as large, granular, nonimmune lymphocytes and are thought to arise from the same lineage as T-lymphocytes.

Lymphocytes participate in mammary gland mucosal immunity by trafficking (homing) to the gland and there is preferential trafficking of certain lymphocyte subpopulations to this tissue site[77,94]. The homing process plays a major role in enhancing the efficiency of immune surveillance and effector responses in potential tissue sites of antigen deposition. Lymphocyte trafficking into different tissue sites, including the mammary gland, can be separated into the following successive steps: tethering (transient adhesion) of the lymphocyte to the endothelium; rolling of the lymphocyte along the endothelium; cellular activation via G-coupled receptors; firm arrest of the lymphocyte to the endothelium; diapedisis of the lymphocyte through the endothelium; and migration of the lymphocyte within the tissue microenvironments[11,12]. Lymphocyte migration is directed by cell-surface molecules known as homing receptors that interact selectively with molecules on endothelial cells and facilitate lymphocyte migration into particular tissue areas[20]. Subsequent entry of the cells into the tissue may require further chemotactic or chemokinetic signals. The underlying mechanisms for shifts in lymphocyte trafficking to the bovine mammary gland have not been elucidated.

Mammary gland and milk lymphocytes are hyporesponsive to mitogenic, antigenic, and allogenic stimulation compared with peripheral blood cells[27,68]. This hyporesponsiveness is speculated to be partially due to distinct lymphocyte subpopulations and to the high proportion of memory T-lymphocytes present in the gland[77]. Under healthy conditions, the mononuclear leukocyte population of the bovine mammary tissue consists of approximately 40-50% T-lymphocytes and 20-25% B-lymphocytes. However, depending on the stage of lactation and tissue location, the percentages and functions of bovine lymphoid cells, especially T-lymphocytes, can vary significantly. Alterations in the proportion and function of lymphocyte subsets, including their ability to migrate to appropriate locations (trafficking), can affect both local and systemic immunity, influence the risk of infection and disease and contribute to disease pathogenesis. In the bovine, shifts in the trafficking patterns and therefore, the ratio of lymphocytes have been correlated to diminished immune functions such as cytokine production, lymphocyte proliferation and cytotoxicity and, thus, disease susceptibility[68]. Although changes in lymphocyte numbers or relative proportions does not necessarily relate to a change in lymphocyte function, alterations in lymphocyte subsets, particularly with the CD4:CD8 ratio, have been associated with immunosuppressive diseases in various species. Reduced secretion of important immunoregulatory cytokines, such as IL-2 and IFN-γ [31,68,73], observed during the postpartum period could be partially explained by the lower relative percentages CD4+ lymphocytes at this time.

T-helper (TH) lymphocytes produce cytokines in response to recognizing specific antigens via MHC II on antigen presenting cells. Recognition of specific antigen by TH lymphocytes stimulates clonal proliferation, differentiation into effector or memory cells and the secretion of cytokines. The repertoire of cytokines produced by CD4+ lymphocytes activate and regulate both humoral and cell-mediated immunity. CD4+ lymphocytes are the predominant T-lymphocyte phenotype in bovine peripheral blood and the supramammary lymph node with a CD4:CD8 ratio > 1 regardless of lactation stage[68,95]. However, the percentage and effector functions of CD4+ lymphocytes is diminished in the peripheral blood, supramammary lymph nodes, and mammary gland during the postpartum compared with the mid to late lactating period[68]. Lowered percentage of these cells partially explains reduced production of IL-2 and IFN-γ observed postpartum[31,68,73]. Because the relative percentages of TH1 or TH2 subsets or responses can greatly influence the type and amount of cytokines produced, a predominance of one subpopulation or response over the other can alter host immune responsiveness. Recently, bovine peripheral blood CD4+ lymphocytes have been shown to function primarily as TH2 compared with TH1 effector cells during the postpartum period[66]. Greater percentages of TH2 effector cells can further explain lower levels of IL-2 and IFN-γ observed at this time.

In contrast to the peripheral blood and supramammary lymph node, CD8+ lymphocytes are the predominant T-lymphocyte phenotype in both the tissue and secretions of the healthy mammary gland. $CD8^+$ lymphocytes comprise about 50-60% of the T-lymphocyte population in the mammary gland with the ratio of CD4 to CD8 ratio constitutively < 1 [3,68,77,94]. This suggests that CD8+ lymphocytes preferentially traffic to the bovine mammary gland under normal conditions. The vast majority of T-lymphocytes in mammary parenchyma and lacteal secretion are CD8+ αβ T-lymphocytes that display characteristics of memory cells[77]. The functional significance for the elevated frequency of CD8+ over CD4+ lymphocytes in both milk and mammary tissue of bovines has not been thoroughly defined.

Researchers demonstrated that $CD8^+$ lymphocytes activated during bacterial infections can suppress important host immune responses[29,30,59]. During the postpartum period, peripheral blood CD8+ lymphocytes mainly produced IL-4 and demonstrated limited cytotoxic activity[67]. These data indicate that CD8+ lymphocytes, in the absence of pathogen-induced activation, can suppress the activity of other immune cells thus, contributing to delayed host immune responsiveness during the early stages of pathogenesis.

T-lymphocytes bearing the γδ form of the TCR are the predominant T-cell phenotype in epithelial tissues such as skin, intestine, and lung[7,39]. In the ruminant system, the prevalence of γδ T-lymphocytes is quite high, with the highest percentage found in calves. Ruminants express greater levels of γδ T-lymphocytes in mammary secretions and mammary parenchyma relative to blood[38]. The finding that γδ T-lymphocyte percentages decrease significantly in the mammary parenchyma during the postpartum period[68] raises interesting questions about the potential effector functions of these lymphoid cells that preferentially home to epithelial surfaces. Although the specific function of these cells in the mammary gland are not well defined, γδ T-lymphocytes can mediate cytotoxicity with variable involvement of MHC and can mediate some NK activity[40]. γδ T-lymphocytes stimulated with IL-2 are capable of recognizing and lysing malignant breast carcinoma cells lines[43]. The cytotoxic ability of these lymphoid cells suggests they are capable of eliminating altered epithelial cells[39]. Additionally, there are indications that γδ T-lymphocytes play a role in antibacterial immunity and may provide a unique barrier function for mucosal microenvironments to bacterial pathogens. The percentage of γδ T lymphocytes in mammary gland parenchyma significantly decreases during periods of heightened susceptibility to bacterial infections[68] suggesting that these lymphocytes may constitute an important antibacterial defense of the bovine mammary gland. Changes in γδ T-lymphocytes percentages during the postpartum period also may influence, via chemokine, production of leukocyte subpopulations that traffic to the gland at this time[6].

Natural killer (NK) cells are large, granular, nonimmune lymphocytes that possess cytotoxic ability in the absence of MHC restriction. NK cells have been shown to play an important role in host defense against bacterial pathogens. NK cells are capable of lysing target cells through a diverse repertoire of mechanisms[12,78]. These pathways include antibody-dependent cell-mediated cytotoxicity (ADCC), granule exocytosis, release of cytolytic factors and receptor-mediated antigen recognition. Their ADCC activity is mediated through expression of CD16 Fc, the receptor for IgG, on the surface of NK cells. Once the NK cell binds an antibody bound target cell via Fc receptors, target-cell destruction is mediated by degranulation of porforin-containing granules. Natural killer cells can also secrete various toxic molecules such as TNF-α that may initiate apoptosis in altered cells. In the bovine mammary gland, the role of NK cells in host defense remains to be elucidated

The primary role of B-lymphocytes is to produce antibodies against invading pathogens. Unlike macrophages and PMN, B-lymphocytes utilize their cell-surface receptors to recognize specific pathogens. B-lymphocytes can internalize, process, and present antigen in the context of major histocompatibility complex (MHC II) to T-helper cells. Upon presentation of the processed antigen to T-helper lymphocytes, IL-2 is secreted by the T cell, which in turn induces proliferation and differentiation of the B-lymphocyte into either antibody producing plasma cells or memory cells . Under certain conditions B-lymphocyte differentiation can be directly stimulated by an antigen such as lipopolysaccharide.

Plasma cells secrete immunoglobulins (Ig) which are important specific soluble effector molecules of the humoral immune response[72]. Antibodies are present in bovine milk and are either synthesized locally or are selectively transported or transducted from serum[5,47,68]. The four different classes of Ig that play a predominant role in mammary gland defense against bacterial pathogens include, $IgG_1$, $IgG_2$, IgM, and IgA, Of these, $IgG_1$ is the predominant isotype in uninfected mammary glands. $IgG_1$, $IgG_2$, and IgM act as opsonins and facilitate PMN and M$\phi$ phagocytosis. IgA is thought to play a role in toxin neutralization, agglutination, and hindering bacterial colonization[47]. In contrast to T-lymphocytes, the percentages of B-lymphocytes remain fairly constant regardless of lactation stage. However during the postpartum period, B-lymphocyte functional capabilities, including antibody production, is diminished[48].

## 4. PMN LEUKOCYTE

The fine structure of the bovine PMN leukocyte has been carefully defined in classic studies[57,58]. The cell is delineated by a plasma membrane that has a number of functionally important receptors. These include L-selectin and ∃2-integrin adhesion molecules associated with the binding of PMN to endothelial cells that are important for migration into sites of infection[21]. Membrane receptors for the Fc component of the $IgG_2$ and IgM classes of immunoglobulins and complement component C3b are necessary for mediating phagocytosis of invading bacteria[21]. Dying or apoptotic PMN express receptors that mark them for quick disposal by macrophages[62].

The most prominent characteristic of the PMN is the multilobulated nucleus. The multilobulated nucleus is important because it allows the PMN to line up its nuclear lobes in a thin line, allowing for rapid migration between endothelial cells. Macrophages on the other hand have a large horseshoe shaped nucleus that makes migration between endothelial cells more difficult. Thus, the PMN is the first newly migrated phagocytic cell to arrive at an infection site. Within the cytoplasm are isles of glycogen that make up 20% of the cell on a dry weight basis and numerous bactericidal granules that are used by the cell in killing bacteria. Like other species, bovine PMN contain azurophilic and specific granules. They also contain a third novel granule that is larger, denser and more numerous than the other two granules. It contains lactoferrin, which is also found in secondary granules, but does not contain constituents common to azurophil granules. Instead, it contains a group of highly cationic proteins and is the exclusive store of powerful oxygen-independent bactericidal compounds[23]. The most important antibacterial mechanism derived from azurophilic granules is the myeloperoxidase-hydrogen peroxide-halide system[35]. Myeloperoxidase in the presence of hydrogen peroxide and halide ions kills bacteria.

The life cycle of bovine PMN is brief. In the bone marrow, this cell requires 10 to 14 days to mature[4]. After maturation, PMN may be stored for a few additional days. The mature PMN leaves the hematopoietic compartment of the bone marrow and enters the vascular sinus by traveling in migration channels through the endothelial cell. The PMN circulate in the blood stream briefly (half-life of 8.9 hr) [16], leave the blood stream by diapedesis between endothelial cells, and enter tissues where they function as phagocytes for 1 - 2 days. Older neutrophils are removed by macrophages as they undergo apoptosis or programmed cell death. The continuous migration of PMN into mammary tissue provides the first immunological line of defense against bacterial invasion. The nursing or milking stimulus induces directed migration of fresh PMN into mammary tissue[51]. Thus, the normal sterile mammary gland is supplied with a constant source of PMN. Furthermore, drainage of newly synthesized milk into milk ducts leads to removal of freshly migrated PMN, and further exudation of PMN into newly formed milk in the alveoli . However, once in the lumen of alveoli, ingestion of fat and casein causes a loss in phagocytic and bactericidal functions and leads to death of PMN[53]. Milking removes compromised PMN, which are replaced by healthy PMN, thus enhancing defense against bacterial infection.

In healthy animals, production and destruction of PMN is tightly regulated, which keeps their number in blood, milk, and tissue constant[32]. Influx of PMN into the mammary gland occurs at a low level for immune surveillance but increases rapidly in response to bacterial invasion. Here, potent chemical messengers guide PMN toward foci of infection. Potent chemoattractants for bovine PMN include C5a, a cleavage product of the fifth component of complement, various lipopolysaccharides (LPS), leukotriene B4, and interleukin-1 (Il-1), Il-2, and Il-8,[19,24,76,96]. It was recently discovered that in the presence of LPS mammary secretory cells secrete Il-8 [8]. While the influx of PMN into the infection site is important for controlling spread of the bacterium, their presence also results in tissue damage that leads to fibrosis and impaired mammary function[2,15,49]. PMN promote tissue injury and disturbed mammary function via 1) reactive oxygen metabolite generation (the respiratory burst) and 2) granular enzyme release (degranulation)[34,44]. It has been shown that vitamin E and the selenium containing enzyme glutathione peroxidase provide protection against damaging effects of reactive oxygen metabolites[70].

The first events to occur in the process of phagocytosis are contact and recognition between the phagocyte and bacterium. This is accomplished by specific antibodies ($IgG_2$ and IgM) that recognize the bacterium through Fab regions on the antibodies and Fc receptors on the phagocyte. Activation of complement components C3b and C3bi, on the surface of bacteria following antibody union, also promote phagocytosis by binding to CR1 and CR3 receptors on PMN[22]. During migration in response to infection, some of the bound antibodies and complement components are removed and new Fc receptors appear on the PMN surface[22,91]. Also, T-lymphocyte derived cytokines secreted in response to inflammation, induce further increases in Fc receptors for $IgG_2$ [92]. Thus, PMN are fully armed to do battle with the invading horde of bacteria, resulting in a more rapid ingestion and elimination of the pathogens. Also, lectin-carbohydrate receptors found on the surface of PMN recognize carbohydrate rich fimbriae of *Escherichia coli* in the absence of specific opsonins[55], and is referred to as non-opsonic phagocytosis. In the absence of specific opsonins, PMN are able to bind and ingest *E. coli*, and may be an important mechanism in controlling intramammary infections by Gram-negative bacteria. After contact and recognition, pseudopods form around the microbe. Fusion of the engulfing pseudopods results in the formation of a phagocytic vacuole or phagosome. Cytoplasmic granules migrate toward the phagosome where the membrane surrounding the granule fuses with the internalized plasma membrane that lines the phagosome, creating the phagolysosome. As a result of this, bactericidal contents of the granule are then emptied into the phagolysosome where digestion of the microbe occurs.

## 5. APOPTOSIS

Apoptosis or programmed cell death is a physiological form of cell death that results in cell removal without inducing an inflammatory reaction. Adequate numbers of circulating leukocytes requires a balance between cell proliferation and apoptosis. In the blood stream PMN seldom die an apoptotic cell death. Apoptosis occurs specifically at the tissue level. The process of apoptosis happens through well defined morphological and biochemical events. First, the cell shrinks and changes in the membrane structure take place. In later stages, the DNA in the nucleus of the cell fragments in pieces of consistent sizes. Finally, the cell breaks up into a number of membrane-bound fragments containing structurally intact organelles. These cell fragments, called apoptotic bodies are phagocytosed by neighboring cells and rapidly degraded. Necrosis is a form of cell death that results in cell swelling and blebbing. Every component is adversely affected and finally the cell reptures. This results in the release of toxic cell components into the surrounding tissue[64]. Many procedures have been developed to measure apoptosis in blood based on the typical features for apoptotic cells as described above. Most procedures apply flow cytometry. Recently, a flow cytometric procedure was adapted to detect apoptosis of bovine PMN in whole blood[79].

Expression of programmed cell death can be altered within the inflammatory microenvironment. Inflammatory mediators modulate PMN apoptosis. In vitro tests on human PMN have shown that bacterial products such as LPS[36] and host derived cytokines such as IL-1, interferon-([17] and granulocyte-macrophage colony stimulating factors[18] retard PMN apoptosis. TNF-$\alpha$[75], IL-6 [1] and the respiratory burst[86] induces PMN apoptosis. The net biologic effect of these competing compounds in vivo is unknown. In vitro studies on bovine PMN of early- and mid-lactating cows have shown that incubating blood with TNF-$\alpha$ accelerates PMN apoptosis. The same effect is observed after incubation of blood with high concentrations of LPS (10 $\mu$g/ml)[80]. This effect is indirect. LPS stimulates macrophages to release cytokines. These cytokines in turn prime PMN. Coincubation of LPS treated blood with TNF-$\alpha$ or actinomycin D, a well known apoptosis inducer of bovine PMN, did not have any additional accelerating effect on apoptosis. This suggests that primed cells are less sensitive to apoptosis following treatments with actinomycin D and TNF-$\alpha$[80]. In vitro studies on human PMN have shown that glucocorticosteroids have an apoptosis retarding effect[42]. Such research has not yet been performed on bovine PMN.

From in vitro studies on human PMN, it is known that apoptotic PMN display a loss of function such as cytoskeletal functions, phagocytosis, granular enzyme release, and respiratory burst activity[87]. Diapedesis of bovine PMN through an in vitro model consisting of a monolayer of epithelial cells isolated from the bovine mammary gland on a collagen coated membrane, caused a decrease in phagocytosis and oxidative burst[71]. These decreases are assumed to be due to PMN apoptosis. Diapedesis of PMN through the collagen coated membrane only, also induced apoptosis of the diapedesed PMN. However, the apoptotic response was negated by a monolayer of endothelial cells[81].

## 6. LPS- CD14- TUMOR NECROSIS (TNF) PATHWAY

The CD14 antigen, which is commonly found on monocytes and macrophages but not on circulating neutrophils and lymphocytes, was recently discovered on bovine mammary PMN and macrophages[55]. This receptor bind LPS-protein complexes and induces the synthesis and release of TNF[93]. TNF up-regulates PMN phagocytosis, adherence, chemotaxis and release of reactive oxygen metabolites[65]. It is recognized that two forms of CD14 exist, a membrane and soluble form[41]. The soluble form results from the shedding of membrane CD14 (mCD14). Soluble CD14 can bind LPS directly and prevent it from binding to mCD14, thus preventing over secretion of TNF that could lead to increased severity of clinical symptoms. Soluble CD14 has been tentatively identified in bovine milk[83] and may play an important role in neutralizing LPS and controlling the clinical symptoms associated with acute coliform mastitis. CD14 has recently been cloned and the protein has been expressed[82]. Intramammary and systemic use of this protein may provide a means of eliminating the potential damaging effects of LPS during acute coliform mastitis .

## 7. MAMMARY EPITHELIAL CELL DAMAGE

A significant negative correlation exists between milk SCC and milk yield[60]. Infection is the primary reason for increased SCC. During experimentally induced *Staphylococcus aureus* mastitis, PMN seem to traverse the mammary epithelium at alveolar regions exhibiting extensive morphological damage[26]. The tissue damage and decline in milk yield may result from damage to secretory tissue by bacterial toxins, as a result of neutrophil migration, or as a consequence of phagocytic activity of neutrophils. Attempts have been made to dissociate these factors to assess their individual potential for damaging the mammary epithelium.

A mammary explant model was developed to assess mammary tissue damage caused by normal PMN function[15]. Mammary tissue from lactating Holsteins was cultured in the presence of intact or lysed PMN, or PMN that were allowed to phagocytose opsonized zymosan. The relative extent of damage to mammary epithelium caused by PMN treatments was as follows: phagocytosing PMN > lysed PMN > intact PMN. Cytological damage observed included cell sloughing from the basement membrane, nuclear pyknosis, cell debris in luminal areas and epithelial vacuolation. During phagocytosis, lysosomes migrate toward the particle being internalized and lysosomal enzymes are released before fusion of the pseudopods is completed, resulting in release of lysosomal contents to the outside of the

PMN[23]. Once within the alveolar lumen, ducts and cisterns of the gland, PMN phagocytose fat globules, casein and , when infection is present, bacteria. Phagocytic activity may in part acçount for the negative correlation between SCC and milk production.

## 8. COMPENSATORY GROWTH OF THE MAMMARY GLAND

Normal quarters within an udder are capable of partially compensating for a dysfunctional quarter(s). In 1938, Swett et al.[74] reported that, in a single cow with non-functional rear quarters, growth and development of the front quarters compensated for parenchymal loss in the rear quarters with partial compensation of milk production. More recently, Woolford[89,90] compared milk yields of 39 cows experimentally infected with *S. aureus* with their infection-free identical twins. Milk production declined approximately 20% in infected quarters. In multiparous cows this loss was compensated for by an increase in milk production in noninfected quarters. In contrast, within-udder compensation did not occur in response to mastitis in heifers. In general clinical mastitis is associated with a decrease in total milk yield[33], the magnitude of this reduction presumably being a reflection of the degree of systemic involvement[10,28]. In contrast to the milk production response to subclinical mastitis[89], a compensatory response to lack of milk removal in adjacent glands was evident in both cows and heifers. Hamann and Reichmuth[25] similarly reported increased compensatory milk production when one to three adjacent quarters were not milked. Capuco and Akers[13] demonstrated that lack of milk removal in two quarters induced increased cell proliferation in the adjacent, compensatory quarters, and also tended to increase metabolic activity per cell. Thus, compensation likely induces both hyperplasia and hypertrophy. The compensatory response of a mammary gland to subclinical mastitis, or milk stasis in adjacent glands may be due to reactivation of quiescent cells, epithelial hyperplasia or hypertrophy. Evidence exists for a combination of effects. In any event, it is clear that division of mammary epithelial cells is not limited to major periods of mammogenesis - the prepubertal period, estrous cycles, and gestation. Cell turnover, and likely the potential for increasing the number of mammary secretory cells, exists throughout the lactation cycle[14]. The hyperplastic and hypertrophic response to frequent milking are clearly mediated by a locally active feedback inhibitor[88]. The mediator(s) of the compensatory proliferation and milk yield response is unknown.

## 9. USE OF MONOCLONAL ANTIBODIES FOR TARGETED LYSIS OF MASTITIS PATHOGENS

Monoclonal antibodies provide useful reagents for the study of PMN surface antigens and receptors, as well as to probe mechanisms of cell activation. A number of anti-bovine neutrophil monoclonal antibodies (MAB) have been produced[61]. Some of these MAB recognize CD11/CD18 integrins[21] and L-selectin adhesion molecules[84]. Others, when binding to antigens on the PMN cell surface regulate chemotactic, phagocytic and oxidative burst activity[61]. Bi-functional MAB were recently created by linking MAB to *Staphylococcus aureus* to MAB to PMN[85]. When injected into the mammary gland, one end hooks up to a *S. aureus* while the other end hooks up to a PMN. When binding to the PMN occurs the MAB causes the cell to undergo oxidative burst killing the bound organism. This new technology represents a new generation of therapeutics for the treatment and prevention of bacterial diseases in food producing animals.

## 10. SUMMARY

During immune surveillance of mammary tissue there is preferential trafficking of certain lymphocyte subpopulations. This homing process plays a major role in enhancing the efficiency of immune surveillance. During diapedesis of PMN into the mammary gland, several functionally important receptors are up-regulated, allowing for a more efficient phagocytosis of invading Gram-positive and Gram-negative mastitis pathogens. In vitro studies revealed that PMN and bacterial toxins produce extensive tissue damage that results in a decrease in milk synthesis. Furthermore, damage to epithelial cells by toxins secreted by *S. aureus* contributes to the pathogenicity of this organism. Advances in biotechnology have made available tools, techniques, and products for use in mastitis research. The study of apoptosis may contribute to an understanding of the impairment in PMN function in cows and its relation to mastitis. The cloning and expression of CD14 may provide a means of neutralizing LPS and minimizing its damaging effect on mammary secretory tissue. Use of bifunctional antibodies for targeted cytotoxicity of mastitis pathogens may prove useful in the treatment and prevention of bovine mastitis.

# 11. REFERENCES


1. Afford, S.C., Pongracz, J., Stockley, R.A., Crocker, J., Burnett, D. The induction by human interleukin-6 of apoptosis in the promonocytic cell line U937 and human neutrophils. Journal of Biological Chemistry, 1992, 267:21612-21616.
2. Akers, R.M., Thompson, W. Effect of induced leukocyte migration on mammary cell morphology and milk component biosynthesis. Journal of Dairy Science, 1987, 70:1685-1695.
3. Asai, K., K. Kai, H. Rikiishi, S. Sugawara, Y. Maruyama, T. Yamaguchi, M. Ohta, Kumagai, K. Variation in CD4+ T and CD8+ T lymphocyte subpopulations in bovine mammary gland secretions during lactating and non-lactating periods. Veterinary Immunology Immunopathology, 1998, 65:51- 61.
4. Bainton, D.F., Ullyot, J.L., Farquahar, M.G. The development of neutrophilic polymorphonuclear leukocytes in human bone marrow. Journal of Experimental Medicine, 1971, 134: 907-934.
5. Bastida-Corcvera, K.F. 1992. The enhancement of mammary gland immunity through vaccination. Page 335 *in* Bovine Medicine: Diseases and Husbandry of Cattle. A.H. Andrews, R.W. Blowey, H. Boyd, and R.C. Eddy, ed. Blackwell Sci Publ., Cambridge, MA.
6. Boismenu, R., L. Feng, Y. Y. Xia, J. C. Chang, Harvan, W. L. Chemokine expression by intraepithelial γδ T cells: implications for the recruitment of inflammatory cells to damage epithelia. Journal of Immunology, 1996, 157:985-992.
7. Boismenu, R.. Havran, W. L. An innate view of γδ T cells. Current Opinions in Immunology, 1997, 9:57- 63.
8. Boudjellab, N., Chan-Tang, H.S., Li, B.S., Zhao, X. Interleukin 8 response by bovine mammary epithelial cells to lipopolysaccharide stimulation. American Journal of Veterinary Research, 1998, 59: 1563-1567.
9. Brittenden, J., S. D. Heys, J. Ross, Eremin, O. Natural killer cells and cancer. Cancer, 1996, 77:1226-1243.
10. Burvenich, C., Heyneman, R., Fabry, J., Vandeputte-Van Messom, G., Massart-Leen, A.M., Roets, E. Possible role for bovine somatotropin (BST) during the recovery of experimentally induced *E. coli* mastitis in cows soon after parturition. Proc. Int. Conference on Mastitis, 1989, St. Georgen/Langsee, Karnten, Austria, 4-9.
11. Butcher, E. C. The regulation of lymphocyte traffic. Current Topics in Microbiology and Immunology, 1986, 128:85-122.
12. Butcher, E. C., Picker, L. J. Lymphocyte homing and homeostasis. Science, 1996, 272:60-66.
13. Capuco, A.V., Akers, R.M. Thymidine incorporation by lactating mammary epithelium during compensatory mammary growth in beef cattle. Journal of Dairy Science, 1990, 3:3094-3103.
14. Capuco, A.V., Byatt, J. Cell trunover in the mammary gland. Journal of Dairy Science, 1998, 81(Suppl.1):224.
15. Capuco, A.V., Paape, M.J., Nickerson, S.C. In vitro study of polymorphonuclear leukocyte damage to mammary tissues of lactating cows. American Journal of Veterinary Research, 1986, 47: 663-668.
16. Carlson, G.P., Kaneko, J.J. Intravascular granulocyte kinetics in developing calves. American Journal of Veterinary Research, 1975, April:421-425.

17. Collota, F., Re, F., Polentarutti, N., Sozzani, S., Mantovani, A. Modulation of granulocyte and programmed cell death by cytokines and bacterial products. Blood, 1992, 80:2012- 2020.
18. Cox, G., Gauldie, J., Jordana, M. Bronchial epithelial cell derived cytokines (G-CSF and GM-CSF) promote the survival of peripheral blood neutrophils in vitro. American Journal of Cell Molecular Biology, 1992, 7:507-517.
19. Daley, M.J., Coyle, P.A., Williams, T.J., Furda, G., Dougherty, R., Hayes, P.W. *Staphylococcus aureus* mastitis: pathogenesis and treatment with bovine interleukin - 1∃ and interleukin - 2. Journal of Dairy Science, 1991, 74:4413-4424.
20. Dailey, M. O. Expression of T lymphocyte adhesion molecules: Regulation during antigen-induced T cell activation and differentiation. Critical Reviews in Immunology, 998,18:153-184.
21. Delcommenne, M., Letesson, J.J., Depelchin, A. Characterization of monoclonal antibodies raised against CD11a, CD11c and CD18 in the bovine species. Proceedings International Conference on Mastitis Physiology and Pathology, University of Ghent, 1990, page 81.
22. DiCarlo, A.L., Paape, M.J. Comparison of C3b binding to bovine peripheral blood and mammary gland neutrophils (PMN). American Journal of Veterinary Research, 1996, 57:151-156.
23. Gennaro, R.B., Dewald, B., Horisberger, U., Gubler, H.U., Baggiolini, M. A novel type of cytoplasmic granule in bovine neutrophils. Journal of Cell Biology, 1983, 96:1651-1661.
24. Gray, G.D., Knight, K.A., Nelson, R.D., Herron, M.J. Chemotactic requirements of bovine leukocytes. American Journal of Veterinary Research, 1982, 43:757-759.
25. Hamann, J., Reichmuth, J. Compensatory milk production within the bovine udder: Effects of short-term non-milking of single quarters. Journal of Dairy Research, 1990, 57:17-22.
26. Harmon, R.J., Heald, C.W. Migration of polymorphonuclear leukocytes into the bovine mammary gland during experimentally induced *Staphylococcus aureus* mastitis. American Journal of Veterinarian Research, 1982, 43:992-998.
27. Harp, J. A., Nonnecke, B. J. Regulation of mitogenic responses by bovine milk lymphocytes. Veterinary Immunology Immunopathology, 1986, 11:215-224.
28. Heyneman, R., Burvenich, C., Vercauteren, R. Interaction between the respiratory burst activity of neutrophil leukocytes and experimentally induced *Escherichia coli* mastitis in cows. Journal of Dairy Science, 1990, 73:985 - 994.
29. Hisatsune, T., A. Enomoto, K. Nishijima, Y. Minai, Y. Asano, T. Tada, Kaminogawa, S. CD8+ suppressor T cell clone capable of inhibiting the antigen- and anti-T cell receptor-induced proliferation of Th clones without cytolytic activity. Journal of Immunology, 1990, 145:2421-2426.
30. Holly, M., Y.S. Lin, Rogers, T.J. Induction of suppressor cells by staphylococcal enterotoxin B: identification of a suppressor cell circuit in the generation of suppressor-effector cells. Immunology,1988, 64:643-648.
31. Ishikawa, H., T. Shirahata, Hasegawa, K. Interferon-γ production of mitogen stimulated peripheral lymphocytes in perinatal cows. Journal Veterinary Medical Science, 1994, 56:735-738.
32. Jain, N.C. Clinical interpretation of changes in leukocyte numbers and morphology. In Schalm's Veterinary Hematology, Lea and Febiger, Philadelphia, Pennsylvania, 1986.
33. Janzen, J.J. Economic losses resulting from mastitis. A review, Journal of Dairy Science, 1970, 53:1151-1160.

34. Kehrli, M.E., Shuster, D.E. Factors affecting milk somatic cells and their role in health of the bovine mammary gland. Journal of Dairy Science, 1994, 77:619-627.
35. Klebanoff, S.J. Myeloperoxidase - mediated antimicrobial systems and their role in leukocyte function. In Biochemistry of the Phagocytic Process, North-Holland Publishing Company, London, 1970, 89-114.
36. Lee, A., Whyte, M.K.B., Haslett, C. Inhibition of apoptosis and prolongation of neutrophil functional longevity by inflammatory mediators. Journal of Leukocyte Biology, 1993, 54: 283-291.
37. Lee, C.W., Wooding, F.B.P., Kemp, P. Identification, properties, and differential counts of cell populations using electron microscopy of dry cows secretion, colostrum, and milk from normal cows. Journal of Dairy Science, 1980, 47:39-50.
38. Machugh, N. D., J.K. Mburu, M.J. Carol, C.R. Wyatt, J.A. Orden, Davis, ,W.C. Identification of two distinct subsets of bovine $\gamma\delta$ T cells with unique cell surface phenotype and tissue distribution. Immunology, 1997, 92:340-345.
39. Mackay, C.R., Hein, Marked variations in gamma delta T cell numbers and distribution throughout the life of sheep. Current Topics in Microbiology and Immunology, 199,173:107-111.
40. Mackay, C.R., W.R. Hein, M.H. Brown, Matzinger, P. Unusual expression of CD2 in sheep: implications for T cell interactions. European Journal of Immunology, 1988,18(11):1681-1688.
41. Maliszewski, C.R. CD14 and immune response to lipopolysaccharide. Science, 1990, 252: 1321 - 1322.
42. Meagher, L.C., Cousin, J.M., Seckl, J.R., Haslett, C. Opposing effects of glucocorticosteroids on the rate of apoptosis in neutrophilic and eosinophilic granulocytes. Journal of Immunology, 1996, 156 : 4422 - 4428.
43. Miescher, S., M. Schreyer, C. Barras, P. Capasso, von Fliedner, V. Sparse distribution of gamma/delta T lymphocytes around human epithelial tumors predominantly infiltrated by primed/memory T cells. Cancer Immunology Immunotherapy, 1990, 32:81-87
44. Miller, J.K., Brzezinska-Slebodzinska, Madsen, F.C. Oxidative stress, antioxidants, and animal function. Journal of Dairy Science, 1993, 76:2812 - 2823.
45. Miller, R.H., Norman, H.D., Wiggans, G.R., Wright, J.R. National survey of herd average somatic cell counts on DHI test days. National Mastitis Council Annual Meeting Proceedings, 1999, 161-162.
46. Miller, R.H., Paape, M.J., Fulton, L.A. The relationship of milk somatic cell count to milk yield for Holstein heifers after first calving. Journal of Dairy Science, 1993, 76:728-733.
47. Musoke, A.J., F.R. Rurangirwa, . Nantulya, V.M. 1987. Biological properties of bovine immunoglobulins and systemic antibody responses. Page 393 *in* The Ruminant Immune System in Health and Disease. W.I. Morrison, ed. Cambridge Univ. Press, Cambridge, England.
48. Nagahata, H., A. Ogawa, Y. Sanada, H. Noda, Yamamoto, S. Peripartum changes in antibody producing capabilities of lymphocytes from dairy cows. Veterinary Quarterly, 1992, 14:39-40.
49. Nickerson, S.C., Heald, C.W. Histopathologic response of the bovine mammary gland to experimentally induced *Staphylococcus aureus* infection. American Journal of Veterinary Research, 1981, 42:1351-1354.
50. Östensson, K., Hageltorn, M., Aström, G. Differential cell counting in fraction-collected milk from dairy cows. Acta Veterinaria Scandinavica, 1988, 29:493-500.

51. Paape, M.J., Guidry, A.J. Effect of milking on leukocytes in the subcutaneous abdominal vein of the cow. Journal of Dairy Science, 1969, 52:998-1002.
52. Paape, M.J., Guidry, A.J., Jain, N.C., Miller, R.H. Leukocytic defense mechanisms in the udder. Flemish Veterinary Journal, 1991, Supplement 1:95-109.
53. Paape, M.J., Guidry, A.J., Kirk, S.T., Bolt, D.J. Measurement of phagocytosis of $^{32}$P-labeled *Staphylococcus aureus* by bovine leukocytes: Lysostaphin digestion and inhibitory effect of cream. American Journal of Veterinary Research, 1975, 36:1737-1743.
54. Paape, M.J., Hafs, H.D., Snyder, W.W. Variation of estimated numbers of milk somatic cells stained with Wright's stain or Pyronin Y-methyl green stain. Journal of Dairy Science, 1963, 46:1211-1216.
55. Paape, M.J., Lillius, E.M., Wiitanen, P.A., Kontio, M.P. Intramammary defense against infections induced by *Escherichia coli* in cows. American Journal of Veterinary Research, 1996, 57:477-482.
56. Paape, M.J., Weinland, B.T. Effect of abraded intramammary device on milk yield, tissue damage and cellular composition. Journal of Dairy Science, 1988, 71:250-256.
57. Paape M.J., Wergin, W.P. The leukocyte as a defense mechanism. Journal of the American Veterinary Association, 1977, 170:1214-1223.
58. Paape, M.J., Wergin, W.P., Guidry, A.J., Pearson, R.E. Leukocytes - the second line of defense against invading mastitis pathogens. Journal of Dairy Science, 1979, 62:135-153.
59. Park, Y. H., L. K. Fox, M. J. Hamilton, Davis, W. C. Suppression of proliferative responses of BoCD8+ T lymphocytes in the mammary gland of cows infected with *Staphylococcus aureus* mastitis. Veterinary Immunology Immunopathology, 1993, 36:137-151.
60. Raubertas, R.F., Shook, G.E. Relationship between lactation measures of somatic cell concentration and milk yield. Journal of Dairy Science, 1982, 65:419-425.
61. Salgar, S.K., Paape, M.J., Alston-Mills, B., Peters, R.R. Modulation of bovine neutrophil functions by monoclonal antibodies. American Journal of Veterinary Research, 1994, 55: 227-233.
62. Savill, J. Recognition and phagocytosis of cells undergoing apoptosis. British Medical Bulletin, 1997, 53:491-508.
63. Schalm, O.W., Carroll, E.J., Jain, N.C. In Bovine Mastitis, Lea and Febiger, Philadelphia, 1971, pages 103-106.
64. Schwartzman, R.A., Cidlowsky, J.A. Apoptosis: the biochemistry and molecular biology of programmed cell death. Endocrinology Review, 1993, 14:133-151.
65. Semnani, M.J., Kabbur, M.B., Jain, N.C. Activation of bovine neutrophil functions by interferon-gamma, tumor necrosis factor alpha, and interleukin - 1 alpha. Comparative Haematology International, 1993, 3:81-88.
66. Shafer-Weaver, K.A., C.M. Corl, Sordillo, L.M. Shifts in bovine CD4+ subpopulations increase T-helper-2 compared with T-helper-1 effector cells during the postpartum period. Journal of Dairy Science, 1999, 82:1696-706
67. Shafer-Weaver, K.A., Sordillo L.M. Bovine CD8+ suppressor lymphocytes alter immune responsiveness during the postpartum period. Veterinary Immunology Immunopathology, 1997, 56:53-64.
68. Shafer-Weaver, K. A., G. M. Pighetti, Sordillo, L. M. Diminished mammary gland lymphocyte functions parallel shifts in trafficking patterns during the postpartum period. Proceedings Society of Experimental Biology and Medicine, 1996, 212:271-280.

69. Sheldrake, R.F., Hoare, R.J.T., McGregor, G.D. Lactation stage, parity, and infection affecting somatic cells, electrical conductivity, and serum albumin in milk. Journal of Dairy Science, 1983, 66:542-547.
70. Smith, K.L., Harrison, J.H., Hancock, D.D., Todhunter, D.A., Conrad, H.R. Effect of vitamin E and selenium supplementation on incidence of clinical mastitis and duration of clinical symptoms. Journal of Dairy Science, 1984, 67:1293-1300.
71. Smits, E., Burvenich, C., Guidry, A.J., Heyneman, R., Massart-Leen, A. Diapedesis across mammary epithelium reduces phagocytic and oxidative burst of bovine neutrophils. Journal of Immunology and Immunopathology, 1999, In press.
72. Sordillo, L.M., Nickerson, S.C. Quantification and immunoglobulin classification of plasma cells in nonlactating bovine mammary tissue. Journal of Dairy Science,1988, 71:84-91.
73. Sordillo, L. M., M. J. Redmond, M. Campos, L. Warren, and L. A. Babiuk. 1991. Cytokine activity in bovine mammary gland secretions during the periparturient period. Canadian Journal of Veterinary Research, 55:298-301.
74. Swett, W.W., Matthews, C.A., Miller, F.W., Graves, R.R. Nature's compensation for the lost quarter of a cow's udder. Journal of Dairy Science, 1938, 21:7-15.
75. Takeda, Y., Watanabe, H., Yonehara, S., Yamashita, T., Saito, S., Sendo, F. Rapid acceleration of neutrophil apoptosis by tumor necrosis factor-∀. International Immunology, 1993, 5:691-694.
76. Taubb, D.D., Pooenheim, J.J. Review of the chemokine meeting: Third International symposium of Chemotactic Cytokines. Cytokine, 1993:175-179
77. Taylor, B.C., Dellinger, J.D., Cullor, J.S., Stott, J.L. Bovine milk lymphocytes display the phenotype of memory T cells and are predominantly CD8+. Cell Immunology, 1994, 156:245-253.
78. TTrinchieri, G. Biology of natural killer cells. Advances in Immunology, 1989, 47:187-376.
79. Van Oostveldt, K., Dosogne, H., Burvenich, C., Paape, M.J., Brochez, V., Van den Eeckhout, E. Flow cytometric procedure to detect apoptosis of bovine polymorphonuclear leukocytes in whole blood. Veterinary Immunology Immunopathology, 1999, In press.
80. Van Oostveldt, K., Burvenich, C., Paape, M.J. Effect of LPS on apoptosis of bovine neutrophils. European Journal of Physiology, 1999, In press.
81. Van Oostveldt, K., Burvenich, C., Paape, M.J., Meyer, E. The effect of diapedesis on the apoptotic response of isolated bovine neutrophils. Cell Biology International, 1999, In press.
82. Wang, Y. Personal communication, 1999.
83. Wang, Y., Paape, M.J., Detection and identification of soluble CD14 in bovine milk. Molecular Biology of the Cell, 1997, Supplement 1:85a.
84. Wang, Y., Paape, M.J., Leino, L., Capuco, A.V., Narva, H. Functional and phenotypic characterization of monoclonal antibodies to bovine L-selectin. American Journal of Veterinary Research, 1997, 58:1392-1401.
85. Wang, Y., Paape, M.J., Segal, D.M., Rainard, P., Poutrel, B., Nakamura, Y. Production of bispecific antibodies to bovine polymorphonuclear neutrophils and to *Staphylococcus aureus* capsular polysaccharide type 5. Journal of Animal Science, 1998, Supplement 1:37.
86. Watson, R.W., Redmond, H.P., Wang, J.H., Condron, C., Bouchier-Hayes, D. Neutrophils undergo apoptosis following ingestion of *Escherichia coli*. Journal of Immunology, 1996, 156:3986-3992.

87. Whyte, M.K.B., Meagher, L.C., MacDermot, J., Haslett, C. Impairment of function in aging neutrophils is associated with apoptosis. Journal of Immunology, 1993, 150:5124-5134.
88. Wilde, C.J., Calvert, D.T., Daily, A., Peaker, M. The effect of goat milk fractions on synthesis of milk constituents by rabbit mammary explants and on milk yield in vivo: Evidence for autocrine control of milk secretion. Biochemistry Journal, 1987, 242:285-288.
89. Woolford, M.W. The relationship between mastitis and milk yield. Kieler Milchwirtschaftliche Forschungsberichte, 1985, 37:224-232.
90. Woolford, M.W., Williamson, J.H., Copeman, P.J.A., Napper A.R., Phillips, D.S.M., Uljee, E. An identical twin study of milk production losses due to subclinical mastitis. Proceedings Ruakura Farmers Conference, 1983, Ruakura, New Zealand, 115-119.
91. Worku, M., Paape, M.J., Filep, R., Miller, R.H. Effect of in vitro and in vivo migration of bovine neutrophils on binding and expression of Fc receptors for $IgG_2$ and IgM. American Journal of Veterinary Research, 1994, 55:221-226.
92. Worku, M, Paape, M.J., Marquardt, W.W. Modulation of Fc receptors for IgG on bovine polymorphonuclear neutrophils by interferon-gamma through de novo RNA transcription and protein synthesis. American Journal of Veterinary Research, 1994, 55:234-238.
93. Wright, S.D., Ramos, R.A., Tobias, P.S., Ulevitch, R.J., Mathison, J.C. CD14, a receptor for complexes of lipopolysaccharide (LPS) and LPS binding protein. Science, 1990, 249: 1431-1436.
94. Yang, T. J., Ayoub, I.A., Rewinski, M. J. Lactation stage-dependent changes of lymphocyte subpopulations in mammary secretions: inversion of CD4+/CD8+ T cell ratios at parturition. American Journal of Reproductive Immunology, 1997, 37:378-383.
95. Yang, T.J., Mather, J.F., Rabinovsky, E.D. Changes in subpopulations of lymphocytes in peripheral blood, and supramammary and prescapular lymph nodes of cows with mastitis and normal cows. Veterinary Immunology Immunopathology, 1988, 18:279-85.
96. Zwahlen, R.D., Roth, D.R. Chemotactic competence of neutrophils from neonatal calves: functional comparison with neutrophils from adult cattle. Inflammation, 1990, 14:109-115.

32.

# Mammary Gland Immunology And Neonate Protection In Pigs

*Homing of lymphocytes into the MG*

Salmon H.
*Laboratoire Lymphocytes et Immunité des Muqueuses, INRA, 37380 Nouzilly*

Key words: mammary gland, lymphocyte, homing, plasma cells, hormone, secretory IgA, lactogenic immunity, entero-mammary link

Abstract: Since placenta of pregnant sows are impermeable to immunoglobulin passage, the neonates are born agammaglobulinemic ; although immunocompetent, they are unable to develop rapidly an immune response which will protect their systemic and mucosal compartments ; thus their survival depend upon the passive acquisition of maternal immunity including at least 3 components: i) a systemic humoral immunity, transmitted through colostrum conveying mainly by IgG; these IgG are transferred from maternal serum via Fcγ receptors on the epithelial cells of mammary gland (MG). ii) a local humoral immunity, especially secretory IgA (IgAs), transmitted mainly by milk (lactogenic immunity) until weaning. IgAs are secreted by MG recruited plasma cells and are excreted in milk via secretory component of epithelial cells : these IgA exhibit a specificity for the antigens present in the maternal digestive tract, the so-called "entero-mammary link"; this link is due to the migration of lymphocytes from the gut to the mammary gland ; they are recruited from the blood via the interaction of their homing receptor (α4β7) with the developmentally regulated mucosal vascular addresin MadCAM-1. In the MG, MadCAM-1 increased in pregnancy (probably under oestrogenic stimulation) but regressed in lactation ; its density is closely related to the T cell numbers in MG; in contrast the increase in plasma cell numbers is not related to MadCAM-1 density. Thus IgA precursor cells (α4β7 B cells) seem to be recruited by a milk B cell chemoattractant. On the other hand, presence of T and B lymphocytes in MG (some of them originating from the systemic compartment), sustains the attempts of MG immunization and the results sustain the view of a true local immune response. iii) possibly but not formally proved, a cellular immunity transmitted via maternal immunocompetent cells present in mammary secretions; the exported lymphocytes may represent a selected population of lymphocytes after their passage through the MG epithelium.

## 1. INTRODUCTION

Pigs (and ruminants too) are artiodactyls that have a epitheliochorial placenta impermeable to immunoglobulins (Ig), so the neonates are born hypo-or agammaglobulinemic ; although immunocompetent at birth, due to the lack of previous antigenic sensitization during embryonic life, neonates are unable to develop rapidly an immune response which will protect their systemic and mucosal compartments so that their survival depend upon the passive acquisition of maternal immunity, and this consists of various components: 1) a systemic humoral immunity, in the form of IgG, transmitted through colostrum within the first 24-36 hours after birth. The maternal serum antibodies are transferred via colostrum which is particularly rich in IgG due to a specific concentrative mechanism in the mammary tissue. 2) a local humoral immunity, especially in the form of secretory IgA (IgAs), transmitted mainly by milk (lactogenic immunity) until weaning. The antibodies present in mammary secretions exhibit a specificity for the antigens and microorganisms present in the maternal digestive tract. After ingestion, these immunoglobulins are not absorbed via the intestinal mucosa of the young, but instead they remain in situ where they provide a local protection against endemic microorganisms and dietary antigens. 3) lastly, a cellular immunity transmitted via maternal immunocompetent cells present in mammary secretions.

## 2. ORIGIN OF IMMUNOGLOBULINS IN COLOSTRUM AND MILK

The maternal antibodies are transferred from blood to colostrum through the mediation of Fcγ receptors on the surface of epithelial cells ; hence, colostrum is particularly rich in IgG due to a specific concentrative mechanism in the mammary tissue. In addition, 40% of colostral IgA and 85% of IgM are also derived from serum[3].

The existence of local antibody synthesis superimposed to transudation is supported by immunohistochemical location of IgA and IgG plasma cells in MG[5]the majority of milk antibodies (70% of IgG and greater than 90% of IgM and IgA) result from local synthesis in the MG[3] by these plasma cells. In sow similarly to mice, very few plasma cells were detected in the MG before parturition but their numbers increase very rapidly throughout lactation[5]This suggests that there may be hormonal control[13]. Mammotropic hormones also seem to influence the binding of antibodies to mammary epithelial cells, as well as their transepithelial transport[2].

## 3. ORIGIN OF MAMMARY GLAND IGA PLASMA CELLS : THE LACTOGENIC IMMUNITY

In many mammals, including humans, the ingestion of maternal milk confers the offspring with a certain degree of immunity against infection of the digestive tract and limits the number of infections in the upper respiratory system. Although numerous nonspecific factors have been identified in milk, this protection seems to be due to secretory antibodies present in mammary secretions[2,20]. Detection of IgA specific for Escherichia coli 083 in the milk of women who were orally immunized against this bacteria and the presence of anti-pneumococcus antibodies in the milk of rabbits exposed to this intestinal antigen, even in the absence of systemic humoral response, suggests the existence of a relationship between lactogenic immunity and intestinal local immunity. In sows, the existence of this gut-mammary link is sustained by the observations of the presence of specific IgA antibodies in milk of individuals, whose gut were stimulated by the corresponding antigens or entero-pathogenic virus of transmissible gastro-enteritis[14,15]. The lack of evidence for dissemination of the intestinal antigen to the mammary gland and the demonstration of a lymphoblast migration in the organism suggests that during pregnancy and lactation, cells of gut-associated lymphoid tissues are capable of migrating, via the mesenteric lymph node (MLN), the thoracic duct and the blood, to mammary tissues. This has been demonstrated in mouse; shortly before parturition and during lactation, the IgA lymphoblasts from the MLN, compared to those from peripheral lymph nodes, preferentially migrate to the MG.

## 4. DUAL ORIGIN OF RECRUITED IMMUNOCYTES INTO THE MG

The number of MG lymphocytes (T, B and null) increases from day 80 of pregnancy, in parallel to the increase of prolactin receptor density onto the epithelial cells[16]. When lymphocytes emigrating from mesenteric and inguinal lymph nodes are labelled distinctly, they were found 24 hours later in MG at the same frequency[12]: this study carried out on the whole lymphocyte populations indicated that a proportion of the MG lymphocytes arose from both the systemic and the mucosal lymphocyte pools, this later including the IgA lymphoblast entero-mammary cycle.

The different migration pathways of lymphocytes in the organism, reflected by the compartmentalisation of the immune system into systemic and mucosal, may be determined by the expression of particular structures on the surface of endothelial cells (Vascular addressin) and complementary structures on the membranes of lymphocytes (Homing receptor[21]). The endothelia of the venules within the lamina propria of the gut and in the lactating MG[22] express a mucosal addressin cell adhesion molecule, MadCAM-1. This molecule interacts with the $\alpha 4\beta 7$ integrin present onto gut derived lymphocytes[11].

Our data in mouse[23], indicate the presence of MadCAM-1 on endothelial cells of MG with a maximal expression at the end of pregnancy (probably under estrogenic stimulation) but regression in lactation ; its density is closely related to the $\alpha 4\beta 7$ T cell numbers in MG; in contrast the increase in plasma cell numbers is not related to MadCAM-1 density, since plasma cell increased when MadCAM-1 decreased. Thus IgA precursor cells ($\alpha 4\beta 7$, IgA B cells) need a milk B cell chemoattractant[6,7], as compared to T cells, to be recruited during the lactation.. We have sought for a similar factor in the sow[1]in the lactoserum, but not in the blood serum sampled at the same time, we have found a peptide with a chemotactic activity[1] towards B lymphocytes from the mesenteric but not inguinal lymph-node ; in addition, a peptide derived from bovine β-casein was found similarly chemoattractant for swine B lymphoblast as the sow lactoserum itself[8].

All cell types involved in the immune response were present in the MG at the different stages of gestation and lactation and nearer the alveolar epithelium as gestation proceeded: T lymphocytes, including CD4 and CD8, B lymphocytes and class II bearing cells (epithelial cells and macrophages). T lymphocytes accumulated early in pregnancy, specifically T helper cells; the specific increase of IgA lymphocytes occurring after this phase could suggest a role for these T cells in the induction of IgA response. The local accumulation of immune cells and the increase in CD8 cells near the epithelium suggests a role in local immune defence[5,13]. Only few functions of the MG lymphocytes have been explored in vitro; mammary lymphocytes had comparable maximum levels of PHA and ConA stimulation as blood lymphocytes, suggesting the presence of virgin lymphocytes[13].

## 5. INTRA-MAMMARY IMMUNISATION

The above observations raise the possibility of a genuine local immune response ; thus to ascertain this hypothesis, a search was made about intra-mammary immunisation, inasmuch as the MG could be a better route than the gut, needing less antigen because of the absence of degradation; in addition, this is reminiscent of the nature experiment where the piglets may inoculate antigen in the MG during suckling.

In sow, ferritin used alone led to an IgM response whilst in presence of complete Freund's adjuvant, which provokes a local granuloma formation and targets the antigen to the secretory site, there was a predominant IgA response[14]. The intra-mammary inoculation of live attenuated gastro-enteritis virus in pregnant sow induced persisting high levels of IgG neutralizing antibodies in milk whereas in lactation IgA antibodies were elicited[10], which fit with the data about the dual origin of lymphocytes in MG.

## 6. TRANSFER OF CELLULAR IMMUNITY BY MILK CELLS TO NEONATE

Depending upon the animal species and animals within species, milk contains $2x10^5$ to $10^7$ cells/ml. There are epithelial cells (31% of total cells in sow's milk) and non-nucleated cell fragments, and granulocytes, mainly represented by neutrophils (47%) and a few eosinophils (1%), lymphocytes (12%) and macrophages (9%)[17].The mean ratio of CD4+:CD8+ T lymphocytes in the peripheral blood and mammary gland secretions was 1.53 and 0.85 respectively. It can be speculated that the high ratio of CD8+ cells in secretion is most likely due to overflow of the cells which are mainly located intra-epithelialy, into milk. Activated CD8+ T lymphocytes may play an important role in the regulation and expression of the local immune response to pathogens[9].

Therefore, milk contains immune components to permit the passive transfer of specific cellular immunity. A number of research studies carried out in humans and rodents demonstrated that there was transmission from mother to offspring, through mammary secretions, of such reactions as cell-mediated hypersensitivity and skin transplant rejection. However, the efficiency of the transfer depends upon the ability of the cells to survive in the digestive tract of the young. That is, unless the passive acquisition by the young of the immunity mediated by maternal cells results from the passage of soluble factors produced by lymphocytes, such as transfer factor, rather than from the actual transfer of lymphocytes themselves. In new-born pigs, intra-epithelial lymphocytes are devoid of NK activity against virally infected cells and transfer from adult pigs of blood mononuclear cells increased the resistance to gastro-enteritis virus[4].

# 7. CONCLUSIONS

Study of lactogenic immunity and of the protection of the new-born demonstrates the compartmentalisation of the immune system into systemic and local systems. There is no problem to enhance the systemic protection of the neonate by a convenient systemic immunisation of the mother ; but the biological importance of the transmission of secretory antibodies from mother to offspring, that ensures the protection of the mucosa which is the first site to experience the antigenic challenge, requires a better understanding of the physiology of IgA humoral immune response. The later can take place at various levels: one step would be to investigate further the mechanisms that induce the expression of the IgA isotype by B cell precursors, for instance those localized in Peyer's patches, without overlooking the part played by the auxiliary T cells, antigen-presenting cells and the nature of the antigen itself. The regulation of the circulation pathways of the IgA lymphoblasts still are unclear, particularly if we take into account the variation due to the hormonal status. Because research has demonstrated existence of organ specific markers in vascular endothelia that determines the homing of immunocompetent cells, it seems important to verify that these findings can be extended to all mammals. Localized at the mucosal level, IgA lymphoblasts differentiate into plasma cells that locally produce specific secretory antibodies. There are few data available of the possible regulation *in situ* of this local humoral response by lymphokines and/or by tissue factors released by adjacent cells.

Lastly, with a better insight into what determines the expression of membrane antigens at the level of the endothelial cell, sites that are specifically recognized by homing receptors present on the surface of lymphocytes, it may be possible to envision the control of lymphocyte circulation in the organism. Understanding this control process may, for example, allow us to induce specific migration of immunocompetent cells to the mammary gland to start or enhance the specific humoral immune response and to improve the immune quality of the milk produced by the mammary gland.

# 8. REFERENCES

1. Abda R, Chevaleyre C, Salmon H: Effect of cryopreservation on chemotaxis of lymphocytes. Cryobiology. 36:184-193, 1998
2. Berthon P, Salmon H, Martinet J, Houdebine LM: Immunological factors in mammary secretions. Biologie de la lactation 389-414; 102
3. Bourne FJ, Curtis J: The transfer of immunoglobins IgG, IgA and IgM from serum to colostrum and milk in the sow. Immunology 24:157-162, 1973
4. Cepica A, Derbyshire JB. The effect of adoptive transfer of mononuclear leukocytes from an adult donor on spontaneous cell-mediated cytotoxicity and resistance to transmissible gastroenteritis in neonatal piglets. Can.J.Comp.Med. 48:360-364, 1984
5. Chabaudie N, Le Jan C, Olivier M, Salmon H: Lymphocyte subsets in the mammary gland of sows. Res.Vet.Sci. 55:351-355, 1993
6. Czinn SJ, Lamm ME: Selective chemotaxis of subsets of B lymphocytes from gut-associated lymphoid tissue and its implications for the recruitment of mucosal plasma cells. J.Immunol. 136:3607-3611, 1986
7. Czinn SJ, Robinson J, Lamm ME: Chemotaxis as a mechanism for recruitment of mucosal plasma cell precursors. Adv.Exp.Med.Biol. 216A:305-311, 1987
8. Fronteau D, Tanneau, G, Henry G, Chevaleyrec C, Leonil J, Salmon, H. Activités chimiotactique de lait d'artiodactyle sur les lymphocytes porcins. 30, 363-367. 1998. Paris, Institut Technique du porc. Journées Rech. Porcine en France. 98.
9. Park YH, Fox LK, Hamilton MJ, Davis WC: Bovine mononuclear leukocyte subpopulations in peripheral blood and mammary gland secretions during lactation. J.Dairy.Sci. 75:998-1006, 1992
10. Saif, LJ., Bohl, EH. Passive, immunity to transmissible gastrœnteritis virus: intramammary viral inoculation of sows. Annals N. Y Acad. Sc., 409,708-722, 1983.
11. Salmi M, Adams D, Jalkanen S: Cell adhesion and migration. IV. Lymphocyte trafficking in the intestine and liver. Am.J.Physiol. 274:G1-G6,1998.
12. Salmon H: Surface markers of swine lymphocytes: application to the study of local immune system in mammary gland and transplanted gut., Swine in biomedical research. Edited by Tumbleson ME. New York, Plenum, 1986, pp 1855-1864.
13. Salmon H: The intestinal and mammary immune system in pigs. Vet.Immunol.Immunopathol. 17:367-388, 1987.
14. Salmon H: Humoral lactogenic immunity in the sow: basis and practice. Pig News and Information 10:151-157, 1989.
15. Salmon H: (Lactogenic immunity and vaccinal protection in swine). Vet.Res. 26:232-237, 1995.
16. Salmon H, Delouis C: (Kinetics of lymphocyte sub-populations and plasma cells in the mammary gland of primiparous sows in relation to gestation and lactation). Ann.Rech.Vet. 13:41-49, 1982.
17. Schollenberger A, Degorski A, Frymus T: Cells of sow mammary secretions. I. Morphology and differential counts during lactation. Zentralblatt. 33:31-38, 1986.
18. Schollenberger A, Frymus T, Degorski A: Cells of sow mammary secretions. II. Characterization of lymphocyte populations. Zentralblatt. 33:39-46, 1986.
19. Schollenberger A, Frymus T, Degorski A: Cells of sow mammary secretions. III. Some properties of phagocytic cells. Zentralblatt. 33:353-359, 1986.
20. Silim A, Rekik MR, Roy RS, Salmon H, Pastoret PP: Immunité chez le foetus et le nouveau-né, Immunologie Animale. Edited by Pastoret PP, Govaerts A, Bazin H. Paris, Flammarion, 1990, pp 197-204

21. Springer TA: Traffic signals for lymphocyte recirculation and leukocyte emigration: the multistep paradigm. Cell 76:301-314, 1994.
22. Streeter PR, Berg EL, Rouse BT, Bargatze RF, Butcher EC: A tissue-specific endothelial cell molecule involved in lymphocyte homing. Nature 331:41-46, 1988.
23. Tanneau G.M Hibrand-Saint Oyant L Chevaleyre C.C Salmon. Differential recruitment of T and IgA B lymphocytes in the developing Mammary Gland in relation to Homing Receptors and Vascular addressins, J. Histochem. Cytochem. 1999 (in press).

33.

# Relationship Between Teat Tissue Immune Defences and Intramammary Infections

Alfonso Zecconi*, Jörn Hamann°, Valerio Bronzo*, Paolo Moroni*, Giulia Giovannini* & Renata Piccinini*

** Faculty of Veterinary Medicine, Institute for Infectious Diseases, Milan Italy, ° Tierärztliche Hochschule Hannover, Zentrum für Lebensmittelwissenschaften, Hannover Germany*

Key words: Teat , teat tissues, immunity, milking machine, intramammary infections

Abstract: The teat is the main entrance for pathogens into the mammary gland. It also acts as a sensory, motor and primary defence organ. This latter function is important in preventing intramammary infections while efficiency in preventing new infections is determined by teat tissue integrity. Machine milking may evoke mechanical and circulatory impairment in teat tissues. These local metabolic disorders may decrease the efficiency of the local immune defence mechanisms. Teat tissue changes can be estimated by measuring teat thickness before and after milking. Experimental and field studies showed a high correlation between changes in thickness and infection risk. Teats with >5% change in thickness have significantly increased teat duct colonisation rates and intramammary infection rates. The link between changes in teat thickness and infections should be found in changes in local immune defences and measurable changes in cytological and biochemical immune factors are expected. Indeed, the application of experimental milking conditions (i.e. no pulsation milking and positive pressure milking) showed to have a significant influence on some non specific immune factors in teat secretion. Positive pressure milking increases PMNs content and decreases macrophages content of teat secretion . Some enzymes such as NAGase and lysozyme were decreased by positive pressure milking, the concentration of the same enzymes were higher after no pulsation milking. A better knowledge on the interaction between the teat apex immune defense mechanisms and the machine milking process is necessary to reduce the new infection rate of the bovine mammary gland.

# 1. INTRODUCTION

The teat is the main entrance for pathogens into the mammary gland and works as the first line of defence against invading pathogens. Under physiological conditions, the teat canal acts as an efficient valve guarding the entrance of the teat cistern. The average length of the teat canal is about 10 mm and ranges from 3 to 18 mm. The teat canal diameter varies markedly depending on the measured location (distal, middle, and proximal). The data for the teat diameter cover a range of 0,35 to 5,0 mm with an average value of around 2 mm[7].

There are two major components of the anatomical defences: the sphincter muscle and the teat keratin. The sphincter muscle is a smooth muscle distributed in spiral form starting from the base of the teat and ending around the teat canal. Its function is to keep the teat canal closed, thus reducing the risk of invasion by pathogens[8]. Pathogens entering the teat canal encounter a mesh-like matrix of keratin. Keratin seals the teat canal between milkings and during dry period. Keratin adsorbs bacteria in mesh-like network, then sheds the adsorbed bacteria during next milking. Keratin also has antibacterial activity[3].

Within teat tissues, there are also cellular and humoral defences. Cellular defences in udder tissue and secretion include: polymorphonuclear leukocytes (PMN), monocytes, macrophages, lymphocytes, plasma cells, and mast-cells. The concentration of these cells, particularly of plasma cells increases from deep parenchyma to Furstenber's rosettes[11]. The humoral defences of the mammary gland include: lactoferrin, lysozyme, immunoglobulins, NAG-ase and basic proteins[3,12].

# 2. FACTORS AFFECTING TEAT TISSUES STATUS

The most important aspect in reducing the contamination risk especially concerning contagious pathogens (e.g. *Staph.aureus*) consists in the maintenance of a healthy skin condition of the teat. As long as the skin is healthy most pathogens have only a limited chance to survive on the teat skin[9,17]. This can only be achieved as long as the integrity and functionality of stratum corneum is ensured. Indeed, several studies showed that after artificial teat chapping the colonisation rate by *Staph.aureus* was significantly increased[5]. Colonisation of the teat skin predisposes the cow to new intramammary infections[13]

## 2.1 Teat canal closure

The closure of the teat canal is an active process related to the function of the smooth muscle in the teat wall. The impulses from the autonomic sympathetic nervous system keep these muscles under a constant tension[14]. Milk accumulation during the intermilking period increases the tone, whereas milk withdrawal decreases the sympathetic tone. Consequently, diameter and penetrability of the teat canal have the highest values just after milking. Therefore, as a physiological phenomenon, the ease of bacterial penetration into the teat canal is increased for approximately 2 hours after milking compared with later stages of the intermilking interval[16]. Improper machine milking function contribute to circulatory impairment which counteracts to maintain the closure of the teat canal between milkings to prevent leakage, and keep the keratin occluding the canal lumen compressed as a barrier in preventing the penetration of bacteria to the teat cistern.

## 2.2 Teat canal keratin lining

The teat canal is lined by a stratified squamous epithelium which shows in comparison with the general bovine skin much greater widths of the stratum granulosum and stratum corneum. The stratum corneum is equivalent to the keratin layer. Physical attributes of the keratin and its chemical, bacteriostatic or bactericidal properties limit or prevent the penetration of pathogens into the teat cistern. Teat canal keratin represents a primary defence mechanism as a physical barrier by occlusion, adsorption of bacteria to the keratin surface, elimination of bacteria by desquamation of corrupted cells during milking. Its components (lipids, basic proteins) have also bactericidal effects. The fundamental importance of the teat canal keratin lining as infection barrier is well documented since more than 30 years[10]. However, different types of mastitis pathogens can growth and survive on the keratin for considerable time periods[4].

## 2.3 Blood and lymph circulation

The teat blood supply is mainly provided by the teat artery (arteria papillae) which runs from the base to the tip of the teat near the inner surface of the wall. Teat veins are characterised by a very thick muscular wall. They drain into the Fürstenberg's venous ring, a circle of veins at the teat base. Based on their anatomic structure the veins are well situated to maintain blood flow during the machine milking process when the pressures are applied to the teat differences. In contrast to the secretory tissue, the teat skin and teat wall blood circulation show no lactation dependent variation[15]. The lymphatic system of the teat consists of a superficial and deeper part. The abundance of lymph vessels in the teat tissue has been characterised as a "lymphatic corpus cavernosus" which becomes turgescent during sucking or milking[1].

Under emotional stress induced by fright and pain or by local impairment, blood flow intensity can be decreased and this results in a lower uptake of oxygen from the circulation and might be able to increase the infection risk. In addition to these processes, teat stimulation may decrease the sympathetic tone of the mammary gland so that blood supply is increased. However, the rate and amplitude of the teat and teat sphincter muscle contractions, which are partly responsible for blood movement in the tissue, are decreased. The integrity of the circulation system providing a sufficient blood supply is the basis to enable optimum functioning of all mechanisms involved in the defence against pathogens. Physiological and pathological changes in the circulatory system associated with milking may decrease the efficacy of the defence systems at least for some time after milking.

## 3. TEAT STATUS AND INFECTION RISK

The major factor affecting teat status is the milking machine. Milking machine influence status by: a.: moving pathogens into the teat cistern during milking; b.: increasing external teat contamination; and c.:decreasing teat tissue defences potential.

A significant higher proportion of bacteriological positive findings has been observed for teat duct swabbing samples compared with foremilk samples. Therefore, teat duct colonization can be assumed to be a source of new intramammary infections (IMI), even though spontaneous elimination of pathogens can often take place[19]. The risk of new IMI increases as much as teat tissues immune defences are impaired. Machine milking may evoke mechanical and circulatory impairment of teat tissue. Frequently this impairment is subclinical and there is no way to restore the physiological conditions. Using teat thickness measurement before and after milking[6] it is possible to evaluate teat tissue status due to the action of machine milking. Several field studies based on this method[19,20] have shown:

a. a significant association between teat thickness changes and IMI;
b. the IMI risk increases in association with both an increase and a decrease of teat thickness;
c. the IMI risk is higher when thickness increases than when a decrease of thickness is observed;

d. within different bacteria species, only coagulase negative staphylococci IMI showed to be significantly associated with teat thickness changes.

# 4. TEAT STATUS AND IMMUNE DEFENCES

Even if the association between teat thickness changes, and more in general, between machine milking action and IMI risk can be demonstrated, information on the immune factors affected by the action of milking machine, are relatively few. To address this issue, two different "extreme" milking systems (positive pressure and no pulsation) were applied on the same group of cows. Different immune parameters were assessed on the milk and on the teat secretion[2]. Figures 1 and 2 summarise the results for some of them.

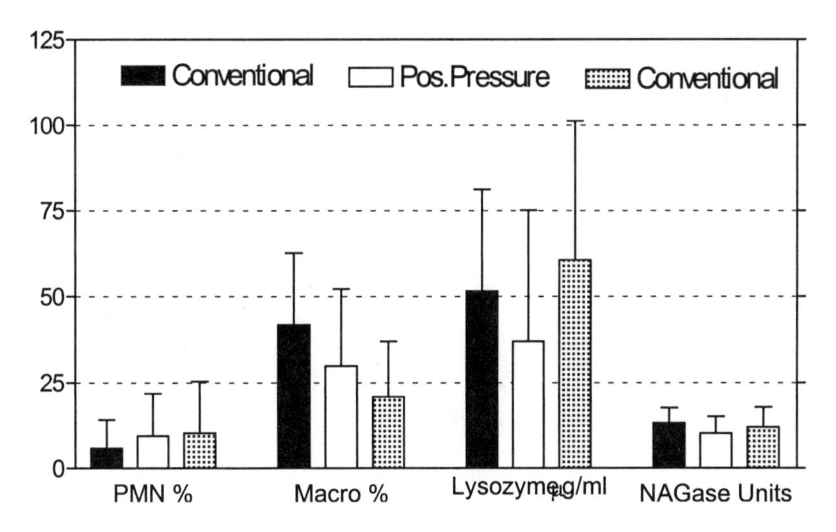


Fig. 1 Cellular ratio and enzyme concentration in teat secretion measured in 32 quarters of 8 primiparous cows milked with a positive pressure milking system for 8 days. Conventional milking system was applied for the same period of time before and after the application of positive pressure milking system.

The results showed that positive pressure milking system increased PMN ratios and decreased macrophage ratios during and after the application of positive pressure milking system. This latter one induced a decrease of lysozyme concentration in teat secretion, while milking system did not significantly influence NAGase concentration.

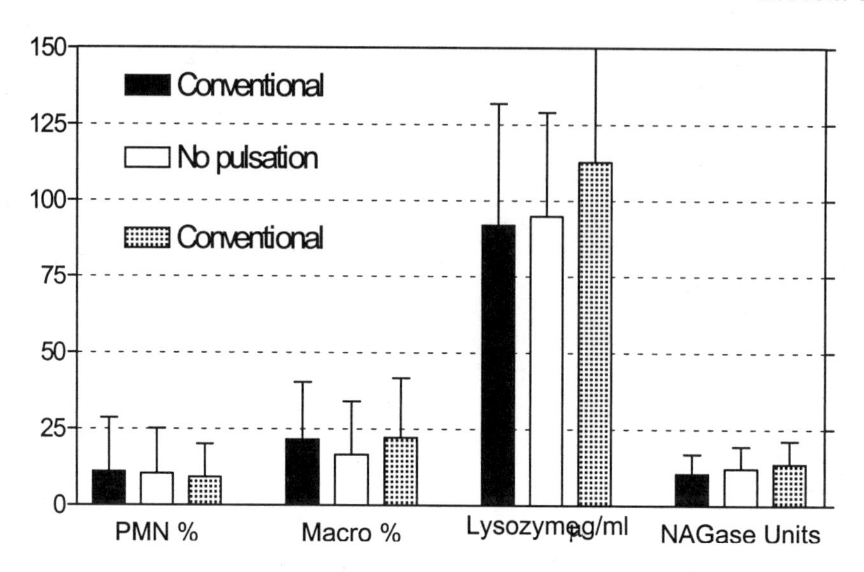


Fig. 2 Cellular ratio and enzyme concentration in teat secretion measured in 32 quarters of 8 primiparous cows milked with a no-pulsation milking system for 8 days. Conventional milking system was applied for the same period of time before and after the application of no pulsation milking system.

The results showed that PMN ratios were not influenced by no pulsation milking system, while a decrease of macrophage ratio was observed. Lysozyme and NAGase concentration showed a slight increase during and after the application of no pulsation milking system.

The results of these experiments confirmed that machine milking could affect teat immune factors. The pattern of factor changes was different depending on the type of milking system applied.

## 5. CONCLUSIONS

The efficiency of bovine defences is one of the most important factors in the epidemiology of intramammary infections. Within the bovine immune defences, teat plays a pivotal role being the main entrance for bacteria to the udder. The maintenance of a physiological blood supply and the avoidance of mechanical impairment resulting from unsuitable milking procedures and malfunctioning milking machines are fundamental to ensure teat tissue integrity.

More knowledge on the influences of milking machine on teat immune defences is needed to avoid the impairment of these latter ones and to ensure that milking machine will milk the cow as much physiologically as possible.

# 6. REFERENCES


1. Blum, J.W., D. Schams And R. Bruckmaier (1989): Catecholamines, oxytocin and milk removal in dairy cows. J. Dairy Res. 56, 167 –177
2. Bronzo, V.,Hamann, J.& Zecconi, A. 1995. Changes in anatomical, cytological and biochemical parameters of the teat associated with the use of three different milking units. Proc. 3rd Int.Mastitis SeminarTel Aviv, 1,110-113 .
3. Craven, N. & Williams, M. R. 1985. Defences of the bovine mammary gland against infection and prospects for their enhancement. Veterinary Immunology & Immunopathology 10 71-127
4. Du Preez, J.H. (1985): Teat canal infections. Kiel. Milchwirtschaftl. Forschungs-berichte, 267 – 273
5. Fox, L. K. and M. S. Cumming (1996): Relationship between thickness, chapping and Staphylococcus aureus colonization of the bovine teat tissue. J. Dairy Res. 63, 369 - 375
6. Hamann, J. 1985. Measurement of machine milking induced teat tissue reactions. Milchwissenschaft 40 16-18.
7. Hamann, J. 1987. Effect of machine milking on teat end condition – a literature review. IDF-Bulletin, No 215, 33 –53
8. Hamann, J. and C. Burvenich 1994. Physiological status of the bovine teat. IDF-Bulletin No 297, 3-12
9. King, J. S. 1981. Streptococcus uberis: A review of its role as a causative organism of bovine mastitis. II. Control of infection. Br. Vet. J. 137, 160 -165
10. Morse, G. E. and J. Platonov 1964. Basic studies of bovine mastitis: I. The role of the teat canal as a barrier to udder infections with Streptococcus agalactiae. J. Dairy Sci. 47, 696-700
11. Nickerson, S. C. & Pankey, J. W. 1983. Cytologic obdervations of the bovine teat end. J. Vet. Res. 44 1433-1441.
12. Outerridge, P. M. & Lee, C. S. 1988 The defence mechanism of the mammary gland of domestic ruminants. Moving Frontiers in Veterinary Immunology. R. Pandey. Basel, Karger: 165-196.
13. Pankey, J. W., R. J. Eberhart, A. L. Cumming, R. D. Daggett, R. J: Farnsworth and C. K. Mcduff 1984. Uptake of postmilking teat antisepsis. J. Dairy Sci. 67, 1336 – 1353
14. Peeters, G., R. Coussens and G. Sierens 1949. Physiology of the nerves in the bovine mammary gland. Arch. Int. Pharmacodyn. Ther. 79: 75 - 82
15. Peeters, G., A. Houvenaghel, E. Roets, A.-M. Massart-Leen, R. Verbeke, G. Dhondt and F. Verschooten 1979. Electromagnetic blood flow recording and balance of nutrients in the udder of the lactating cow. J. Anim. Sci. 48: 1143 - 1153
16. SCHULTZE, W.D. and S.C. BRIGHT 1983. Changes in penetrability of bovine papillary duct to endotoxin after milking. Am. J. Vet. Res. 44: 2373 – 2375
17. Smith,A., H.G.J. Coetzee 1979. The survival of Staphylococcus aureus on teats and orifices of cows in the dry period. S. Afr. J. Dairy Technol. 11 79 – 81
18. Tagand, R. 1932. Anatomie des vaisseaux mammaires. Lait XII: 881 – 893
19. Zecconi, A.,Hamann, J.,Bronzo, V.& Ruffo, G. 1992. Machine-induced teat tissue reaction and infection risk in a dairy herd from contagious mastitis pathogens. J.Dairy Research 59 265-271.
20. Zecconi, A.,Bronzo, V.,Piccinini, R.,Moroni, P.& Ruffo, G. 1996. Field study on the relationship between teat thickness changes and intramammary infections

34.

# Immunological Aspects of Pregnancy-Associated Glycoproteins

Dosogne H., A.M. Massart-Leën and C.Burvenich
*University of Ghent, Faculty of Veterinary Medicine, Department of Physiology, Biochemistry and Biometrics, Merelbeke, Belgium*

Key words: Pregnancy associated glycoproteins

Abstract: The incidence of severe cases of acute *E. coli* mastitis in dairy cows is highest during early lactation. This phenomenon has been associated with a decreased function and decreased numbers of circulating polymorphonuclear neutrophil leukocytes (PMN). The cause of this impaired function and decreased number is poorly understood. Stress, hormonal and metabolic alterations around parturition and the onset of lactation may play a role in this phenomenon. Several molecules, such as cortisol and beta-hydroxybutyrate have been found to alter the oxidative burst activity of circulating PMN around parturition. Pregnancy-Associated Glycoprotein (bPAG) could also be involved. The theory of immunosuppression by bPAG was investigated because analogous glycoproteins produced by the placenta of other species exert local immuno-suppression in order to maintain the histoincompatible feto-maternal unit. The production and subsequent release into the maternal circulation of bPAG is ensured by the binucleate cells from the trophoblast and starts already at implantation. However, peak levels are only reached 1 week before parturition. Due to the long half-life time of this molecule, high levels are found in plasma until 2 weeks after calving. The co-occurrence of the impairment of PMN oxidative burst activity in the early postpartum period and a peak in plasma bPAG concentrations might support the hypothesis of an immunosuppressive effect of PAG. Moreover, an inhibitory effect of bPAG on the proliferation of bovine bone marrow progenitor cells has been found recently in our laboratory. bPAG occurs in colostrum, but its effect on milk cells has not been clarified. It is concluded that interaction between the physiology of reproduction and lactation on the one side and immune function on the other side in dairy cattle requires further research.

# 1. INTRODUCTION

The mammalian placenta represents the contact area between foetus and mother. Besides being a barrier, it plays a role in nutrient supply to and waste removal from the foetus. Moreover, it has an important endocrinological-immunological function. It can affect the maternal metabolism and plays a role in mammogenesis and lactogenesis. Certain glycoproteins are an important aspect of this endocrinological-immunological function. These molecules are typical for gestation and indicate the development of a trophoblast. They are secreted into the maternal circulation soon after implantation. Presently, about 100 different of these glycoproteins are identified. In humans and horses, chorion gonadotrophins (hCG and eCG, respectively) are commonly used for early pregnancy detection. In cows, the bovine pregnancy-associated glycoprotein (PAG) is determined for these purposes. A luteotropic hormone (LH) activity is attributed to CG, but for PAG there is no clear function discovered until now.

# 2. Physiology of placental glycoproteins

## 2.1 Chorion Gonadotrophin

The CG is a glycoprotein that consists of an alpha and a beta subunit. It is produced during pregnancy in several species, but not in ruminants. Both human hCG and equine eCG (also called pregnant mare serum gonadotropin) have a clear luteinizing hormone (LH)-activity. When administered to other species, its effect is predominantly that of the follicle stimulating hormone (FSH). Whereas hCG is produced by the syncytiotrophoblast, eCG is produced by the endometrial cups. Human CG would also postpone luteal regression. Equine CG is responsible for maintenance of primary corpora lutea and for induction of secondary corpora lutea[24]. Equine CG, which has a high percentage of sialic acid, would realise an immunological isolation of the trophoblast by absorption of destructive factors produced by the maternal immune system. It would also play a role in the accumulation of leukocytes around the endometrial cups[17].

## 2.2 Placental lactogen

Placental lactogen (PL) is a single-chain polypeptide that is produced by trophoblast cells in humans, sheep and cows. Human PL has a prolactin and growth hormone activity, because it stimulates milk secretion on the one hand and plays a role in proliferation of the mammary gland and maternal intermediary metabolism on the other hand. In sheep, oPL would also have prolactin activity. In this species, a growth hormone activity has not been identified. Horses and pigs do not produce PL.

## 2.3 Other pregnancy-associated glycoproteins in humans

The pregnancy-specific B1 glycoprotein is produced by the syncytiotrophoblast. It would exert an immunosuppressive effect with prevention of foetal rejection. Pregnancy-associated plasma protein A (PAPP-A) is also produced by the syncytiotrophoblast in early pregnancy[21]. PAPP-A is a strong inhibitor of leukocyte elastase, a protease that can affect the chorionvilli of the trophoblast. Pregnancy-associated α2-glycoprotein is a high molecular weight (360 kDa) glycoprotein with immunosuppressive properties both in vitro and in vivo[9].

## 2.4 Interferon-γ

Interferon-γ (IFN-() in ruminants is produced by the mononuclear cells of trophoblast epithelium even before implantation. It inhibits the production of prostaglandins by the endometrium and prevents luteolysis in sheep and cows. IFN-γ has the typical features of interferons: antiviral activity, modulation of the immune system and inhibition of cell proliferation[6].

## 2.5 Uterine milk proteins

In sheep, uterine milk proteins (UTMP) are produced by uterine epithelium under the influence of progesterone. They probably have an immunoregulatory function because both in vitro and in vivo they suppress a number of immune responses, in particular the proliferation of lymphocytes. UTMP show an amino acid sequence homology with the family of serine proteinase inhibitors, but a functional activity as serine proteinase has not been demonstrated so far[14]. In the bovine, a progesterone-induced UTMP has also been demonstrated, but possible immunoregulatory effects have not been investigated[13].

## 2.6 Bovine pregnancy-associated glycoproteins

### 2.6.1 Identification of pregnancy-associated glycoproteins

In 1982, Butler and co-workers isolated the pregnancy-specific glycoprotein B from rabbit antisera after injection with extracts from bovine cotyledons. It appeared to be a new molecule. PSPB was a group of analogous proteins with a molecular weight varying between 47 and 90 kDa. Later, the pregnancy-associated glycoprotein was isolated from bovine fetal cotyledons[1,28]. From amino acid sequence analysis, it appeared to be the same molecule as PSPB[18]. Besides in cows, PAG has been identified in sheep[27], deer[25], goats[15], bisons and moose[12]. Recently, new members of the PAG family were discovered in the porcine[23] and equine[10] placenta.

### 2.6.2 Biochemistry of pregnancy-associated glycoproteins

PAG belongs to the family of aspartic proteinases. From analysis of amino acid sequence homology, PAG demonstrated a surprising 60% homology with the family of pepsinogens. The aspartic proteinase family is a highly conserved family of which also pepsin, chymosine, cathepsin E and D and renine are members. Because of some mutations in active sites and because of their inactivity in proteolytic assays, it has been concluded that both bovine and ovine PAG are probably not active as proteïnases[11,26]. Bovine PAG has a bilobed macromolecular structure with a cleft that can bind peptides 7 to 8 amino acids long. The cDNA of bPAG-1 coded for a polypeptide of 380 amino acids long. From cDNA sequencing, it has been concluded that the theoretical molecular weight of bPAG should be 36 kDa. However, its actual molecular weight is 67 kDa. Addition of carbohydrates to the 4 possible sites of N-linked glycosylation can lead to a substantial increase in size and heterogeneity. From these data, it is rather likely that PAG functions as a carrier of bioactive peptides than as a proteolytic agent. An alternative hypothesis is that they are hormones which bind to specific receptor molecules[20].

### 2.6.3 Function of bovine pregnancy-associated glycoprotein

Until now there has been no clear function attributed to PAG. However, their evolutionary survival and abundant presence at the foeto-maternal site suggest that they are not simply a curiosity. Since the isolation and characterization of PAG, different hypotheses for its function have been presented. One is that they are placental hormones directed towards the mother, perhaps with an action on the corpus luteum. Indeed, PAG1 stimulated the induction of prostanoids in both cultured luteal cells and endometrial explants[4]. However, reports on the action of PAG on progesterone production have generated variable results and more research is required to elucidate their possible role as hormones.

Several lines of evidence have lead to the hypothesis that PAG could have an immunosuppressive effect in dairy cows. The incidence of acute *E. coli* mastitis is very high shortly after calving[2]. Impairment of neutrophil function plays an important role in the pathogenesis of this disease. Although bPAG is produced and released at implantation, peak values of this glycoprotein in the maternal circulation are only found around calving[29]. Concentrations in maternal plasma increase up to 2,500 ng/ml at the day of calving. Also, other pregnancy-associated glycoproteins have been found to exert immunosuppressive effects. For example, pregnancy-associated plasma protein A is a strong inhibitor of leukocyte elastase[21]. Moreover, higher levels of PAG were found in the plasma of cows during their first pregnancy than in following pregnancies[29] (Dosogne et al., unpublished results). In first pregnancies higher titers of antibodies are found against some trophoblast antigens that are recognized as foreign by the maternal immune system. This is consistent with the concept of an immunosuppressive effect of PAG.

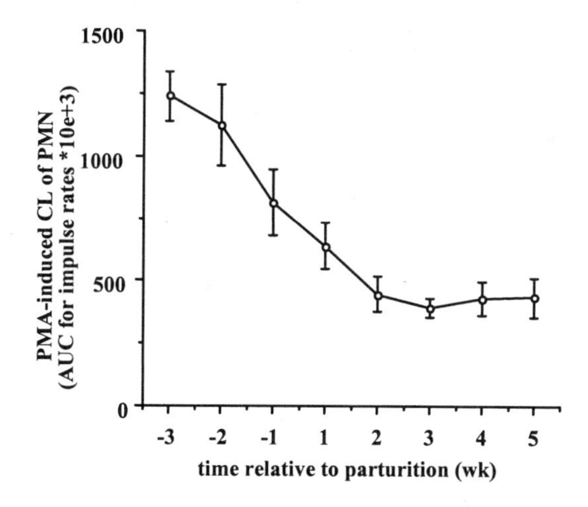


Fig. 1. Phorbol myristate acetate (PMA)-induced chemiluminescence (CL) activity of polymorphonuclear leukocytes (PMN) isolated from blood of cows before and during the early postpartum period. Values are means ± SEM of 11 cows.

The hypothesis of immunosuppression in periparturient cows by PAG was investigated in studies with healthy cows and in cows with mastitis. The phagocytosis and oxidative burst activity of polymorphonuclear neutrophil leukocytes (PMN) isolated from blood and PAG concentrations in plasma were evaluated in 2 longitudinal studies in dairy cows from 3 weeks before until 5 weeks after calving, carried out in the United States and in Europe. Phagocytosis of *Escherichia coli* was not changed in the early postpartum period. In both studies, a significant decrease in oxidative burst activity of PMN was observed between 1 and 3 weeks after calving (Fig. 1). Figure 1 presents the results from the European study. Results from the USA study were similar. In all cows, a very significant increase in plasma bPAG concentration was found between 1 week before and 2 weeks after calving (Fig. 2). The peak of bPAG concentration in plasma immediately preceded the alterations of blood PMN functions. These results suggested that bPAG could be associated with inhibition of PMN function of dairy cows during the early postpartum period[5].

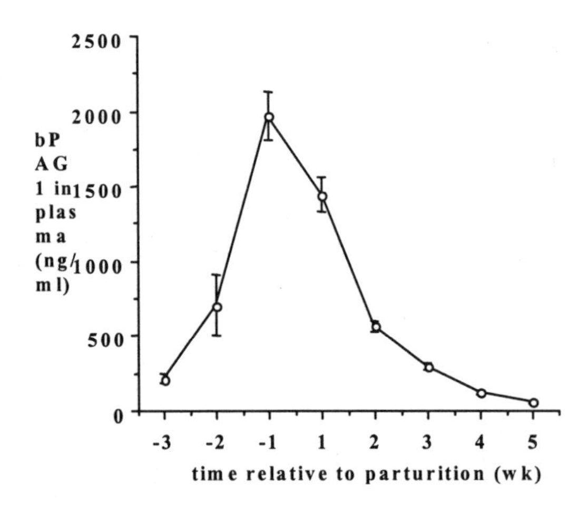


Fig. 2. Pregnancy-associated glycoprotein (PAG) concentrations in plasma of cows before and during the early postpartum period. Values are means ± SEM of 11 cows.

The mechanism of action of bPAG on PMN function is still poorly understood. Binding studies between the PAG molecule and bovine PMN isolated from blood and milk have been unsuccessful so far (Dosogne et al., unpublished results). It has been found that PAG inhibits the proliferation of bone marrow cells that are the precursors of mature neutrophils[8]. This could play a role the presence of high numbers of immature neutrophils in the circulation around calving. However, an important condition for acceptance of this hypothesis, is that bPAG has to penetrate the blood-bone marrow barrier. This is not evident for such a large molecule. At this moment, not much is known about the penetrability of the blood-bone marrow barrier. A blood-bone marrow barrier to macrophage colony-stimulating factor, a glycoprotein of only 23 kDa, has been demonstrated in mice[22]. The penetrability of the blood-bone marrow barrier to bPAG and other molecules is an interesting subject for further research.

A second part of the investigations of a role of PAG in bovine mastitis entails its possible effects on neutrophil function during mastitis. For this purpose, 6 early lactating dairy cows were experimentally infected with 20 ml $1.10^4$ cfu *E. coli* P4:O32. Both neutrophil functions and the concentration of PAG in plasma were investigated at several times before and during mastitis. An increased number of immature neutrophils was found in plasma with a peak of 72% immature neutrophils per 100 neutrophils at 12 h after infection (Fig. 3). This was associated with an observed decrease of detoxification of endotoxins by the neutrophil enzyme acyloxyacyl hydrolase (Fig. 3).

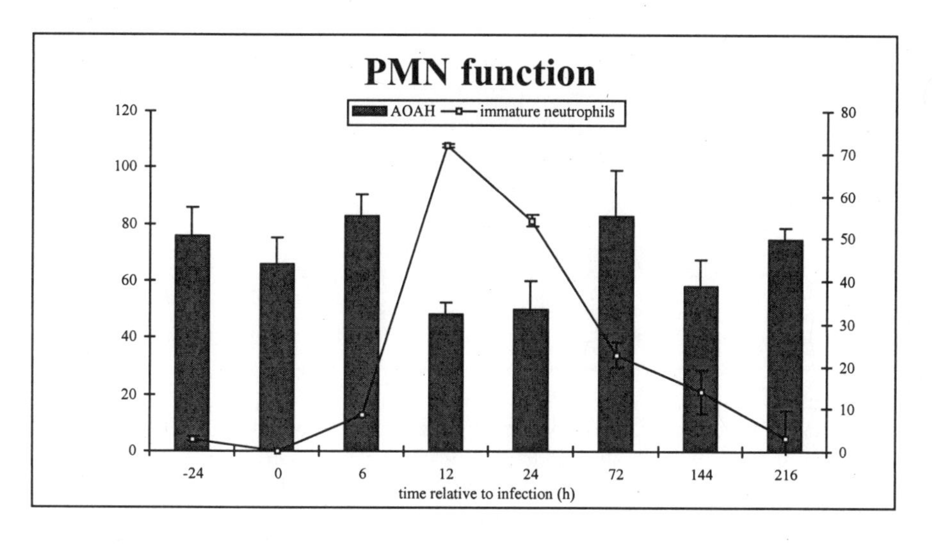


Fig. 3. Percentage of immature neutrophils and acyloxyacyl hydrolase activity of bovine polymorphonuclear leukocytes isolated from blood before and during experimental infection of the mammary gland with *E. coli*. Values are means ± sem of 6 cows.

During inflammation, cortisol is released into the circulation. This results in an impairment of protein synthesis and a stabilization of lysosomal membranes. This may affect the breakdown of the PAG molecule in pathological conditions because both the production and the release of proteolytic enzymes is inhibited. Alternatively, acute-phase reactant proteins are synthesized by the liver and released into the circulation during inflammation[19]. Increased serum levels of $\alpha$1-acid glycoprotein have been found in cattle following intravenous *Fusobacterium necrophorum*-induced inflammation[16]. Moreover, the severity and outcome of experimentally induced mastitis in pregnant heifers was related to the amount of acid-soluble glycoproteins in plasma[7]. In analogy with these results, the concentration of PAG in plasma during mastitis was investigated. In our experiment, a gradual decrease of PAG plasma levels was found before and during experimentally induced *E. coli* mastitis in early lactation cows (Fig. 4).

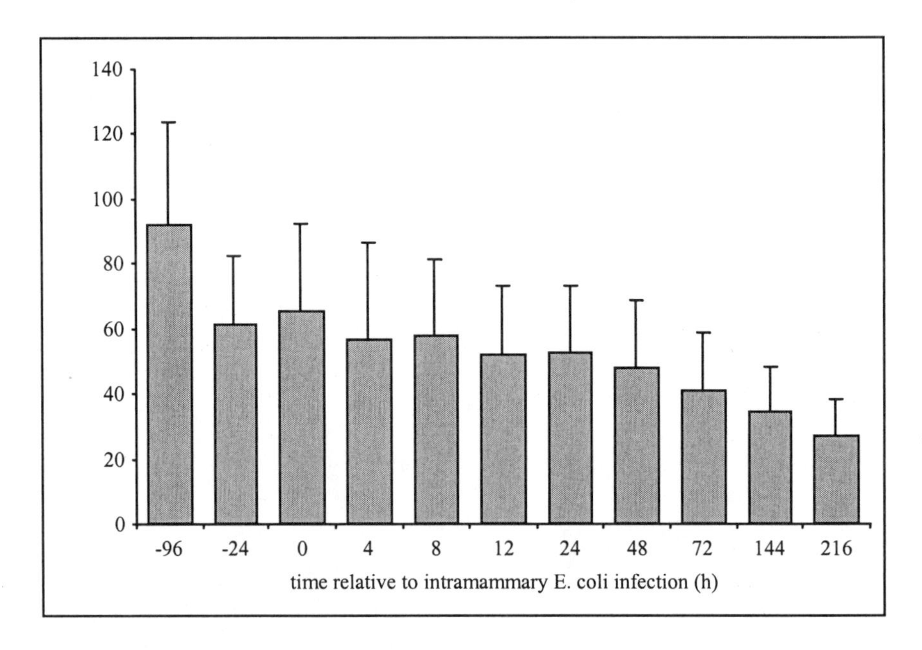


Fig. 4. Concentration of PAG in plasma from cows before and during experimental infection of the mammary gland with *E. coli*. Values are means ± sem of 6 cows.

This decrease, however, had already started before the induction of mastitis in these early lactation cows. It was not different from the decrease of PAG in plasma in healthy cows (Fig 2) when the data were arranged in function of the day relative to parturition. Therefore, the breakdown of the PAG molecule was not changed during mastitis.

## 3. CONCLUSION

It appears that most glycoproteins that are produced by the mammalian placenta, are involved in the maintenance of pregnancy. In the case of chorion gonadotrophins, this is ensured by a luteinizing hormone effect. For other glycoproteins, a suppressive effect towards the maternal immune system has been demonstrated. In the case of bovine PAG, effects have not been clearly demonstrated. However, its high concentrations in maternal plasma around calving suggests that its possible immunosuppressive effects may be extended beyond the maintenance of pregnancy. It may play a role in the pathogenesis of acute *E. coli* mastitis by impairment of neutrophil function. Until now, the only known mechanism of action of this molecule on the immune system is the inhibition of proliferation of bone marrow cells *in vitro*. The role of PAG in periparturient immunosuppression requires further research.

## 4. ACKNOWLEDGMENTS

The research on immunomodulatory effects of bPAG has been supported by the Belgian Ministry of Small Enterprises and Agriculture D1/2-5741A. The authors greatly acknowledge the scientific and technical assistance of Prof. J.-F. Beckers and Dr. J. Sulon from the University of Liege, Faculty of Veterinary Medicine.

## 5. REFERENCES

1. Beckers J.F., Wouters-Ballman P., Ectors F. (1988). Isolation and radioimmunoassay of a bovine pregnancy-specific protein. Theriogenology 29, 219 (abstract).
2. Burvenich, C., Paape, M.J., Hill, A.W., Guidry, A.J., Miller, R.H., Heyneman, R., Kremer, W.D.J. & Brand, A. (1994). Role of the neutrophil leukocyte in the local and systemic reactions during experimentally induced *E. coli* mastitis in cows immediately after calving. *The Veterinary Quarterly*, **16**, 45-50.
3. Butler J.E., Hamilton W.C., Sasser R.G., Ruder C.A., Hass G.M., Williams R.J. (1982). Detection and partial characterization of two bovine pregnancy-specific proteins. Biology of Reproduction 26, 925-933.
4. Del Vecchio R.P., Sutherland W.D., Sasser R.G. (1995). Effect of pregnancy-specific protein B on luteal cell progesterone, prostaglandin, and oxytocin production during two stages of the bovine estrous cycle. J. Anim. Sci. 73, 2662-2668.
5. Dosogne H., Burvenich C., Freeman A.E., Kehrli M.E., Detilleux J.C., Sulon J, Beckers J.-F., Hoeben D. (1999). Pregnancy-associated glycoprotein and decreased polymorphonuclear leukocyte function in early postpartum dairy cows. Veterinary Immunology and Immunopathology 67, 47-54.
6. Ealy A.D., Green J.A., Alexenko A.P., Keisler D.H., Roberts R.M. (1998). Different ovine interferon-tau genes are not expressed identically and their protein products display different activities. Biol. Reprod. 58, 566-573.
7. Hirvonen J., Pyörälä S., Jousimies-Somer H. (1996). Acute phase response in heifers with experimentally induced mastitis. J. Dairy Res. 63, 351-360.
8. Hoeben D., Burvenich C., Massart-Leën A.M., Lenjou M, Nijs G., Van Bockstaele D., Beckers J.-F. (1999). In vitro effect of ketone bodies, glucocorticosteroids and bovine

pregnancy-associated glycoprotein on cultures of bone marrow progenitor cells of cows and calves. Vet. Immunol. Immunopathol. (in press).

9. Horne C.H.W., Armstrong S.S., Thomson A.W., Thomson W.D. (1983). Detection of pregnancy associated $\alpha$2-glycoprotein, an immunosuppressive agent, in IgA producing plasma cells and in body secretions. Clin. Exp. Immunol. 51, 631-638.
10. Green J.A., Xie S., Newman A., Szafranska B., Roberts R.M., Baker C.B., McDowell K. (1995). Pregnancy-associated glycoproteins of the horse. Biology of Reproduction 50 (Suppl.1), Abstract 152.
11. Guruprasad K., Blundell T.L., Xie S., Green J., Szafranska B., Nagel R.J., McDowell K., Baker C.B., Roberts R.M. (1996). Comparative modelling and analysis of amino acid substitutions suggests that the family of pregnancy-associated glycoproteins includes both active and inactive aspartic proteinases. Protein engineering 9, 849-856.
12. Haigh J.C., Dalton W.J., Ruder C.A., Sasser R.G. (1993). Diagnosis of pregnancy in moose using a bovine assay for PSPB. Theriogenology 40, 905-911.
13. Hansen P.J. (1997). Interactions between the immune system and the bovine conceptus. Theriogenology 47, 121-130.
14. Hansen P.J., Liu W.-J. (1997). Biology of progesterone-induced uterine serpins. In: Church et al. (ed.). Chemistry and Biology of Serpins. Plenum Press, New York., 143-154.
15. Houston B.D., Robbins C.T., Ruder C.A., Sasser R.G. (1986). Pregnancy detection in mountain goats by assay for PSPB. Journal of Wildlife Management 50, 740-742.
16. Itoh H., Motoi Y., Haritani M., Kobayashi M., Tamura K., Takase K., Oikawa S. (1997). Immunohistochemical localization of $\alpha$1-acid glycoprotein in liver tissues of bovine fetuses, newborn calves, and sich or healthy adult cattle. Am. J. Vet. Res. 58, 725-728.
17. Koets A.P. (1995). The equine endometrial cup reaction: a review. Vet.Quart. 17, 21-29.
18. Lynch R.A., Alexander B.M., Sasser R.G. (1992). The cloning and expression of the bovine pregnancy-specific protein B (bPSPB) gene. Biology of Reproduction 46 (Suppl.), Abstract 89.
19. Ricca G.A., Hamilton R.W., McLean J.W. (1981). Rat $\alpha$1-glycoprotein mRNA. J. Biol. Chem. 256, 1032-1038.
20. Roberts R.M., Xie S., Nagel R.J., Low B., Green J., Beckers J.F. (1995). Glycoproteins of the aspartyl proteinase gene family secreted by the developing placenta. In: Aspartic proteinases: structure, function, biology, and biomedical implications. Takahashi K. (ed.) Plenum Press, New York, 231-240.
21. Rosen S.W. (1986). New placental proteins: chemistry, physiology and clinical use. Placenta 7, 575-594.
22. Shadduck R.K., Waheed A., Wing E.J. (1989). Demonstration of a blood-bone marrow barrier to macrophage colony-stimulating factor. Blood 73, 68-73.
23. Szafranska B., Xie S., Green J., Roberts R.M. (1995). Porcine pregnancy-associated glycoproteins: new members of the aspartic proteinase gene family expressed in trophectoderm. Biology of Reproduction 53, 21-28.
24. Talamantes F., Ogren L. (1994) The placenta as an endocrine organ: polypeptides. The Physiology of reproduction, edited by E. Knobil and J. Neill et al., Raven Press, Ltd, New York ©, Chapter 52, 2093-2144.
25. Wood A.K., Short R.E., Darling A.E., Dusek G.L., Sasser R.G., Ruder C.A. (1986). Serum assays for detecting pregnancy in mule and white-tailed deer. Journal of Wildlife Management 50, 684-687.
26. Xie S., Low B.G., Kramer K.K., Nagel R.J., Anthony R.V., Zoli A.P., Beckers J.F. (1991). Identification of the major pregnancy-specific antigens in cattle and sheep as

inactive members of the aspartic proteinase family. Proceedings of the National Academy of Science USA 88, 10247-10251.

27. Zoli A.P., Beckers J.F., Ectors F. (1995). Isolement et caractérisation partielle d'une glycoprotéine associée à la gestation chez la brebis. Annales de Médecine Vétérinaire 139, 177-184.
28. Zoli A.P., Beckers J.F., Wouters-Ballman P., Closset J., Falmagne P., Ectors F. (1991). Purification and characterization of a bovine pregnancy-associated glycoprotein. Biology of Reproduction 45, 1-10.
29. Zoli A.P., Guilbault L.A., Delahaut P., Ortiz W.B., Beckers J.F. (1992). Radioimmunoassay of a bovine pregnancy-associated glycoprotein in serum: its application for pregnancy diagnosis. Biology of Reproduction 46, 83-92.

# 35.

# Vaccines Against Bovine Mastitis due to *Streptococcus uberis* Current Status and Future Prospects.

James A. Leigh
*Institute for Animal Health, Compton Laboratory, Compton, Newbury, Berks UK.*

Key words: Vaccines, mastitis

Abstract The prevalence of bovine mastitis in the UK has been reduced over the past twenty five years due to the implementation of a five-point control plan aimed at reducing exposure, duration and transmission of intramammary infections by bacteria. This has markedly reduced the incidence of bovine mastitis caused by bacteria which show a contagious route of transmission but has had little effect on the incidence of mastitis due to bacteria which infect the gland from an environmental reservoir. Streptococcus uberis is one such bacterium which is responsible for a significant proportion of clinical mastitis worldwide. The inadequacies of the current methods of mastitis control have led to the search for additional measures to prevent intramammary infection by this bacterium. A live vaccine in combination with an intramammary administration of a soluble cell surface extract was shown to induce protection of the mammary gland from experimental challenge with S. uberis. Protection was strain specific, but was achieved in the absence of opsonic activity and without a large influx of neutrophils. One hypothesis is that protection was achieved by reducing the rate of bacterial growth in vivo. This view has led to the identification and exploitation of a novel plasminogen activator as a vaccine antigen. Vaccines containing this antigen conferred cross strain protection.

## 1. INTRODUCTION

Three species of *Streptococcus* are intimately associated with intramammary infection in the dairy cow. Two, *Streptococcus dysgalactiae* and *Streptococcus agalactiae* are transmitted largely from animal to animal by contagious routes. The third, *Streptococcus uberis*, is also transmitted to the bovine mammary gland, at a significant frequency, via contact with the organism in the environment. Due to the ubiquitous nature of *S. uberis* in the environment of the dairy cow teat end contamination by this organism poses a constant threat of infection. At present, mastitis due to *S. uberis* remains a persistent problem which impacts on the economic production of milk and on the welfare of the dairy cow [1].

## 2. CURRENT STATUS

### 2.1 Live vaccines

Immunisation with live *S. uberis* by the sub-cutaneous route, along with the administration of a soluble preparation of bacterial cell wall antigens by the intramammary route, conferred some protection against subsequent experimental challenge. All challenged quarters on the non-immunised animals shed high numbers of bacteria (~$10^7$ cfu / ml of milk) and developed clinical mastitis. In contrast, quarters of animals which received the vaccine shed considerably fewer bacteria (<$10^2$ cfu / ml of milk) and only two out of eight quarters, on four animals, developed clinical mastitis[2]. Subsequent experiments have confirmed the protective effect of vaccination with live *S. uberis* but also demonstrated that this was less effective against strains other than that administered as the immunising antigen[3]. Neither milk nor serum from immunised animals promoted an increase in the uptake and killing of *S. uberis* by bovine neutrophils. The control of infection was not associated with a marked influx of neutrophils into the mammary gland. One possible explanation for the protective effect was that vaccination had decreased the rate at which *S. uberis* was able to colonise the gland. Consequently, it was possible that bacterial numbers had been controlled due to the dilution and elimination of *S. uberis* within the milk.

Due to the strain specific nature of the protection obtained and the wide diversity of strains which are able to induce clinical disease even within a single herd[4,5], it is unlikely that further vaccine development will be centred on a live antigen. However, it is possible, but as yet untested, that a similar regime may protect against a small number of related strains in the field.

This series of experiments demonstrated, for the first time, that protection from intramammary infection by *S. uberis* and the consequent mastitis could be achieved in the absence of a marked inflammatory response and the production of opsonic activity. This challenged the view that killing of *S. uberis* by neutrophils was essential for the control of intramammary infection. Furthermore, the interpretation of these data offered an alternative hypothesis, namely the inhibition of bacterial colonisation of the mammary gland, around which strategic research on *S. uberis* should be centred.

## 2.2 Sub-unit vaccines

The possession of the biochemical machinery necessary to obtain adequate nutrition from the host is a prerequisite for maintenance of a bacterial infection. Following experimentally induced infection of the lactating mammary gland, *S. uberis* is found predominantly in the luminal areas of secretory alveoli and ductular tissue[6], indicating that much of the bacterial growth occurs in residual and newly synthesised milk. This environment is likely to be deficient in free and peptide associated amino acids[7].

*S. uberis* is auxotrophic for between 10 and 13 amino acids and 8 were commonly required by all strains[8]. It has been postulated that early in pathogenesis, prior to the induction of an inflammatory response, growth of *S. uberis* would be facilitated by the ability to hydrolyse host proteins[9]. However, *S. uberis* does not hydrolyse protein directly[9] and in a chemically defined medium in which a single, essential amino acid was omitted, the inclusion of intact alpha, beta or kappa bovine caseins failed to restore growth[8].This species has been shown to activate bovine and ovine plasminogen [9] to the serine protease, plasmin. Plasminogen occurs naturally in bovine milk[10]. In the absence of certain essential amino acids, growth of *S. uberis* can be restored by the inclusion of plasmin-hydrolysed caseins[8] thus demonstrating that acquisition of some essential nutrients may be achieved by this route.

Despite the multiplicity of bacterial activities which must be employed by *S. uberis* in the colonisation the bovine mammary gland, the plasminogen activator (PauA) is currently the only molecule to be assigned a putative role in this process. Sub-cutaneous immunisation with concentrated culture supernatant containing PauA from one strain conferred between 37.5% - 62.5% protection from clinical disease following experimental challenge with a different strain[11]. The mean bacterial recovery from immunised animals did not exceed $10^3$ cfu / ml of milk whereas that from non vaccinated animals was around $10^7$ cfu / ml of milk. Protection correlated with the production of neutralising activity towards PauA and was achieved without a marked inflammatory response. The mean SCC from immunised animals remained below 300,000 cells / ml of milk whereas that from control animals exceeded 5,000,000 cells /ml of milk. Furthermore, immunisation with a similar preparation from which the plasminogen activator had been removed using an immobilised monoclonal antibody showed no protective effect[11].

PauA is produced by the vast majority of strains isolated from clinical cases of bovine mastitis; it appears to be the major bovine plasminogen activator produced by *S. uberis*, and it can be neutralised by specific antibody. It has been determined that the *pau*A genes from two strains, one from the UK and the other from the USA, shared over 99% sequence identity (Rosey *et al.*, 1995 unpublished) suggesting the antigen is highly conserved between distinct strains over a wide geographical area. A subsequent survey of ten strains from the UK, USA and Denmark revealed similar levels of homology in nine[12]. The other strain (an isolate from Denmark), which has now been confirmed as *S. uberis* by analysis of the V2 region of the 16S rDNA sequence (L. Johnsen personal communication), did not have the *pau*A gene but was able to activate plasminogen through the action of a distinct protein of 45kda. The frequency of this gene rather than *pau*A within the population of *S. uberis* would appear low, but has not yet been determined precisely.

## 3. FUTURE PROSPECTS

Due to wide variety of strains of *S. uberis* which are associated with clinical disease any vaccine aimed at this organism must show broad, cross strain protection. The live vaccine which showed very good protection against the vaccine strain clearly did not protect against one other strain that was tested. The precise mode of protection induced by the live vaccine has not yet been determined and as a consequence the importance of particular antigens in not known. However, the possibility remains that such a regime may confer cross protection against specific clusters of related strains. Until such data is available such an approach is unlikely to be taken forward commercially. A sub unit vaccine based on the plasminogen activator, PauA, has shown cross protective effects following experimental challenge, its efficacy in field trial is yet to be shown. Most, if not all, strains of *S. uberis* isolated from clinical disease are able to activate bovine plasminogen. In only one case has this been shown to be mediated by a molecule other than PauA. It is important that the frequency at which this alternate gene occurs within *S. uberis* from clinical samples is determined, as such strains are unlikely be controlled by a vaccine directed solely at PauA.

To date, research has not led to the commercial production of an effective vaccine against any of the streptococcal species involved in mastitis. However, the prospects of an effective vaccine against *S. uberis* are significantly better than they have been at any time in the past. If successful, the approach of a growth retarding immunity may be extended to confer greater protection and may be extrapolated to other bacterial, and in particular streptococcal, species.

# 4. REFERENCES


1. Leigh, J. A. (1999) *Streptococcus uberis*: A permanent barrier to the control of bovine mastitis? The Veterinary Journal. 157. 225-238.
2. Hill, A. W., Finch, J. M., Leigh, J. A. & Field, T. R. (1994) Immune modification of the pathogenesis of *Streptococcus uberis* mastitis in the dairy cow. *FEMS Immunology & Medical Microbiology*. 8, 109-118
3. Finch, J.M, Winter, A., Walton, A. & Leigh, J. A. (1997) Further studies on the efficacy of a live vaccine against mastitis caused by *Streptococcus uberis*. *Vaccine*.15, 1138-1143
4. Hill, A. W. & Leigh, J. A. (1989) DNA-fingerprinting of *Streptococcus uberis*: A useful tool for the epidemiology of bovine mastitis. *Epidemiology and Infection*. 103, 165-171
5. Jayarao, B. M., Oliver, S. P., Tagg, J. R. & Matthews, K. R. (1991) Genotypic and phenotypic analysis of *Streptococcus uberis* isolated from bovine mammary secretions. *Epidemiology & Infection*. 107, 543-555
6. Thomas, L. H., Haider, W., Hill, A. W. & Cook, R. S. (1994) Pathologic findings of experimentally induced *Streptococcus uberis* infection in the mammary gland of cows. *American Journal of Veterinary Research*. 55, 1723-1728
7. Aston, J. W. (1975) Amino acids in milk. Their determination by gas-liquid chromotography and their variation due to mastitic infection. *Australian Journal of Dairy Technology*. 30, 55-59
8. Kitt, A. J. & Leigh, J. A. (1997) The auxotrophic nature of *Streptococcus uberis*: The acquisition of essential amino acids from plasmin derived casein peptides. *Advances in Experimental Medicine & Biology*. 418, 647-650
9. Leigh, J. A. (1993) Activation of bovine plasminogen by *Streptococcus uberis*. *FEMS Microbiology Letters*. 114, 67-72
10. Kaminogawa, S & Yamauchi, K (1972) Decomposition of beta casein by milk protease. Similarity of the decomposed products to temperature sensitive and R-caseins. Agricultural and Biological Chemistry. 36, 255-260
11. Leigh, J. A., Finch, J. M., Field, T. R., Real, N. C., Winter, Walton, A. W. & Hodgkinson S. M. (1999) Vaccination with the plasminogen activator from *Streptococcus uberis* induces an inhibitory response and protects against experimental infection in the dairy cow. Vaccine. 17, 851-857
12. Johnsen, L. B., Poulsen, K., Kilian, M & Petersen, T. E. (1999) Purification and cloning of a streptokinase from *Streptococcus uberis*. Infection & Immunity. 67, 1072-1078

# Index